Symmetry: A Mathematical Exploration

Symmetry: A Mathematical Exploration

Edited by Finn Morris

Clanrye International
www.clanryeinternational.com

Clanrye International,
750 Third Avenue, 9th Floor,
New York, NY 10017, USA

ISBN: 978-1-64726-654-7

Cataloging-in-Publication Data

Symmetry : a mathematical exploration / edited by Finn Morris.
p. cm.
Includes bibliographical references and index.
ISBN 978-1-64726-654-7
1. Symmetry (Mathematics). 2. Automorphisms. 3. Group theory. I. Morris, Finn.
QA174.7.S96 S96 2023
516.1--dc23

For information on all Clanrye International publications
visit our website at www.clanryeinternational.com

Clanrye International

Contents

Preface

Over the recent decade, advancements and applications have progressed exponentially. This has led to the increased interest in this field and projects are being conducted to enhance knowledge. The main objective of this book is to present some of the critical challenges and provide insights into possible solutions. This book will answer the varied questions that arise in the field and also provide an increased scope for furthering studies.

Symmetry can be defined as a type of invariance. It refers to the property of a mathematical object remaining unchanged under a set of operations or transformations. Symmetry deals with the identification and use of invariants for any of the various transformations for any paired dataset and characterizations associated with it. In mathematics, all kinds of structures have their own kind of symmetry. For instance, a matrix is symmetric if the original matrix is equal to its transposed version. An integration of applied mathematics with symmetry may work as a powerful instrument for reducing and solving varied problems. There are several applications of symmetry in mathematics. Some of these include matrices, groups, tensors, quantum mechanics, probability theory, and differential equations. This book investigates the role of symmetry in mathematics in detail. Through it, we attempt to further enlighten the readers about the new concepts in this area of study. Students, researchers, experts, and all associated with the application and study of symmetry will benefit alike from this book.

I hope that this book, with its visionary approach, will be a valuable addition and will promote interest among readers. Each of the authors has provided their extraordinary competence in their specific fields by providing different perspectives as they come from diverse nations and regions. I thank them for their contributions.

Editor

1

Modified Born–Infeld-Dilaton-Axion Coupling in Supersymmetry

Yermek Aldabergenov [1,2]* and Sergei V. Ketov [3,4,5,]*

1 Department of Physics, Faculty of Science, Chulalongkorn University, Thanon Phayathai, Pathumwan, Bangkok 10330, Thailand
2 Institute of Experimental and Theoretical Physics, Al-Farabi Kazakh National University, 71 Al-Farabi Avenue, Almaty 050040, Kazakhstan
3 Department of Physics, Tokyo Metropolitan University, Minami-ohsawa 1-1, Hachioji-shi, Tokyo 192-0397, Japan
4 Research School of High Energy Physics, Tomsk Polytechnic University, 30 Lenin Avenue, Tomsk 634050, Russia
5 Kavli Institute for the Physics and Mathematics of the Universe (IPMU), The University of Tokyo, Chiba 277-8568, Japan
* Correspondence: Yermek.A@chula.ac.th (Y.A.); ketov@tmu.ac.jp (S.V.K.)

Abstract: We propose the supersymmetric extension of the modified Born–Infeld-axion-dilaton non-linear electrodynamics that has confined static abelian solutions used for describing the electromagnetic confinement in the presence of axion and dilaton fields, as well as charged matter. The supersymmetric extension also has the non-trivial scalar potential that implies the upper bounds on the matter fields.

Keywords: born–infeld model; supersymmetry; dilaton; axion; electromagnetic; confinement; abelian

1. Introduction

Born–Infeld (BI) theory [1] is the profound extension of Maxwell electrodynamics, distinguished by its several theoretical features: (i) the Lagrangian as a density; (ii) Lorentz invariance; (iii) the resolution of the Coulomb singularity of the static electric field of a point-like charge; (iv) the upper limits on the values of electric and magnetic fields; (v) causal propagation of waves (no shock waves); (vi) the electric-magnetic self-duality; (vii) the origin in string theory as the low-energy effective theory of open strings and D-branes (see [2–9] for all these known aspects of the BI theory).

The supersymmetric extensions of the BI theory exhibit the additional highly-non-trivial and unexpected features, such as (viii) the extra hidden (non-linearly) realized supersymmetries allowing (ix) the interpretation of the supersymmetric BI actions as the Maxwell–Goldstone actions associated with partial spontaneous supersymmetry breaking [10–14] and (x) adding Fayet–Iliopoulos (FI) terms [15].

The BI theory and its supersymmetric extensions are the very special and remarkable examples of the non-linear electrodynamics, whose applicability is not limited to string theory. Moreover, for the sake of phenomenological applications, one may sacrifice some of the distinguished features (i–x) mentioned above. The phenomenological applications may also include the coupling to axion and dilaton fields to the non-linear electrodynamics, beyond the standard (minimal) coupling to the charged fields.

One such interesting application was proposed in [16], where the natural extension of the Born–Infeld action was used, which preserves the properties (i) and (ii) above. The authors of [16]

discovered the existence of confined solutions to the modified BI equations of motion, i.e., the fields with finite support in spacetime, in the presence of axion and dilaton fields. This feature is not shared by the original BI theory and represents the electromagnetic confinement that may be of great interest to astrophysics and cosmology because it may imply the existence of new forms of matter (in the form of charged or uncharged stars invisible to us, like black holes).

It is, therefore, of interest, to construct the supersymmetric extension of the modified BI theory proposed in [16] and couple it to charged matter and a dilaton-axion superfield. Our paper is devoted to this purpose.

The paper is organized as follows. In Section 2, we briefly review the Born–Infeld theory and its $N = 1$ supersymmetric extension in four spacetime dimensions. In Section 3, we also review the standard coupling of the BI theory to dilaton and axion, together with its supersymmetric extension. Sections 2 and 3 provide the foundation for our new results given in Sections 4 and 5. Our conclusion is Section 6.

2. Supersymmetric Born–Infeld Theory

The standard Born–Infeld (BI) Lagrangian reads [1]:

$$\mathcal{L}_{\rm BI} = \frac{1}{b^2}\left(1 - \sqrt{-\det(\eta_{mn} + bF_{mn})}\right) , \quad F_{mn} = \partial_m A_n - \partial_n A_m \, , \tag{1}$$

with the BI coupling constant b. In four spacetime dimensions, it takes the form:

$$\mathcal{L}_{\rm BI} = \frac{1}{b^2}\left(1 - \sqrt{1 + \frac{b^2}{2}F^2 - \frac{b^4}{16}(F\tilde{F})^2}\right) = -\frac{F^2}{4} + \frac{\frac{b^2}{16}[(F^2)^2 + (F\tilde{F})^2]}{1 + \frac{b^2}{4}F^2 + \sqrt{1 + \frac{b^2}{2}F^2 - \frac{b^4}{16}(F\tilde{F})^2}} , \tag{2}$$

where we have used the notation $\tilde{F}^{mn} = \frac{1}{2}\epsilon^{mnkl}F_{kl}$, $F^2 \equiv F_{mn}F^{mn}$, and $F\tilde{F} \equiv \frac{1}{2}\epsilon^{mnkl}F_{mn}F_{kl}$ with Levi–Civita ϵ.

The equations of motion (EOM) and Bianchi identities (BIs) of the BI theory can be written as follows:

$$\partial^n G_{mn} = 0 \, , \tag{3}$$

$$\partial^n \tilde{F}_{mn} = 0 \, , \tag{4}$$

respectively, where we have used the notation:

$$G_{mn} = -2\frac{\partial \mathcal{L}}{\partial F^{mn}} \, . \tag{5}$$

The EOM and BIs are invariant under the $SO(2)$ electric-magnetic duality rotations:

$$\begin{aligned} F_{mn} &\to \cos\gamma F_{mn} + \sin\gamma \tilde{G}_{mn} \, , \\ G_{mn} &\to \cos\gamma G_{mn} + \sin\gamma \tilde{F}_{mn} \, , \end{aligned} \tag{6}$$

where we have used the notation $\tilde{G}^{mn} = \frac{1}{2}\epsilon^{mnkl}G_{kl}$. As regards a generic non-linear electrodynamics with the Lagrangian $\mathcal{L}(F)$, the condition of the electric-magnetic duality reads [5]:

$$G\tilde{G} = F\tilde{F} \, . \tag{7}$$

The Lagrangian (2) can be expanded as:

$$\mathcal{L}_{\rm BI} = -\frac{1}{4}F^2 + \frac{b^2}{32}\left[(F^2)^2 + (F\tilde{F})^2\right] + \frac{b^4}{128}\left[(F^2)^3 + F^2(F\tilde{F})^2\right] + \mathcal{O}(b^6) \, , \tag{8}$$

whose leading term is the standard Lagrangian of Maxwell electrodynamics.

The form (2) of the BI theory is most convenient for its (rigid) supersymmetrization in superspace. In $N = 1$ superspace, the supersymmetric BI theory is described by the Lagrangian [10,17] (for a review of the Born–Infeld theory and its supersymmetric extensions, see, e.g., [9,14]):

$$\mathcal{L}_{\rm sBI} = \frac{1}{4}\left(\int d^2\theta W^2 + \text{h.c.}\right) + \frac{b^2}{4}\int d^4\theta \frac{W^2\overline{W}^2}{1+\frac{b^2}{2}(\omega+\bar{\omega})+\sqrt{1+b^2(\omega+\bar{\omega})+\frac{b^4}{4}(\omega-\bar{\omega})^2}}\,, \tag{9}$$

in terms of the superfield strength $W_\alpha = -\frac{1}{4}\overline{D}^2 D_\alpha V$ of the real gauge superfield V with $W^2 \equiv W^\alpha W_\alpha$ and:

$$\omega \equiv \frac{1}{8}D^2W^2 = \frac{1}{4}(F^2 - 2D^2 - iF\tilde{F}) + \ldots\,, \tag{10}$$

where the dots stand for the fermionic and the higher order (in θ) terms.

The Lagrangian (9) is invariant under the $U(1)$ gauge transformations:

$$V \to V - \Lambda - \overline{\Lambda} \tag{11}$$

with the chiral superfield gauge parameter Λ. In addition to the manifest $N = 1$ supersymmetry (SUSY), the SUSY BI action is also known to be invariant under the second (non-linearly-realized) supersymmetry [11,12].

The bosonic part of the Lagrangian (9) reads:

$$\mathcal{L}_{\rm sBI} = -\frac{A}{4} + \frac{\frac{b^2}{16}(A^2+B^2)}{1+\frac{b^2}{4}A+\sqrt{1+\frac{b^2}{2}A-\frac{b^4}{16}B^2}} = \frac{1}{b^2}\left(1-\sqrt{1+\frac{b^2}{2}A-\frac{b^4}{16}B^2}\right)\,, \tag{12}$$

where $A \equiv F^2 - 2D^2$ and $B \equiv F\tilde{F}$. The real auxiliary field D can be eliminated by its equation of motion, $D = 0$, then the resulting Lagrangian coincides with Equations (1) and (2).

The SUSY BI theory (9) can be further generalized by adding matter chiral superfields Φ_i (charged under the $U(1)$ gauge symmetry with charges q_i), together with a Fayet–Iliopoulos (FI) term:

$$\mathcal{L} = \mathcal{L}_{\rm sBI} + \int d^4\theta \left[K(\overline{\Phi}_i e^{q_i V}\Phi_i) + \xi V\right]\,, \quad \Phi_i \to e^{q_i\Lambda}\Phi_i\,, \tag{13}$$

where we have introduced the arbitrary function K (we do not demand renormalizability) and the real constant FI parameter ξ. In terms of the field components, we find:

$$\mathcal{L} = \mathcal{L}_{\rm sBI} - XD + \ldots\,, \tag{14}$$

where we have introduced the notation:

$$X \equiv -\frac{1}{2}(K_V + \xi) \quad \text{and} \quad K_V \equiv \frac{\partial K}{\partial V}|_{\theta=0}\,, \tag{15}$$

and the dots stand for the terms that are irrelevant in the D equation of motion.

When the matter fields are absent, $K = 0$, the BI theory with the FI term (13) retains its hidden $N = 2$ SUSY [15], although the FI term spontaneously breaks the linear $N = 1$ SUSY. When the charged matter fields are present, the second non-linear SUSY is explicitly broken.

Given the Lagrangian (14), the EOM of D reads:

$$D = X\sqrt{1+\frac{b^2}{2}(F^2-2D^2)-\frac{b^4}{16}(F\tilde{F})^2}\,, \tag{16}$$

and its solution is given by:

$$D = \frac{X}{\sqrt{1+b^2X^2}} \sqrt{1+\frac{b^2}{2}F^2 - \frac{b^4}{16}(F\tilde{F})^2}\,. \tag{17}$$

Substituting the solution back into the Lagrangian yields:

$$\mathcal{L} = \frac{1}{b^2}\left(1 - \sqrt{1+b^2X^2}\sqrt{1+\frac{b^2}{2}F^2 - \frac{b^4}{16}(F\tilde{F})^2}\right)\,. \tag{18}$$

Therefore, we obtain the scalar potential:

$$V_D = \frac{1}{b^2}\left(\sqrt{1+b^2X^2}-1\right) = \frac{1}{2}X^2 - \frac{b^2}{8}X^4 + \frac{b^4}{16}X^6 + \mathcal{O}(b^6)\,. \tag{19}$$

3. BI Coupling to Dilaton-Axion and Its Supersymmetrization

The standard coupling of Born–Infeld theory to dilaton field ϕ and axion field C reads (the full theory also includes the dilaton and axion kinetic terms):

$$\begin{aligned}\tilde{\mathcal{L}}_{\mathrm{BI}} &= \frac{1}{b^2}\left(1 - \sqrt{-\det(\eta_{mn} + be^{\phi/2}F_{mn})}\right) + \frac{C}{4}F\tilde{F} \\ &= \frac{1}{b^2}\left(1 - \sqrt{1+\frac{b^2}{2}e^{-\phi}F^2 - \frac{b^4}{16}e^{-2\phi}(F\tilde{F})^2}\right) + \frac{C}{4}F\tilde{F}\,. \end{aligned} \tag{20}$$

In this case, the $SO(2)$ electromagnetic self-duality can be extended to the $SL(2,\mathbb{R})$ self-duality [5] (in quantum theory (superstrings), the $SL(2,\mathbb{R})$ is broken into its discrete subgroup $SL(2,\mathbb{Z})$). After introducing the complex dilaton-axion field and its vacuum expectation value (VEV),

$$\tau \equiv C + ie^{-\phi}\,, \quad \tau_0 = \frac{\Theta}{2\pi} + \frac{4\pi i}{e^2}\,, \tag{21}$$

respectively, where Θ is the vacuum theta-angle and e is the $U(1)$ electric charge, the $SL(2,\mathbb{R})$ transformations read:

$$\tau \to \frac{a\tau + b}{c\tau + d}\,, \quad \text{with} \quad a,b,c,d \in \mathbb{R}\,, \quad ad - bc = 1\,. \tag{22}$$

The SUSY extension of the action (20) takes the form [18]:

$$\tilde{\mathcal{L}}_{\mathrm{sBI}} = \tfrac{1}{4i}\left(\int d^2\theta\tau W^2 - \text{h.c.}\right) + \tfrac{b^2}{16}\int d^4\theta \frac{|\tau-\bar{\tau}|^2 W^2\overline{W}^2}{1+\frac{b^2}{2}(\omega'+\bar{\omega}')+\sqrt{1+b^2(\omega'+\bar{\omega}')+\frac{b^4}{4}(\omega'-\bar{\omega}')^2}}\,, \tag{23}$$

where we have used the notation:

$$\omega' \equiv \frac{1}{16i}(\tau-\bar{\tau})D^2W^2 = \frac{1}{4}e^{-\phi}(A - iB) + \ldots\,, \tag{24}$$

and have promoted τ to the chiral superfield. The bosonic part of the Lagrangian (23) reads:

$$\tilde{\mathcal{L}}_{\mathrm{sBI}} = \frac{1}{b^2}\left(1 - \sqrt{1+\frac{b^2}{2}e^{-\phi}(F^2 - 2D^2) - \frac{b^4}{16}e^{-2\phi}(F\tilde{F})^2}\right) + \frac{C}{4}F\tilde{F}\,. \tag{25}$$

The auxiliary field D is eliminated by its EOM, $D = 0$, and the resulting Lagrangian coincides with (20). In the parametrization (21), the Lagrangian should be rescaled as $\mathcal{L} \to \frac{e^2}{4\pi}\mathcal{L}$, in order to obtain the canonical kinetic term of F_{mn}.

4. Modified BI Theory with Dilaton-Axion-Like Couplings

As was argued in the Introduction, let us consider the modified coupling of the dilaton-axion field to BI theory, which was proposed in [16]:

$$\begin{aligned}\mathcal{L}_{\mathrm{BI}'} &= \frac{1}{b^2}\left(1-\sqrt{-\det(\eta_{mn}+b\alpha F_{mn}+b\beta\tilde{F}_{mn})}\right)\\ &= \frac{1}{b^2}\left(1-\sqrt{1+\frac{b^2}{2}\left(e^{-\phi}F^2-CF\tilde{F}\right)-\frac{b^4}{16}\left(e^{-\phi}F\tilde{F}+CF^2\right)^2}\right),\end{aligned} \tag{26}$$

where α and β are related to ϕ and C as:

$$\alpha^2-\beta^2=e^{-\phi}\,, \quad -2\alpha\beta=C\,. \tag{27}$$

The Lagrangian (26) is obtained from the original BI theory (1) by the substitution $F_{mn}\to\alpha F_{mn}+\beta\tilde{F}_{mn}$. As is clear from a comparison of Equations (20) and (26) (see also Equation (28) below), the two theories differ in their C-dependence, but their leading (b-independent) terms coincide.

We refer to the modified theory (26) as the BI$'$ theory and still associate the fields ϕ and C with the dilaton and axion, respectively, as in [16], because they regain their original meaning in the weak coupling limit $b\to 0$. Indeed, in the BI$'$ theory, the shift symmetry $C\to C+const.$, defining the axion, is lost (together with the electromagnetic self-duality, since $G\tilde{G}\neq F\tilde{F}$), while C directly affects the equations of motion for the electromagnetic field A_m. However, after expanding the action (26),

$$\begin{aligned}\mathcal{L}_{\mathrm{BI}'} = &-\frac{1}{4}e^{-\phi}F^2+\frac{1}{4}CF\tilde{F}+\frac{b^2}{32}(e^{-2\phi}+C^2)\left[(F^2)^2+(F\tilde{F})^2\right]-\\ &-\frac{b^4}{128}(e^{-2\phi}+C^2)(e^{-\phi}F^2-CF\tilde{F})\left[(F^2)^2+(F\tilde{F})^2\right]+\mathcal{O}(b^6)\,,\end{aligned} \tag{28}$$

in the leading order with respect to b, the BI$'$ theory coincides with the Maxwell theory coupled to the dilaton and axion. Hence, for the weak F_{mn} field, the axionic shift symmetry approximately holds (as well as the approximate $SL(2,\mathbb{R})$ duality).

A supersymmetrization of the BI$'$ theory is straightforward and results in:

$$\mathcal{L}_{\mathrm{sBI}'}=\tfrac{1}{4i}\left(\int d^2\theta\tau W^2-\mathrm{h.c.}\right)+\tfrac{b^2}{4}\int d^4\theta\frac{|\tau|^2W^2\overline{W}^2}{1+\frac{b^2}{2}(\omega'+\bar{\omega}')+\sqrt{1+b^2(\omega'+\bar{\omega}')+\frac{b^4}{4}(\omega'-\bar{\omega}')^2}}\,, \tag{29}$$

where we have used the notation:

$$\omega'\equiv\frac{\tau}{8i}D^2W^2=-\frac{\tau}{4}(B+iA)+\ldots \tag{30}$$

together with $A\equiv F^2-2D^2$, $B\equiv F\tilde{F}$, and $\tau=C+ie^{-\phi}+\mathcal{O}(\theta)$.

After expanding the Lagrangian in components, we find:

$$\begin{aligned}\mathcal{L}_{\mathrm{sBI}'} &= -\frac{1}{4}(e^{-\phi}A-CB)+\frac{b^2}{16}\cdot\frac{|\tau|^2(A^2+B^2)}{1+\frac{b^2}{4}(e^{-\phi}A-CB)+\sqrt{1+\frac{b^2}{2}(e^{-\phi}A-CB)-\frac{b^4}{16}(e^{-\phi}B+CA)^2}}\\ &= \frac{1}{b^2}\left(1-\sqrt{1+\frac{b^2}{2}(e^{-\phi}A-CB)-\frac{b^4}{16}(e^{-\phi}B+CA)^2}\right)\\ &= \frac{1}{b^2}\left(1-\sqrt{1+\frac{b^2}{2}\left(e^{-\phi}F^2-2e^{-\phi}D^2-CF\tilde{F}\right)-\frac{b^4}{16}\left(e^{-\phi}F\tilde{F}+CF^2-2CD^2\right)^2}\right).\end{aligned} \tag{31}$$

In the absence of matter fields and FI terms, the auxiliary field D is eliminated by its EOM, $D = 0$, and the resulting Lagrangian coincides with (26).

4.1. Adding Supersymmetric Matter and the FI Term

Adding to the BI′ theory the charged matter and the FI term along the lines of Equations (13) and (14), with:

$$\mathcal{L}' = \mathcal{L}_{\text{sBI}'} - XD + \ldots\,, \tag{32}$$

yields the EOM for D in the form:

$$2b^2\psi D - \frac{b^4}{2}CD\left(\psi F\tilde{F} + CF^2 - 2CD^2\right) = \\ = 2b^2X\sqrt{1 + \frac{b^2}{2}\left(\psi F^2 - 2\psi D^2 - CF\tilde{F}\right) - \frac{b^4}{16}\left(\psi F\tilde{F} + CF^2 - 2CD^2\right)^2}\,, \tag{33}$$

where we have introduced the notation $\psi \equiv e^{-\phi}$. Since $D = 0$ is no longer a valid solution for its EOM, this generates a scalar potential.

Let us study solutions to EOM of D. After setting $F_{mn} = 0$, there are three of them (in terms of D^2),

$$D_1^2 = \frac{1}{3b^2C^2}\left[-4\psi - b^2X^2 + \frac{(2\psi - b^2X^2)^2}{\lambda} + \lambda\right], \tag{34}$$

$$D_{2,3}^2 = \frac{1}{3b^2C^2}\left[-4\psi - b^2X^2 - \frac{\frac{1}{2}(1 \pm i\sqrt{3})(2\psi - b^2X^2)^2}{\lambda} - \frac{1}{2}(1 \mp i\sqrt{3})\lambda\right], \tag{35}$$

where we have used the notation:

$$\lambda \equiv \left(6b|X|\sqrt{3\sigma(\psi^2 + C^2)} + \sigma + 27(\psi^2 + C^2)b^2X^2\right)^{\frac{1}{3}}, \tag{36}$$

$$\sigma \equiv 8\psi^3 + (15\psi^2 + 27C^2)(bX)^2 + 6\psi(bX)^4 - (bX)^6\,. \tag{37}$$

The two real solutions are $D_1 = \pm\sqrt{D_1^2}$. For consistency, the sign of the square roots should be correlated with that of X, i.e., $\pm\sqrt{D_1^2}$ should correspond to $\pm|X|$, respectively.

Since σ appears under the square root in (36), when $\sigma < 0$, the D_1^2 becomes complex, while $D_{2,3}^2$ generically stays complex as well (however, for certain values of X, ψ, and C, the imaginary parts of $D_{1,2,3}$ may vanish). Furthermore, D_1^2 has to be positive, i.e., the condition:

$$\frac{(2\psi - b^2X^2)^2}{\lambda} + \lambda - 4\psi - b^2X^2 \geq 0 \tag{38}$$

must hold. It can be easily checked that (38) is always satisfied for $\sigma \geq 0$.

As is clear from the Lagrangian (31) or EOM (33), the solution (34) does not guarantee that the expression under the square root is positive. This means we have to require that (for $F_{mn} = 0$):

$$1 - b^2\psi D^2 - \frac{b^4}{4}C^2D^4 \geq 0\,, \tag{39}$$

which yields (also demanding the D^2 to be positive):

$$D^2 \leq 2\frac{-\psi + \sqrt{\psi^2 + C^2}}{b^2C^2}\,. \tag{40}$$

Substituting the solution (34) for D^2 leads to a complicated inequality involving X, ψ, and C.

We conclude that the supersymmetric BI′ theory leads to restrictions on the values of $\psi = e^{-\phi}$, C, and $X \equiv -\frac{1}{2}(K_V + \xi)$, in contrast to the standard supersymmetric BI theory, namely Equation (40) together with the condition:

$$\sigma \geq 0 \,. \tag{41}$$

4.2. The Case of a Constant τ

To gain more insight, let us consider the particular case of $\tau = \tau_0$, i.e.:

$$\tau = \frac{\Theta}{2\pi} + \frac{4\pi i}{e^2} \,. \tag{42}$$

We find that the scalar potential of the supersymmetric BI′ theory has a restricted domain in terms of the values of X. In order to get numerical results, as an example, let us set $\Theta = 2\pi$ and $e^2 = 4\pi$, so that $\tau = 1 + i$. We first derive the upper limit on X, using the inequality $\sigma \geq 0$ (with the definition (37)):

$$-(bX)^6 + 6(bX)^4 + 42(bX)^2 + 8 \geq 0 \,. \tag{43}$$

The inequality (43) is respected by:

$$|X| \equiv \frac{1}{2}|K_V + \xi| \leq \sqrt{5 + 3\sqrt{3}}b^{-1} \approx 3.19\, b^{-1} \,. \tag{44}$$

Note that the BI parameter b has the dimension $(\mathrm{mass})^{-2}$ and can be written in terms of the BI mass scale as $b = M_{\mathrm{BI}}^{-2}$, whereas the FI parameter has the dimension of $(\mathrm{mass})^2$.

Figure 1 illustrates the solution D_1 acquiring the non-vanishing imaginary part for $X > X_{\max} \equiv \sqrt{5 + 3\sqrt{3}}\, b^{-1}$.

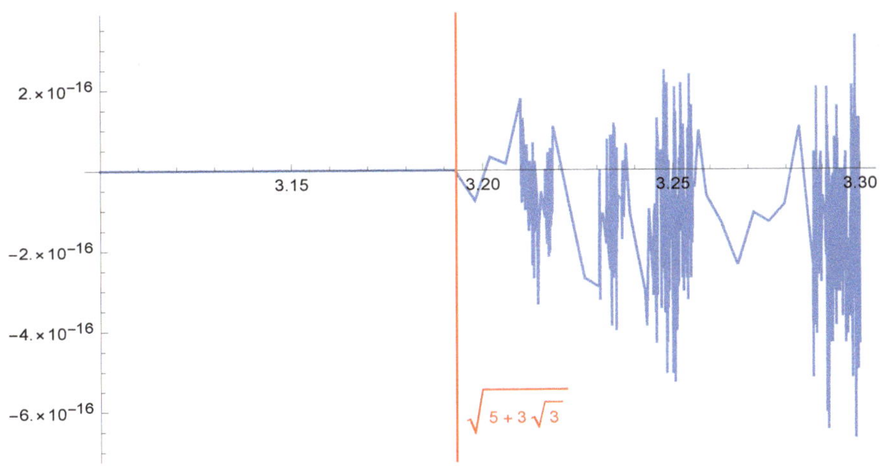

Figure 1. ImD_1 (vertical) versus X (horizontal). For $|X| \leq \sqrt{5 + 3\sqrt{3}}$ (with $b = 1$), the imaginary part of D_1 is zero, so the solution is valid in this region.

If we ignore the matter fields, $K = 0$, we arrive at the upper limit on the absolute value of the FI parameter:

$$|\xi| \leq 2\sqrt{5 + 3\sqrt{3}}\, b^{-1} \,. \tag{45}$$

On the other hand, when the charged matter is present, the condition (44) gives the upper limit on K_V for a given ξ. For instance, with the canonical choice $K = \Phi e^{2V}\overline{\Phi}$ and $K_V = 2|\varphi|^2$, where φ is the leading component of Φ, the charged scalar φ is bounded from above. Setting $\xi = 0$ for simplicity, we find:

$$|\varphi|^2 \leq \sqrt{5 + 3\sqrt{3}}\, b^{-1} \,. \tag{46}$$

Note that the upper limit of $|X|$ is of the same order $(1/b)$ as that for F_{mn}.
Next, let us consider the condition (40). When $\psi = C = 1$, it takes the form:

$$|D| \leq \sqrt{2\sqrt{2} - 2}\, b^{-1} \,. \tag{47}$$

After a substitution of the solution (34) into (47), the resulting inequality can be numerically solved for X, yielding the conditions:

$$|X| \lessapprox 2.02\, b^{-1} \quad \text{and} \quad |X| \gtrapprox 3.04\, b^{-1} \,. \tag{48}$$

5. The Scalar Potential

The scalar potential of the supersymmetric matter-coupled BI′ theory (32) can be obtained by using the solutions (34) for D and setting $F_{mn} = 0$. The exact potential is very complicated, and it seems impossible to obtain its perturbative expansion in terms of b, because the solution to D has the overall factor of b^{-1}.

Though the exact features of the potential depend on the choice of X, it is nevertheless possible to get some general results. For instance, the potential is always non-negative, and its minimum is at $X = 0 \to D = 0$, if it exists, with unbroken SUSY and Minkowski vacuum. This is similar to the ordinary D-term potentials proportional to D^2. With the positive FI term and no matter, the minimum is of the de Sitter type with the D-term SUSY breaking.

The main difference against the ordinary D-term potentials is the existence of restrictions on the values of $X \equiv \frac{1}{2}(K_V + \xi)$ for given values of $\psi = e^{-\Phi}$ and C. These restrictions arise from the conditions (41) and (47) for general ψ and C. If we set $\psi = C = 1$, we obtain the specific results:

$$|X| \lessapprox 2.02\,, \quad 3.04 \lessapprox |X| \lessapprox 3.19, \tag{49}$$

in units of $b = 1$.

We provide the plot of the scalar potential as the function of X in Figure 2. For negative values of X, the negative root of D_1^2 should be used as the solution to D, so that the potential is symmetric under $X \to -X$.

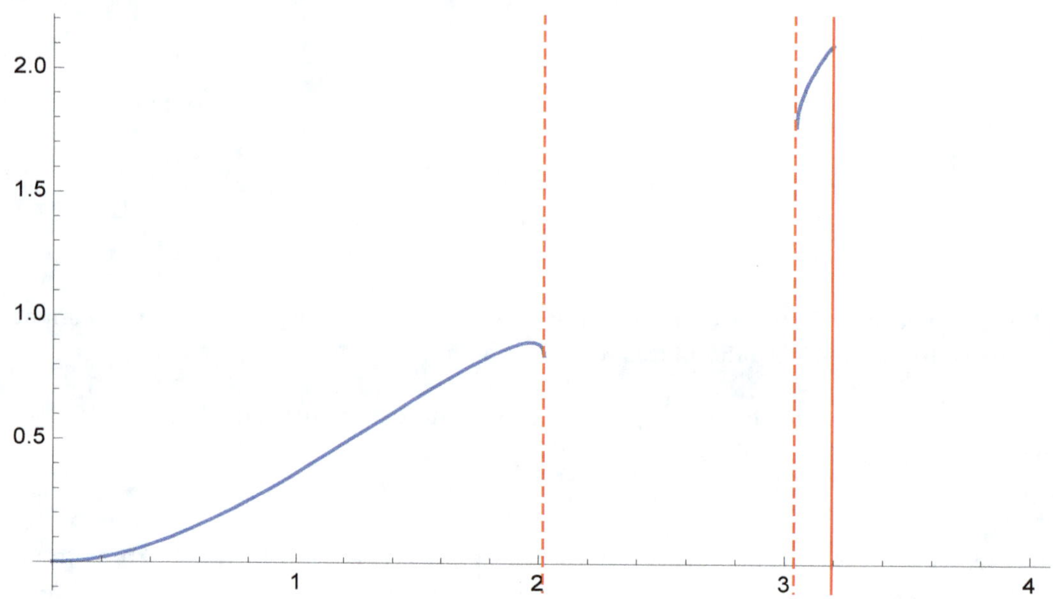

Figure 2. The scalar potential (vertical) versus X (horizontal). The dashed red lines mark the values $X \approx 2.02$ and $X \approx 3.04$, while the solid red line stands at $X \approx 3.19$, the upper limit on X.

6. Conclusions

Our main results are given in Sections 4 and 5.

It is of further interest to explore whether the confining mechanism of [16] also exists in non-abelian extensions of the (modified) Born–Infeld theory and/or their supersymmetric extensions, in the presence of the dilaton-axion superfield. For example, a non-abelian (and non-supersymmetric) Born–Infeld (NBI) theory with a Θ-term, having a similar structure to the BI′ theory, was considered in [7], where it was found that this NBI theory (together with the Higgs sector) admits monopole solutions and has Witten's effect (i.e., a shift of the electric charge of the monopole induced by the presence of the Θ-term [19]), like that in the usual non-abelian gauge theory. Unfortunately, a non-abelian extension of the BI theory and, hence, a supersymmetric NBI theory, is not unique [9].

Author Contributions: Conceptualization, Y.A. and S.V.K.; methodology, Y.A. and S.V.K.; software, Y.A.; validation, S.V.K.; formal analysis, Y.A. and S.V.K.; investigation, Y.A. and S.V.K.; resources, Y.A. and S.V.K.; data curation, Y.A. and S.V.K.; writing–original draft preparation, Y.A. and S.V.K.; writing–review and editing, Y.A. and S.V.K.; visualization, Y.A.; supervision, S.V.K.; project administration, Y.A. and S.V.K.; funding acquisition, Y.A. and S.V.K.

Acknowledgments: Y.A. was supported by the CUniverse research promotion project of Chulalongkorn University under the grant reference CUAASC and the Ministry of Education and Science of the Republic of Kazakhstan under the grant reference AP05133630. S.V.K. was supported by a Grant-in-Aid of the Japanese Society for Promotion of Science (JSPS) under No. 26400252, the Competitiveness Enhancement Program of Tomsk Polytechnic University in Russia, and the World Premier International Research Center Initiative (WPI Initiative), MEXT, Japan.

References

1. Born, M.; Infeld, L. Foundations of the new field theory. *Proc. R. Soc. Lond.* **1934**, *A144*, 425–451. doi:10.1098/rspa.1934.0059. [CrossRef]
2. Boillat, G. Nonlinear electrodynamics—Lagrangians and equations of motion. *J. Math. Phys.* **1970**, *11*, 941–951. doi:10.1063/1.1665231. [CrossRef]
3. Plebanski, J. *Lectures on Non-Linear Electrodynamics*; NORDITA: Copenhagen, Denmark, 1970.
4. Fradkin, E.S.; Tseytlin, A.A. Nonlinear Electrodynamics from Quantized Strings. *Phys. Lett.* **1985**, *163B*, 123–130. doi:10.1016/0370-2693(85)90205-9. [CrossRef]
5. Gaillard, M.K.; Zumino, B. Duality Rotations for Interacting Fields. *Nucl. Phys.* **1981**, *B193*, 221–244. doi:10.1016/0550-3213(81)90527-7. [CrossRef]
6. Ketov, S.V. Solitons, monopoles and duality: From Sine-Gordon to Seiberg-Witten. *Fortschritte der Physik* **1997**, *45*, 237–292. [CrossRef]
7. Grandi, N.E.; Pakman, A.; Schaposnik, F.A.; Silva, G.A. Monopoles, dyons and theta term in Dirac-Born–Infeld theory. *Phys. Rev.* **1999**, *D60*, 125002, doi:10.1103/PhysRevD.60.125002. [CrossRef]
8. Olive, D.I. Aspects of electromagnetic duality. *PoS* **2000**, *tmr2000*, 021. doi:10.22323/1.006.0021. [CrossRef]
9. Ketov, S.V. Many faces of Born–Infeld theory. In Proceedings of the 7th International Wigner Symposium (Wigsym 7), College Park, MD, USA, 24–29 August 2001.
10. Cecotti, S.; Ferrara, S. Supersymmetric Born–Infeld Lagrangians. *Phys. Lett.* **1987**, *B187*, 335–339. doi:10.1016/0370-2693(87)91105-1. [CrossRef]
11. Bagger, J.; Galperin, A. A New Goldstone multiplet for partially broken supersymmetry. *Phys. Rev.* **1997**, *D55*, 1091–1098, doi:10.1103/PhysRevD.55.1091. [CrossRef]
12. Rocek, M.; Tseytlin, A.A. Partial breaking of global D = 4 supersymmetry, constrained superfields, and three-brane actions. *Phys. Rev.* **1999**, *D59*, 106001, doi:10.1103/PhysRevD.59.106001. [CrossRef]
13. Ketov, S.V. A Manifestly N = 2 supersymmetric Born–Infeld action. *Mod. Phys. Lett.* **1999**, *A14*, 501–510, doi:10.1142/S0217732399000559. [CrossRef]
14. Tseytlin, A.A. Born–Infeld action, supersymmetry and string theory In *The Many Faces of the Superworld: Yuri Golfand Memorial*; World Scientific Publishing: Singapore, 1999; pp. 417–452, doi:10.1142/9789812793850_0025.

15. Kuzenko, S.M. The Fayet-Iliopoulos term and nonlinear self-duality. *Phys. Rev.* **2010**, *D81*, 085036, doi:10.1103/PhysRevD.81.085036. [CrossRef]
16. Burton, D.A.; Dereli, T.; Tucker, R.W. Born–Infeld axion–dilaton electrodynamics and electromagnetic confinement. *Phys. Lett.* **2011**, *B703*, 530–535, doi:10.1016/j.physletb.2011.08.039. [CrossRef]
17. Deser, S.; Puzalowski, R. Supersymmetric Nonpolynomial Vector Multiplets and Causal Propagation. *J. Phys.* **1980**, *A13*, 2501. doi:10.1088/0305-4470/13/7/031. [CrossRef]
18. Ketov, S.V. More on the gauge fixed D3-brane action with dilaton axion coupling from N = 1 superspace. *Mod. Phys. Lett.* **2003**, *A18*, 1887–1894, doi:10.1142/S0217732303011733. [CrossRef]
19. Witten, E. Dyons of Charge e theta/2 pi. *Phys. Lett.* **1979**, *B86*, 283–287. doi:10.1016/0370-2693(79)90838-4. [CrossRef]

2

On a $\mathbb{Z}_2^n$-Graded Version of Supersymmetry

Andrew James Bruce

Mathematics Research Unit, University of Luxembourg, Maison du Nombre 6, Avenue de la Fonte, L-4364 Esch-sur-Alzette, Luxembourg; andrewjamesbruce@googlemail.com

Abstract: We extend the notion of super-Minkowski space-time to include $\mathbb{Z}_2^n$-graded (Majorana) spinor coordinates. Our choice of the grading leads to spinor coordinates that are nilpotent but commute amongst themselves. The mathematical framework we employ is the recently developed category of $\mathbb{Z}_2^n$-manifolds understood as locally ringed spaces. The formalism we present resembles $\mathcal{N}$-extended superspace (in the presence of central charges), but with some subtle differences due to the exotic nature of the grading employed.

Keywords: $\mathbb{Z}_2^n$-manifolds; supersymmetry; superspace methods; parastatistics

---

1. Introduction

Supersymmetry was independently discovered by three groups of authors: Gervais & Sakita [1], Gol'fand & Lichtman [2] and Volkov & Akulov [3] in the early 1970s. The first example of an interacting supersymmetric quantum field theory in four dimensions was due to Wess & Zumino [4]. This non-classical symmetry is more than just a way of relating bosonic and fermionic degrees of freedom as supersymmetric field theories can have some remarkable properties. For example, supersymmetry can lead to milder divergences and even 'non-renormalisation theorems'; offer a solution to the hierarchy problem in Grand Unified Theories; remove the tachyon from the spectrum of string theories and naturally leads to a theory of gravity when promoted to a local gauge theory. However, we stress that no experimental evidence that Nature utilises supersymmetry has yet been found. More than this, current experimental data suggest that the (constrained) minimal supersymmetric standard model is disfavoured as a potential realistic model (see Autermann [5]). It is even said that "supersymmetry is a solution looking for a problem" (see for example Kaku [6] (p. 101)).

Alongside the vast wealth of results from theoretical physics, there has been steady growth in the mathematical aspects of supersymmetry since its conception. This knowledge accumulates to the theory of supermanifolds and super Lie groups, with the definition of a supermanifold being due to Berezin & Leĭtes [7]. Much of the fundamental work was carried out between 1965 and 1975 by Berezin and his collaborators. At its 'bare bones' supermathematics is the study of $\mathbb{Z}_2$-graded structures. The study of such things is well motivated by physics via the fact that quasi-classical descriptions of fermions and Faddeev–Popov ghosts both require anticommuting fields. From a pure mathematical perspective, various 'sign rules' appear naturally in algebraic topology and homological algebra (see Mac Lane [8]).

Some very powerful methods for building supersymmetric actions come under the umbrella of superspace methods as first developed (operationally) by Salam & Strathdee [9]. These methods allow for a more geometric picture of supersymmetry as well as giving practical methods of constructing theories. However, the formalism can become clumsy, and not all supersymmetric theories have a superspace formulation. For instance, it is well known that there are no off-shell formulations of $d = 4$, $\mathcal{N} = 4$ theories with a finite number of auxiliary fields. One has to resort to projective (see [10,11]) or harmonic superspaces (see [12]) in order to cover this situation .

At the most basic level, one postulates that standard four-dimensional Minkowski space-time is extended by appending four anticommuting Majorana spinors. Thus, $\mathcal{N} = 1$ super-Minkowski space-time is the supermanifold that admits (global) coordinates

$$(x^{\mu}, \theta_{\alpha}) .$$

The supersymmetry transformations on this supermanifold are

$$x^{\mu} \mapsto x^{\mu} + \frac{1}{4}\epsilon_{\beta}\theta_{\alpha}(C\gamma^{\mu})^{\alpha\beta}, \qquad \theta_{\alpha} \mapsto \theta_{\alpha} + \epsilon_{\alpha}, \tag{1}$$

where ϵ_{α} is an anticommuting Majorana spinor-valued parameter. For a detailed introduction to supersymmetry we suggest West [13], and Wess & Bagger [14]. For an introduction to superspace and methods of quantising supersymmetric field theories one may consult Gates et al. [15]. One can also consult the encyclopedia edited by Duplij, Siegel & Bagger [16] for many details pertaining to supersymmetry and related mathematics, including historical remarks in the section "SUSY Story" written by the founders of supersymmetry.

Extended supersymmetries are then constructed by appending more and more anticommuting Majorana spinors. For example, $\mathcal{N} = 2$ super-Minkowski space-time comes with local coordinates $(x^{\mu}, \theta^{i}_{\alpha})$, $i = 1,2$. In particular, we still have a supermanifold and thus $\theta^1\theta^2 = -\theta^2\theta^1$. The case for $\mathcal{N} > 2$ is analogous. Moreover, $\mathcal{N}$-extended supersymmetry transformations similar to (1) exist.

However, mathematically there is no reason not to consider appending sets of Majorana spinors that do not necessarily anticommute. Indeed, non-anticommuting superspaces have long been studied in the physics literature, see for example [17–22]. The inspiration for many of these works is the well-known fact that various background fields in string theory lead to noncommutative deformations. For example, R-R field backgrounds lead to '$\theta - \theta$' deformations and gravitino backgrounds lead to '$x - \theta$' deformations. We must also mention the work of Frydryszak who studied nilpotent commuting variables and the corresponding generalisation of supersymmetry [23–25].

We will consider the very special instance of Majorana spinors that are $\mathbb{Z}_2^n$-graded commutative. We view this situation as a very mild form of '$\theta - \theta$' non-anticommutativity. This should, of course, be compared with Green's notion of a parafermion (see Green [26], and also Volkov [27] who independently introduced the concept). Because of how we will assign the $\mathbb{Z}_2^n$-grading the supersymmetry algebra '$[Q, Q] \sim P + Z$' where we allow a central charge Z, is not given in terms of just an anticommutator, some terms will be given in terms of a commutator, collectively we use $\mathbb{Z}_2^n$-graded commutators. This will complicate the understanding of Bogomol'nyi-Prasad-Sommerfield (BPS) states, for example. At this juncture, we should also mention the work of Zheltukhin [28] who defined a 'para-Fermionic superspace' in relation to a generalisation of the Neveu–Ramond–Schwarz superstring. As far as we know, Zheltukhin is the first to consider a 'superspace' formed by replacing standard Grassmann coordinates with a $\mathbb{Z}_2^2$-graded version. We must also mention the work of Vasil'iev [29], in which a $\mathbb{Z}_2^2$-Lie algebra structure was uncovered in supergravity with a positive cosmological constant by modifying spinor conjugation. Moreover, $\mathbb{Z}_2^n$-gradings appear in the context of parasupersymmetry, see for example Yang & Jing [30] and Yang, Jing & Ping [31]; can be found behind the symmetries of the Lévy–Leblond equation, see Aizawa, Kuznetsova, Tanaka & Toppan [32]; and Tolstoy's alternative super-Poincaré symmetries [33].

Technically we enter the world of $\mathbb{Z}_2^n$-manifolds (or coloured supermanifolds) which informally are 'manifolds' for which the structure sheaf has a $\mathbb{Z}_2^n$-grading. The $n = 1$ case is just the standard theory of supermanifolds. One can view supermanifolds as a mild form of noncommutative geometry, and similarly one can view $\mathbb{Z}_2^n$-manifolds as a mild form of 'noncommutative supergeometry'. However, as almost all of the differential geometry of supermanifolds and $\mathbb{Z}_2^n$-manifolds can be treated as interpreted using the methods from classical differential geometry and algebraic geometry. Thus, one should view $\mathbb{Z}_2^n$-manifolds as a starting place for more general noncommutative supergeometries where more abstract algebraic methods are needed (see for example de Goursac [34], Grosse & Reiter [35] and Schwarz [19]).

In this paper, we will build, via a $\mathbb{Z}_2^n$-graded generalisation of the super-Poincaré algebra, a $\mathbb{Z}_2^n$-graded version of super-Minkowski space-time. The methods employed are minor modifications of the standard methods of coset spaces as applied to standard supersymmetry. The aim of this paper is to point out that a rather direct generalisation of extended supersymmetry to the setting of $\mathbb{Z}_2^n$-graded geometry exists. Our methodology is to start from a $\mathbb{Z}_2^n$-graded generalisation of a Haag-Łopuszański-Sohnius type algebra [36] and formally integrate it to construct a $\mathbb{Z}_2^n$-manifold (see for example [14] (Section IV) for the standard case). The constructions formally look very similar to the standard case of $\mathcal{N}$-extended supersymmetry. In part, this is why the starting $\mathbb{Z}_2^n$-Lie algebra was chosen as it is.

The organisation of this paper is as follows: In Section 2 we review the locally ringed space approach to $\mathbb{Z}_2^n$-manifolds and establish our conventions with spinors. In Section 3 we start from a $\mathbb{Z}_2^n$-extended Poincaré algebra, build the corresponding $\mathbb{Z}_2^n$-Minkowski space-time, and consider some basic consequences of these constructions. Due to the fact that for the $n = 2$ case some of the expressions greatly simplify, we present further details of $\mathbb{Z}_2^2$-Minkowski space-time including its canonical $\mathbb{Z}_2^2$-SUSY structure. We end this paper with some concluding remarks in Section 4.

2. Preliminaries

2.1. $\mathbb{Z}_2^n$-Manifolds and Their Basic Geometry

We bring the reader's attention to [37–40] where details of the locally ringed space approach to $\mathbb{Z}_2^n$-manifolds can be found. Prior to these works is [41], where the functor of points approach was used to define $\mathbb{Z}_2^n$-manifolds or coloured supermanifolds. We also draw the reader's attention to Marcinek [42]. We will draw upon these works heavily and not present proofs of any formal statements. Moreover, we will largely follow standard notation from the theory of supermanifolds. We restrict our attention to real structures and do not consider the complex analogues at all.

Definition 1 ([37]). *A* locally $\mathbb{Z}_2^n$-ringed space, $n \in \mathbb{N} \setminus \{0\}$, *is a pair* $X := (|X|, \mathcal{O}_X)$ *where* $|X|$ *is a second-countable Hausdorff topological space and a* $\mathcal{O}_X$ *is a sheaf of* $\mathbb{Z}_2^n$*-graded,* $\mathbb{Z}_2^n$*-commutative associative unital* $\mathbb{R}$*-algebras, such that the stalks* $\mathcal{O}_{X,p}$*,* $p \in |X|$ *are local rings.*

In this context, $\mathbb{Z}_2^n$-commutative means that any two sections $a, b \in \mathcal{O}_X(|U|)$, $|U| \subset |X|$ open, of homogeneous degree $\deg(a)$ and $\deg(b) \in \mathbb{Z}_2^n$, respectively, commute with the sign rule

$$ab = (-1)^{\langle \deg(a), \deg(b) \rangle} ba,$$

where $\langle\, , \,\rangle$ is the standard scalar product on $\mathbb{Z}_2^n$.

One should, of course, be reminded of the definition of a superspace, and indeed standard superspaces are examples of locally $\mathbb{Z}_2^n$-ringed space for $n = 1$. To pass from a general superspace to a supermanifold one needs the standard local model of $\mathbb{R}^{p|q}$. The definition of a $\mathbb{Z}_2^n$-manifold is similar to that of a supermanifold, but with some important subtle differences.

First, we need to fix a convention on how we fix the order of elements in $\mathbb{Z}_2^n$, we do this lexicographically by filling in zeros from the left and ones from the right. We will refer to this ordering as the standard ordering. For example, with this choice of ordering

$$\begin{aligned}
\mathbb{Z}_2^2 &= \{(0,0),\ (0,1),\ (1,0),\ (1,1)\}, \\
\mathbb{Z}_2^3 &= \{(0,0,0),\ (0,0,1),\ (0,1,0),\ (1,0,0),\ (0,1,1),\ (1,0,1),\ (1,1,0),\ (1,1,1)\}, \\
\mathbb{Z}_2^4 &= \{(0,0,0,0),\ (0,0,0,1),\ (0,0,1,0),\ (0,1,0,0),\ (1,0,0,0), \\
&\quad (0,0,1,1),\ (0,1,0,1),\ (0,1,1,0),\ (1,0,0,1),\ (1,0,1,0),\ (1,1,0,0), \\
&\quad (0,1,1,1),\ (1,0,1,1),\ (1,1,0,1),(1,1,1,0),\ (1,1,1,1)\}.
\end{aligned}$$

Note that slightly different conventions with the ordering have been employed in the literature. A tuple $\mathbf{q} = (q_1, q_2, \cdots, q_N)$, where $N = 2^n - 1$ provides all the information about the non-zero degree coordinates, which we collectively write as ξ. For example, specifying $\mathbf{q} = \{2, 3, 6\}$ means that we have two coordinates of degree $(0,1)$, three of degree $(1,0)$, and six of degree $(1,1)$.

Definition 2 ([37]). *A (smooth) $\mathbb{Z}_2^n$-manifold of dimension $p|\mathbf{q}$ is a locally $\mathbb{Z}_2^n$-ringed space $M := (|M|, \mathcal{O}_M)$, which is locally isomorphic to the $\mathbb{Z}_2^n$-ringed space $\mathbb{R}^{p|\mathbf{q}} := (\mathbb{R}^p, C^\infty_{\mathbb{R}^p}[[\xi]])$. Here $C^\infty_{\mathbb{R}^p}$ is the structure sheaf on the Euclidean space $\mathbb{R}^p$. Local sections of $\mathbb{R}^{p|\mathbf{q}}$ are formal power series in the $\mathbb{Z}_2^n$-graded variables ξ with smooth coefficients, i.e.,*

$$C^\infty(\mathcal{U}^p)[[\xi]] = \left\{ \sum_{\alpha \in \mathbb{N}^N}^{\infty} \xi^\alpha f_\alpha \mid f_\alpha \in C^\infty(\mathcal{U}^p) \right\},$$

for any $\mathcal{U}^p \subset \mathbb{R}^p$ open. Morphisms *between $\mathbb{Z}_2^n$-manifolds are morphisms of $\mathbb{Z}_2^n$-ringed spaces, that is, pairs $(\phi, \phi^*) : (|M|, \mathcal{O}_M) \to (|N|, \mathcal{O}_N)$ consisting of a continuous map $\phi : |M| \to |N|$ and sheaf morphism $\phi^* : \mathcal{O}_N(|V|) \to \mathcal{O}_M(\phi^{-1}(|V|))$, where $|V| \subset |N|$ is open.*

Example 1 (The local model). *The locally $\mathbb{Z}_2^n$-ringed space $\mathcal{U}^{p|\mathbf{q}} := (\mathcal{U}^p, C^\infty_{\mathcal{U}^p}[[\xi]])$, where $\mathcal{U}^p \subseteq \mathbb{R}^p$, is naturally a $\mathbb{Z}_2^n$-manifold, we refer to such $\mathbb{Z}_2^n$-manifolds as $\mathbb{Z}_2^n$-*superdomains *of dimension $p|\mathbf{q}$. We can employ (natural) coordinates (x^a, ξ^i) on any $\mathbb{Z}_2^n$-superdomain, where x^a form a coordinate system on $\mathcal{U}^p$ and the ξ^i are formal coordinates.*

Canonically associated with any $\mathbb{Z}_2^n$-graded algebra $\mathcal{A}$ is the homogeneous ideal J of $\mathcal{A}$ generated by all homogeneous elements of $\mathcal{A}$ having non-zero degree. If $f : \mathcal{A} \to \mathcal{A}'$ is a morphism of $\mathbb{Z}_2^n$ -graded algebras, then $f(J_\mathcal{A}) \subseteq J_{\mathcal{A}'}$. The J-adic topology plays a fundamental rôle in the theory of $\mathbb{Z}_2^n$-manifolds. In particular, these notions can be 'sheafified'. That is, for any $\mathbb{Z}_2^n$-manifold M, there exists an ideal sheaf $\mathcal{J}$, defined by $\mathcal{J}(|U|) = \langle f \in \mathcal{O}_M(|U|) \mid f \text{ is of non-zero degree} \rangle$. The $\mathcal{J}$-adic topology on $\mathcal{O}_M$ can then be defined in an obvious way (see [37] for details).

Many of the standard results from the theory of supermanifolds pass over to $\mathbb{Z}_2^n$-manifolds. For example, the topological space $|M|$ comes with the structure of a smooth manifold of dimension p, hence our suggestive notation. There exists a canonical projection $\epsilon : \mathcal{O}_M(|M|) \to C^\infty(|M|)$, and it can be shown that $\mathcal{J} = \ker \epsilon$.

The immediate problem with $\mathbb{Z}_2^n$-manifolds is that $\mathcal{J}$ is *not* nilpotent—for standard supermanifolds the ideal sheaf is nilpotent and this is a fundamental property that makes the theory of supermanifolds so well-behaved. However, this loss of nilpotency is replaced by the Hausdorff completeness of the $\mathcal{J}$-adic topology.

Proposition 1 ([37]). *Let M be a $\mathbb{Z}_2^n$-manifold. Then $\mathcal{O}_M$ is $\mathcal{J}$-adically Hausdorff complete as a sheaf of $\mathbb{Z}_2^n$-commutative rings, i.e., the morphism*

$$\mathcal{O}_M \to \lim_{\leftarrow k \in \mathbb{N}} \mathcal{O} \setminus \mathcal{J}^k$$

naturally induced by the filtration of $\mathcal{O}_M$ by the powers of $\mathcal{J}$ is an isomorphism.

The presence of formal power series in the coordinate rings of $\mathbb{Z}_2^n$-manifolds forces one to rely on the Hausdorff-completeness of the $\mathcal{J}$-adic topology. This completeness replaces the standard fact that on supermanifolds, functions of Grassmann odd variables are always polynomials—a result that is often used in extending results from manifolds to supermanifolds.

What makes $\mathbb{Z}_2^n$-manifolds a very workable form of noncommutative geometry is the fact that we have well-defined local models and that the structure of morphisms can (locally) be described at the level of coordinates. That is, we have the chart theorem ([37] (Theorem 7.10)) that basically says that morphisms between the local coordinates can be extended uniquely to morphisms of locally $\mathbb{Z}_2^n$-ringed spaces.

Let M be a $\mathbb{Z}_2^n$-manifold, then an *open $\mathbb{Z}_2^n$-submanifold* of M is a $\mathbb{Z}_2^n$-manifold defined as

$$U := (|U|, \mathcal{O}_{M,|U|}),$$

where $|U| \subseteq |M|$ is an open subset. From the definition of a $\mathbb{Z}_2^n$-manifold, we know that for 'small enough' $|U|$ we have an isomorphism

$$(\varphi, \varphi^*) : U \xrightarrow{\sim} \mathcal{U}^{p|\mathbf{q}} .$$

This local identification allows us to construct a local coordinate system. Dropping explicit reference to the local isomorphism, and so via minor abuse of notation, we write the local coordinates as $x^A = (x^a, \xi^i)$. The commutation rules for these coordinates is determined by the scalar product on $\mathbb{Z}_2^n$ inherited by the standard (Euclidean) scalar product on $\mathbb{R}^n$, i.e.,

$$x^A x^B = (-1)^{\langle \deg(A), \deg(B) \rangle} x^B x^A .$$

Changes of coordinates, i.e., different choices of the local isomorphisms, can be written (using standard abuses of notation) as $x^{A'} = x^{A'}(x)$, where we understand the changes of coordinates to respect the $\mathbb{Z}_2^n$-grading. The subtlety here is that these changes of coordinates need not be polynomial in coordinates of non-zero degree, generically we have a formal power series. We will refer to global sections of the structure sheaf of a $\mathbb{Z}_2^n$-manifold as functions and employ the standard notation $C^\infty(M) := \mathcal{O}_M(|M|)$.

Remark 1. *It is clear that, much like standard supermanifolds, a $\mathbb{Z}_2^n$-manifold is not 'simply' a collection of topological points. The only topological points of M are the points of the reduced manifold $|M|$. To regain some classical intuition one can employ Grothendieck's functor of points. However, we will not need to do so in this work. Or rather, we avoid using the functor of points and work formally in the rare instances we need to—principally in defining group structures on $\mathbb{Z}_2^n$-manifolds.*

Example 2 ($\mathbb{Z}_2^n$-graded Cartesian spaces). *Directly from the definition, it is clear that $\mathbb{R}^{p|\mathbf{q}} := (\mathbb{R}^p, C^\infty(\mathbb{R}^p)[[\xi]])$ is a $\mathbb{Z}_2^n$-manifold. Global coordinates (x^a, ξ^i) can be employed, where the coordinate map is just the identity. In this paper, we will only meet $\mathbb{Z}_2^n$-manifolds that are globally isomorphic to $\mathbb{R}^{p|\mathbf{q}}$ for the appropriate p and* $\mathbf{q}$.

Example 3 (Manifolds and supermanifolds). *Rather trivially, smooth manifolds are $\mathbb{Z}_2^0$-manifolds. Similarly, supermanifolds are $\mathbb{Z}_2^1$-manifolds. Note that for supermanifolds, Grassmann odd coordinates are nilpotent, and thus functions are polynomial in odd variables.*

Example 4 (Double vector bundles). *A double vector bundle D is manifold that has a pair of commuting actions of the multiplicative monoid of reals: $\mathrm{h}_i : \mathbb{R} \times D \to D$ $(i = 1, 2)$. This amounts to being able to find homogeneous (with respect to the natural bi-weight) local coordinates (we suppress indices)*

$$(\underbrace{x}_{(0,0)}, \underbrace{y}_{(0,1)}, \underbrace{z}_{(1,0)}, \underbrace{w}_{(1,1)}),$$

see [43,44]. Admissible changes of homogeneous local coordinates are of the form

$$x' = x'(x), \qquad y' = yS(x), \qquad z' = zT(x), \qquad w' = wU(x) + zyV(x).$$

Given any double vector bundle, one can canonically associate, in a functorial way, a $\mathbb{Z}_2^2$-manifold by declaring the commutation rules for the local coordinates to be determined by the bi-weight. The changes of coordinates remain as given above. For more details consult [37] (Section 6), where the resulting $\mathbb{Z}_2^2$-manifold is denoted ΠD.

The notion of a vector field on a $\mathbb{Z}_2^n$-manifold is naturally given as a $\mathbb{Z}_2^n$-derivation acting on sections of the structure sheaf, i.e., acting on (global) functions. A (homogeneous) vector field is understood as a linear map $X : C^\infty(M) \to C^\infty(M)$, that satisfies the $\mathbb{Z}_2^n$-graded Leibniz rule

$$X(fg) = X(f)g + (-1)^{\langle \deg(X), \deg(f) \rangle} fX(g),$$

for any f and $g \in C^\infty(M)$. It is a simple exercise to see that we have a (left) module structure over $C^\infty(M)$. We denote the module of vector fields as $\mathsf{Vect}(M)$. Any vector field can be 'localised' (see [39] (Lemma 2.2)) in the sense that given $|U| \subset |M|$ there always exists a unique derivation

$$X|_{|U|} : \mathcal{O}_M(|U|) \to \mathcal{O}_M(|U|),$$

such that $X(f)|_{|U|} = X|_{|U|}(f_{|U|})$. Because of this local property, it is clear that one has a sheaf of $\mathcal{O}_M$-modules formed by the local derivations—this defines the *tangent sheaf* of a $\mathbb{Z}_2^n$-manifold (see [39] (Definition 5)). Moreover, this sheaf is locally free and so admits a local basis. The upshot of these considerations is that we can always locally write a vector field as one would on a manifold using a choice of local coordinates, i.e.,

$$X = X^A(x) \frac{\partial}{\partial x^A},$$

where the partial derivatives are defined as standard for the coordinates of degree zero and are defined algebraically for the coordinates of degree non-zero. In the above expression, we drop the explicit reference to the required restriction as is common in standard differential geometry. As for the case of standard supermanifolds, the order of taking partial derivatives matters, but only up to signs,

$$\frac{\partial}{\partial x^A} \frac{\partial}{\partial x^B} = (-1)^{\langle \deg(A), \deg(B) \rangle} \frac{\partial}{\partial x^B} \frac{\partial}{\partial x^A}.$$

Under the commutator

$$[X, Y] := X \circ Y - (-1)^{\langle \deg(X), \deg(Y) \rangle} Y \circ X,$$

$\mathsf{Vect}(M)$ becomes a $\mathbb{Z}_2^n$-Lie algebra (see [45–47]). The grading and symmetry of the Lie bracket are clear, and one can check that the Jacobi identity (written here in Loday–Leibniz form)

$$[X, [Y, Z]] = [[X, Y], Z] + (-1)^{\langle \deg(X), \deg(Y) \rangle} [Y, [X, Z]],$$

holds.

Any geometric object—say a function, vector field or tensor field—on a $\mathbb{Z}_2^n$-manifold carries a natural $\mathbb{Z}_2^n$-grading. We will say that a homogeneous geometric object is even if the total degree is even, and odd if the total degree is odd. Note that this does *not* determine the sign factors in general expressions.

Remark 2. *It is known that the structure sheaf of a $\mathbb{Z}_2^n$-manifold is a nuclear Fréchet sheaf of $\mathbb{Z}_2^n$-graded $\mathbb{Z}_2^n$-commutative associative unital algebras. Moreover, like real manifolds and supermanifolds, the structure* *sheaf can be fully recovered from the algebra of global functions. For example, it can be shown that two $\mathbb{Z}_2^n$-manifolds are diffeomorphic if and only if their algebras of global functions are isomorphic. Let us also mention that the category of $\mathbb{Z}_2^n$-manifolds admits finite products—an essential fact in any rigours treatment of $\mathbb{Z}_2^n$-Lie groups. None of this will feature in the later sections of this paper. For details, the reader should consult Bruce & Poncin [48,49].*

2.2. A Toy $\mathbb{Z}_2^2$-Superspace

Before giving a more careful description of $\mathbb{Z}_2^n$-Minkowski space-time, we proceed with a simpler toy superspace in which we have an underlying one-dimensional space-time. This will allow us to sketch the ideas without the clutter of many indices and in particular spinor indices. Let us consider the $\mathbb{Z}_2^2$-graded Lie algebra with generators P, Q_1, Q_2 and Z of $\mathbb{Z}_2^2$-degrees $(0,0)$, $(0,1)$, $(1,0)$ and $(1,1)$, respectively, given by

$$[Q_1, Q_1] = [Q_2, Q_2] = \frac{1}{2}P, \qquad [Q_2, Q_1] = \frac{1}{2}Z,$$

where all other Lie brackets vanish. Up to conventions, this is essentially two copies of supersymmetry in one-dimension and a central extension. However, one must take care as we have $\mathbb{Z}_2^2$-graded Lie brackets, the first two brackets are 'anticommutators' while the third is a 'commutator' in 'non-graded language'. We want to realise these generators as vector fields on some $\mathbb{Z}_2^2$-manifold, or more correctly we are looking at the associated $\mathbb{Z}_2^2$-Lie group and the left-invariant vector fields. In this case, we can more or less see by eye, i.e., without resorting to coset methods, what the $\mathbb{Z}_2^2$-manifold and left-invariant vector fields should be. With this in mind, consider the $\mathbb{Z}_2^2$-manifold, which we denote as $\mathcal{M}_1^{[2]} \simeq \mathbb{R}^{1|1,1,1}$, that comes with global coordinates

$$(\underbrace{t}_{(0,0)}, \underbrace{\theta^1}_{(0,1)}, \underbrace{\theta^2}_{(1,0)}, \underbrace{z}_{(1,1)}),$$

where again we have used the standard ordering. In comparison with the case of standard superspace, note the unusual commutation relations

$$\theta^1\theta^2 = +\theta^2\theta^1, \qquad z\theta^1 = -\theta^1 z, \qquad z\theta^2 = -\theta^2 z\,.$$

The reader can easily check that the vector fields

$$P = \frac{\partial}{\partial t}, \qquad Q_1 = \frac{\partial}{\partial \theta^1} + \frac{1}{4}\theta^1\frac{\partial}{\partial t} + \frac{1}{4}\theta^2\frac{\partial}{\partial z}, \qquad Q_2 = \frac{\partial}{\partial \theta^2} + \frac{1}{4}\theta^2\frac{\partial}{\partial t} - \frac{1}{4}\theta^1\frac{\partial}{\partial z}, \qquad Z = \frac{\partial}{\partial z},$$

satisfy the given $\mathbb{Z}_2^2$-Lie algebra under standard $\mathbb{Z}_2^2$-commutators. From these vector fields, we can then 'read off' the $(d=1)$ $\mathbb{Z}_2^2$-supersymmetry transformations

$$t \mapsto t + \frac{1}{4}(\epsilon^1\theta^1 + \epsilon^2\theta^2), \qquad \theta^1 \mapsto \theta^1 + \epsilon^1, \qquad \theta^2 \mapsto \theta^2 + \epsilon^2, \qquad z \mapsto z + \frac{1}{4}(\epsilon^1\theta^2 - \epsilon^2\theta^1),$$

where ϵ^1 and ϵ^2 are $\mathbb{Z}_2^2$-graded parameters of degree $(0,1)$ and $(1,0)$, respectively. It is a direct calculation to derive the covariant derivatives, which can easily be shown to be given by:

$$D_1 = \frac{\partial}{\partial \theta^1} - \frac{1}{4}\theta^1\frac{\partial}{\partial t} - \frac{1}{4}\theta^2\frac{\partial}{\partial z}, \qquad D_2 = \frac{\partial}{\partial \theta^2} - \frac{1}{4}\theta^2\frac{\partial}{\partial t} + \frac{1}{4}\theta^1\frac{\partial}{\partial z}\,.$$

As fully expected, we have $[D_1, D_1] = [D_2, D_2] = -\frac{1}{2}P$ and $[D_2, D_1] = -\frac{1}{2}Z$, and all the other possible $\mathbb{Z}_2^2$-commutators vanish.

In order to construct more 'realistic' $\mathbb{Z}_2^n$-superspaces we need to take the underlying space to be Minkowski space-time and promote the odd coordinates to be spinors. We will discuss the specific example of $d = 4$ $\mathbb{Z}_2^2$-Minkowski space-time in Section 3.6.

To understand the 'field content' encoded here let us first remove the central charge, i.e., we set $z = 0$, as this will simplify things—otherwise we will have to deal with a formal power series in z. A (scalar) superfield of degree $(0,0)$ is then of the form

$$\Phi(t,\theta) = q(t) + \theta^1\chi_1(t) + \theta^2\chi_2(t) + 4\theta^2\theta^1 b(t),$$

where we have included a numerical factor for convenience. Superfields require the use of internal Homs in order to define them rigorously, i.e., we need to consider some 'external parametrisations' so that the degrees can be properly attributed. However, we will simply proceed formally as is common in physics. Specifically, the degrees of the components are

$$\deg(q) = (0,0), \qquad \deg(\chi_1) = (0,1), \qquad \deg(\chi_2) = (1,0), \qquad \deg(b) = (1,1).$$

Clearly, the induced commutation rules between these components are different to what is encountered in standard supersymmetric mechanics, see for example Bellucci & Krivonos [50]. In particular, we have odd degrees of freedom that commute with each other and anticommute with one of the even degrees of freedom. At this stage the reader should be reminded of parastatistics in the sense of Green [26]. We define the component form of the $\mathbb{Z}_2^2$-supersymmetry transformations via $\delta\Phi = (\epsilon^1 Q_1 + \epsilon^2 Q_2)\Phi$. It is a straightforward exercise to deduce the following:

$$\delta q(t) = \epsilon^1\chi_1(t) + \epsilon^2\chi_2(t), \qquad \delta\chi_1(t) = -\frac{1}{4}\left(\epsilon^1\dot{q}(t) - \epsilon^2 b(t)\right),$$

$$\delta\chi_2(t) = -\frac{1}{4}\left(\epsilon^2\dot{q}(t) - \epsilon^1 b(t)\right), \qquad \delta b(t) = -\left(\epsilon^1\dot{\chi}_2(t) + \epsilon^2\dot{\chi}_1(t)\right),$$

where 'dot' means the derivative with respect to t. These transformations should be compared to the (off-shell) $d = 1$ $\mathcal{N} = 2$ supersymmetry transformations, see for example [50] (Section 2.2). The reader should note that we work in a manifestly real setting and so no factors of the imaginary unit appear in any of our mathematical expressions.

Via dimensional reduction, for $d = 0$ the $\mathbb{Z}_2^2$-graded Lie algebra we should consider is simply

$$[Q_1, Q_1] = [Q_2, Q_2] = 0, \qquad [Q_2, Q_1] = \frac{1}{2}Z,$$

where the $\mathbb{Z}_2^2$-degrees are assigned as previously. Thus, $\mathcal{M}_0^{[2]} \simeq \mathbb{R}^{0|1,1,1}$ comes with vector fields

$$Q_1 = \frac{\partial}{\partial\theta^1} + \frac{1}{4}\theta^2\frac{\partial}{\partial z}, \qquad Q_2 = \frac{\partial}{\partial\theta^2} - \frac{1}{4}\theta^1\frac{\partial}{\partial z}, \qquad Z = \frac{\partial}{\partial z},$$

that realise the $d = 0$, $\mathbb{Z}_2^2$-supersymmetry algebra. Note that in this case Q_1 and Q_2 are homological, i.e., they square to zero. We do not obtain a bi-complex in this way as the two homological vector fields do not commute, unless the central charge $Z = 0$.

2.3. *Majorana Spinors*

We take the Minkowski metric to be $\eta = \mathrm{diag}(-1, +1, +1, +1)$ and denote 3 + 1 dimensional Minkowski space-time as $\mathcal{M}_4 := (\mathbb{R}^4, \eta)$. That is, we will use the relativists' conventions. We wish to work with *real* objects rather than complex ones. As such we will choose the following real Majorana representation of the Clifford algebra $Cl(3,1)$

$$\gamma^0 = \begin{pmatrix} 0 & +1 & 0 & 0 \\ -1 & 0 & 0 & 0 \\ 0 & 0 & 0 & -1 \\ 0 & 0 & +1 & 0 \end{pmatrix}, \qquad \gamma^1 = \begin{pmatrix} 0 & +1 & 0 & 0 \\ +1 & 0 & 0 & 0 \\ 0 & 0 & 0 & +1 \\ 0 & 0 & +1 & 0 \end{pmatrix},$$

$$\gamma^2 = \begin{pmatrix} +1 & 0 & 0 & 0 \\ 0 & -1 & 0 & 0 \\ 0 & 0 & +1 & 0 \\ 0 & 0 & 0 & -1 \end{pmatrix}, \qquad \gamma^3 = \begin{pmatrix} 0 & 0 & 0 & +1 \\ 0 & 0 & -1 & 0 \\ 0 & -1 & 0 & 0 \\ +1 & 0 & 0 & 0 \end{pmatrix}.$$

These gamma matrices—by definition—satisfy the Clifford-Dirac relation

$$\{\gamma^\mu, \gamma^\nu\} := \gamma^\mu\gamma^\nu + \gamma^\nu\gamma^\mu = 2\eta^{\mu\nu}\mathbb{1}.$$

These 4×4 real matrices act on $\mathbb{R}^4$ in the usual way. With this in mind, we adopt the convention of spin indices (and for matrices more generally) of $(\gamma^\mu)_\alpha{}^\beta$. We then define Majorana spinors as the "things" that these gamma matrices act on. We thus write $\mathbb{M} \simeq \mathbb{R}^4$ for the space of all Majorana spinors. We will conventionally choose the components if a Majorana spinor to have lower spin indices, and so we write u_α for a Majorana spinor. We know that Spin$(3,1)$ acts on Majorana spinors in a linear way, i.e.,

$$u_\alpha \mapsto u'_\alpha = S_\alpha{}^\beta u_\beta.$$

Given an infinitesimal Lorentz transformation

$$\Lambda_\mu{}^\nu = \delta_\mu{}^\nu + \omega_\mu{}^\nu,$$

where $\omega_{\mu\nu} = -\omega_{\nu\mu}$, we know that a Majorana spinor transforms as

$$u'_\alpha = u_\alpha + \frac{1}{4}\omega_{\mu\nu}(\gamma^{\mu\nu})_\alpha{}^\beta u_\beta,$$

where $\gamma^{\mu\nu} := \gamma^{[\mu}\gamma^{\nu]} = \frac{1}{2}(\gamma^\mu\gamma^\nu - \gamma^\nu\gamma^\mu)$. In particular, $\Sigma^{\mu\nu} = \frac{1}{2}\gamma^{\mu\nu}$ give a representation of the Lorentz algebra

$$[\Sigma^{\mu\nu}, \Sigma^{\rho\sigma}] = \eta^{\nu\rho}\Sigma^{\mu\sigma} - \eta^{\mu\rho}\Sigma^{\nu\sigma} + \eta^{\mu\sigma}\Sigma^{\nu\rho} - \eta^{\nu\sigma}\Sigma^{\mu\rho}.$$

By taking the exponential we obtain a finite Lorentz transformation (which by construction is in the proper Lorentz group)

$$u'_\alpha = \exp(\frac{1}{4}\omega_{\mu\nu}\gamma^{\mu\nu})_\alpha{}^\beta u_\beta.$$

As defined here, the Majorana spinor as commuting objects: they are described by the coordinates on $\mathbb{M}$ considered as a linear manifold. In order to have anticommuting Majorana spinors—which are more common in physics—we need to employ the parity reversion functor. Thus, we consider anticommuting Majorana spinors to be described by the coordinates on $\Pi\mathbb{M}$. The manifold $\mathbb{M}$ (and the supermanifold $\Pi\mathbb{M}$) comes equipped with the charge conjugation tensor—this operator exchanges particles and antiparticles. The defining property is

$$C\gamma^\mu C^{-1} = -(\gamma^\mu)^{\mathrm{t}}.$$

Or written more explicitly

$$C^{\alpha\gamma}(\gamma^\mu)_\gamma{}^\delta C_{\delta\beta} = -(\gamma^\mu)^\alpha{}_\beta,$$

Thus, considered as a matrix, in our chosen representation, $C = -\gamma^0$. To set some useful notation, we define $(C\gamma^\mu)^{\alpha\beta} := C^{\alpha\delta}(\gamma^\mu)_\delta{}^\beta$ which a direct computation shows is symmetric in α and β.

Remark 3. *In practice, we will not need the explicit representation of $Cl(3,1)$ that we have chosen. The key point is that we have a completely real representation. Furthermore, real representations are only possible in the 'mostly pluses' convention for the Minkowski metric.*

Remark 4. Dirac spinors *are understood as the complexification of the Majorana spinors, that is we define the space of Dirac spinors as* $\mathbb{D} := \mathbb{M} \otimes_{\mathbb{R}} \mathbb{C} \simeq \mathbb{C}^4$. Weyl spinors *are then defined in terms of a decomposition of the Dirac spinors into a direct sum* $\mathbb{D} = \mathbb{W}_+ \oplus \mathbb{W}_-$ *defined by chirality. More explicitly, we first define* $\gamma_5 := \gamma_0\gamma_1\gamma_2\gamma_3$ *(where we have used the Minkowski metric to lower the indices). The decomposition is then defined in terms of the eigenvalues of* γ_5*. That is,* $u \in \mathbb{W}_\pm$ *if* $\gamma_5 u = \pm \mathrm{i}\, u$*. Via this decomposition, we see that* $\mathbb{W}_\pm \simeq \mathbb{C}^2$*. We will only consider Majorana spinors in this paper.*

2.4. $\mathbb{Z}_2^n$-Graded Majorana Spinors

We can construct $\mathbb{Z}_2^n$-graded Majorana spinors quite directly. From the Batchelor–Gawędzki theorem for $\mathbb{Z}_2^n$-manifolds [38] (Theorem 3.2) we know that any (real) $\mathbb{Z}_2^n$-manifold is noncanonically isomorphic to a $\mathbb{Z}_2^n \setminus \{\underline{0}\}$-graded vector bundle. In particular, any $\mathbb{Z}_2^n \setminus \{\underline{0}\}$-graded manifold is noncanonically isomorphic to the direct sum of the appropriate number of vector spaces—where we assign the appropriate $\mathbb{Z}_2^n \setminus \{\underline{0}\}$-grading to the linear coordinates on each vector space. Each vector space is then considered as a linear $\mathbb{Z}_2^n$-manifold.

Example 5. *Any* $\mathbb{Z}_2^2 \setminus \{\underline{0}\}$*-graded manifold is noncanonically isomorphic to a direct sum of vector spaces* $E = E_{(0,1)} \oplus E_{(1,0)} \oplus E_{(1,1)}$*, where the labels indicate the assignment of the degree to the coordinates on each vector space. Note that as coordinate changes are linear, the assignment of the degree is well defined.*

Then to define the space of $\mathbb{Z}_2^n$-graded Majorana spinors we simply take the appropriate number of copies of $\mathbb{M}$, assign the degrees to the linear coordinates, and then take the direct sum. To set notation, we denote the resulting $\mathbb{Z}_2^n$-manifold as $\mathbb{M}^{[n]}$.

Example 6. *The* $n = 1$ *case is the classical case of anticommuting spinors,* $\mathbb{M}^{[1]} = \Pi\mathbb{M}$.

Example 7. *The* $n = 2$ *case is as follows:* $\mathbb{M}^{[2]} = \mathbb{M}_{(0,1)} \oplus \mathbb{M}_{(1,0)} \oplus \mathbb{M}_{(1,1)}$*. In terms of coordinates, we have three 'species' of Majorana spinors* $(\psi_\alpha, \theta_\beta, u_\gamma)$ *of degree* $(0,1)$*,* $(1,0)$ *and* $(1,1)$ *respectively. The self-commutation rules are*

$$\psi_\alpha\psi_\beta = -\psi_\beta\psi_\alpha, \qquad \theta_\alpha\theta_\beta = -\theta_\beta\theta_\alpha, \qquad u_\alpha u_\beta = +u_\beta u_\alpha,$$

while across the 'species' we have

$$\psi_\alpha\theta_\beta = +\theta_\beta\psi, \qquad \psi_\alpha u_\beta = -u_\beta\psi_\alpha, \qquad \theta_\alpha u_\beta = -u_\beta\theta_\alpha.$$

In the next section, we will be interested in *odd Majorana spinors*. Note that we have the natural decomposition

$$\mathbb{M}^{[n]} = \mathbb{M}_0^{[n]} \oplus \mathbb{M}_1^{[n]},$$

into even and odd spinors defined by the total degree. We will exclusively be interested in $\mathbb{M}_1^{[n]}$ in the rest of this paper. It is a simple counting exercise to show that $\mathbb{M}_1^{[n]}$ has $4 \times 2^{n-1}$ coordinates.

3. $\mathbb{Z}_2^n$-Extended Supersymmetry

3.1. A $\mathbb{Z}_2^n$-Extended Poincaré Algebra

We extend the standard Poincaré algebra by adjoining $\mathbb{Z}_2^n$-graded odd Majorana spinor generators Q_I^α and $\mathbb{Z}_2^n$-graded even central charges Z_{IJ}. Experimentally we know that the underlying real manifold of the resulting $\mathbb{Z}_2^n$-manifold must, locally at least, be Minkowski space-time. Thus, the resulting algebra must contain the Poincaré algebra as the degree zero component. The $\mathbb{Z}_2^n$-graded Lie algebra we consider is a generalisation of the $N = 2^{n-1}$ Haag-Łopuszański-Sohnius type algebra [36], i.e., we allow a central extension:

$$\begin{aligned}
&[P_\mu, J^{\lambda\sigma}] = (\delta_\mu^{\ \lambda}\eta^{\sigma\rho} - \delta_\mu^{\ \sigma}\eta^{\lambda\rho})P_\rho, \\
&[J^{\mu\nu}, J^{\rho\sigma}] = \eta^{\nu\rho}J^{\mu\sigma} - \eta^{\mu\rho}J^{\nu\sigma} - \eta^{\nu\sigma}J^{\mu\rho} + \eta^{\mu\sigma}J^{\nu\rho}, \\
&[J^{\mu\nu}, Q_I^\alpha] = -\frac{1}{4}Q_I^\beta(\gamma^{\mu\nu})_\beta^{\ \alpha}, \\
&[Q_I^\alpha, Q_J^\beta] = \frac{1}{2}\delta_{IJ}(C\gamma^\mu)^{\alpha\beta}P_\mu + C^{\alpha\beta}Z_{IJ}.
\end{aligned}$$

All other Lie brackets are zero. In particular, Z is central. Here P and J are the generators of translations and rotations on Minkowski space-time. To be clear, the degree of the generators is as follows:

$$\begin{aligned}
&\deg(P) = \deg(J) = \underline{0}, && \text{(even)} \\
&\deg(Q_I) = \deg(I) \in \mathbb{Z}_2^n \setminus \{\underline{0}\}, && \text{(odd)} \\
&\deg(Z_{IJ}) = \deg(I) + \deg(J) \in \mathbb{Z}_2^n \setminus \{\underline{0}\}. && \text{(even)}
\end{aligned}$$

From the symmetry of $\mathbb{Z}_2^n$-Lie brackets and the fact that $C^{\alpha\beta} = -C^{\beta\alpha}$, it is clear that $Z_{IJ} = (-1)^{\langle\deg(I),\deg(J)\rangle} Z_{JI}$ for $I \neq J$, and that $Z_{II} = -Z_{II} = 0$. Thus, as Z_{IJ} and Z_{JI} are *not* linearly independent as elements of the $\mathbb{Z}_2^n$-graded Lie algebra.

To see that we do indeed have a $\mathbb{Z}_2^n$-Lie algebra we need to establish that the necessary Jacobi identities are satisfied. First, note that all possible Jacobi identities involving just one Q_I^α hold as this situation essentially reduces to the classical $\mathcal{N} = 1$ super-Poincaré algebra.

The Jacobi identities involving Q_I^α and Q_J^β require checking. Any nested commutators involving two Q and one P are identically zero due to $[Q, Q] = P + Z$ and $[Q, P] = [Q, Z] = 0$ (we suppress the indices and coefficients). The only non-trivial identity to check is

$$[Q_I^\alpha, [J^{\mu\nu}, Q_J^\beta]] \stackrel{?}{=} [[Q_I^\alpha, J^{\mu\nu}], Q_J^\beta] + [J^{\mu\nu}, [Q_I^\alpha, Q_J^\beta]].$$

If $I = J$ then we are back to the standard $N = 1$ super-Poincaré algebra and there is nothing to check. So we assume $I \neq J$. Directly from the algebra we have

$$[Q_I^\alpha, [J^{\mu\nu}, Q_J^\beta]] = -\frac{1}{4}C^{\alpha\gamma}(\gamma^{\mu\nu})_\gamma^{\ \beta}Z_{IJ} = -\frac{1}{4}(C\gamma^{\mu\nu})^{\alpha\beta}Z_{IJ}.$$

From the other side

$$\begin{aligned}
[[Q_I^\alpha, J^{\mu\nu}], Q_J^\beta] + [J^{\mu\nu}, [Q_I^\alpha, Q_J^\beta]] &= -[[J^{\mu\nu}, Q_I^\alpha], Q_J^\beta] + [J^{\mu\nu}, C^{\alpha\beta}Z_{IJ}] \\
&= \frac{1}{4}C^{\gamma\beta}(\gamma^{\mu\nu})_\gamma^{\ \alpha}Z_{IJ} \\
&= -\frac{1}{4}(C\gamma^{\mu\nu})^{\beta\alpha}Z_{IJ},
\end{aligned}$$

where we have used $C^{\alpha\beta} = -C^{\beta\alpha}$. Then as $(C\gamma^{\mu\nu})^{\beta\alpha} = (C\gamma^{\mu\nu})^{\alpha\beta}$ we obtain the desired result.

The Jacobi identity involving Q_I^α, Q_J^β and Q_K^γ is trivially satisfied as $[Q, Q] = P + Z$ and $[Q, P] = [Q, Z] = 0$ implies that $[Q, [Q, Q]] = 0$. Thus, the $\mathbb{Z}_2^n$-super translation algebra, i.e, the part not involving the Lorentz generator $J^{\mu\nu}$, is two-step nilpotent. This is not unexpected given the classical case of the $\mathcal{N} = 1$ super-Poincaré algebra. It is important to note that, by construction, the Poincaré algebra is the degree $\underline{0} = (0, 0, \ldots, 0)$ part of the $\mathbb{Z}_2^n$-extended Poincaré algebra and not just the even part.

Remark 5. *The structures we study are different from the Poincaré parasupergroup of Beckers & Debergh [51] (also see Nikitin & Galkin [52] where central charges were included) and Jarvis [53], though we will later comment on the similarities with parasupersymmetry in Section 3.5. In particular, the (extended)* Poincaré parasuperalgebra *is defined in terms of a double commutator for the spinor-valued generators. The structures we study are $\mathbb{Z}_2^n$-Lie algebras.*

Example 8. *If $n = 1$ then we recover (up to conventions) the standard $\mathcal{N} = 1$ super-Poincaré algebra. Note that we cannot have a non-vanishing central extension in this case.*

Example 9. *Less obvious than the previous example is the case of $n = 2$. On the face of it, this looks like the $\mathcal{N} = 2$ super-Poincaré algebra with a central charge. However, the subtlety is in the assignment of the degree, which are as follows:*

$$\deg(P) = \deg(J) = (0,0), \qquad \deg(Z_{12}) = (1,1),$$
$$\deg(Q_1) = (0,1), \qquad \deg(Q_2) = (1,0).$$

Example 10. *More complicated again—though this example highlights the general features—is the case of $n = 3$. Forgetting the obvious assignment of degree to the generators of the Poincaré algebra, we have*

$$\deg(Q_1) = (0,0,1), \qquad \deg(Z_{12}) = (0,1,1),$$
$$\deg(Q_2) = (0,1,0), \qquad \deg(Z_{13}) = (1,0,1),$$
$$\deg(Q_3) = (1,0,0), \qquad \deg(Z_{14}) = (1,1,0),$$
$$\deg(Q_4) = (1,1,1), \qquad \deg(Z_{23}) = (1,1,0),$$
$$\deg(Z_{24}) = (1,0,1),$$
$$\deg(Z_{34}) = (0,1,1).$$

Remark 6. *We could also include a further modification to the extended Poincaré algebra by including a further term,*

$$[Q_I^\alpha, Q_J^\beta] = \frac{1}{2}\delta_{IJ}(C\gamma^\mu)^{\alpha\beta}P_\mu + C^{\alpha\beta}Z_{IJ} + (C\gamma_5)^{\alpha\beta}Y_{IJ},$$

where $\gamma_5 := \gamma_0\gamma_1\gamma_2\gamma_3$. It was shown by Witten & Olive [54] that in the standard case of $\mathcal{N} \geq 2$ extended supersymmetry that Z and Y have the interpretation as electric *and* magnetic charges *respectively. Such central charges can arise as the boundary terms in supersymmetric field theories. One can think of Z in a more general context as colour charge, or even as a more exotic force charge. In the case of the $\mathbb{Z}_2^n$-extended Poincaré algebra, we have a further problem with interpreting this charge due to the assignment of non-trivial $\mathbb{Z}_2^n$-degree. In $(1+1)$ dimensions, a related supersymmetry algebra where only Y appears in the anticommutator of the supersymmetry generators was studied by Duplij, Soroka & Soroka [55].*

3.2. Direct Consequences of the Algebra

The standard consequences of supersymmetry upon a quantum theory hold in this higher graded setting. For immediate simplicity let us drop the central term (not that this actually makes much difference). Let us suppose that the $\mathbb{Z}_2^n$-super translation algebra (minus central term) can be realised in terms of Hermitian operators acting on some Hilbert space—the Hilbert space of states of some theory. We stress that at this point, it is an assumption that such representations exists. Then the algebraic structure here implies the following:

1. Positivity of energy: a direct computation shows $E := P_0 = \frac{1}{2}\sum_\alpha [Q_I^\alpha, Q_I^\alpha]$ (no sum over I). Then, passing to the representation as Hermitian operators allow us to write $\hat{E} = \sum_\alpha (Q_I^\alpha)^\dagger Q_I^\alpha$ and thus for any state
$$\langle\psi|\hat{E}|\psi\rangle = \langle\psi|\sum_\alpha (Q_I^\alpha)^\dagger Q_I^\alpha|\psi\rangle = ||\sum_\alpha Q_I^\alpha|\psi\rangle||^2 \geq 0.$$
2. Irreducible representations of supersymmetry carry the same value of $P^\mu P_\mu = -m^2$: this follows from $[P, Q] = 0$, which implies that P^2 is a Casimir.
3. The spin of each state in a multiplet varies in steps of 1/2: this follows from $[Q, J] \sim Q$.

Thus, there is no difference between the standard super and $\mathbb{Z}_2^n$-super case with regards to the positivity of energy, Casimirs and the spin of states in each multiplet. However, the notion of BPS states seems more complicated, see Section 3.6.

3.3. $\mathbb{Z}_2^n$-Minkowski Space-Time

Armed with the theory of $\mathbb{Z}_2^n$-manifolds, we can more or less follow the standard methods from supersymmetry to construct the $\mathbb{Z}_2^n$-Lie group associated with the $\mathbb{Z}_2^n$-super translation algebra. That is, we can exponentiate the $\mathbb{Z}_2^n$-Lie algebra to obtain a $\mathbb{Z}_2^n$-Lie group. To make rigorous mathematical sense of all these constructions, one needs the functor of points—we will avoid a discussion of this here and work formally when needed.

With the above comments in mind, we take two even elements of the $\mathbb{Z}_2^n$-super translation algebra
$$A = x^\mu P_\mu + \theta_\alpha^I Q_I^\alpha + \frac{1}{2} z^{JK} Z_{KJ}, \qquad B = x'^\mu P_\mu + \theta'^I_\alpha Q_I^\alpha + \frac{1}{2} z'^{JK} Z_{KJ},$$
where the factor of one half is included to remove the overcounting—we understand z and Z to be graded symmetric. Then using the Campbell–Baker–Hausdorff formula we obtain
$$\begin{aligned} A \circ B = &\left(x^\mu + x'^\mu + \frac{1}{2}\theta'^I_\beta \theta^J_\alpha \delta_{JI}(C\gamma^\mu)^{\alpha\beta}\right) P_\mu \\ &+ (\theta^I_\alpha + \theta'^I_\alpha) Q^\alpha_I \\ &+ \frac{1}{2}\left(z^{JK} + z'^{JK} + \theta'^{(K}_{\beta}\,\theta^{J)}_{\alpha} C^{\alpha\beta}\right) Z_{JK}, \end{aligned} \tag{2}$$
where $\theta'^{(K}_{\beta}\,\theta^{J)}_{\alpha} := \frac{1}{2}\left(\theta'^K_\beta \theta^J_\alpha + (-1)^{\langle \deg(K), \deg(J)\rangle}\, \theta'^J_\beta \theta^K_\alpha\right)$.

Definition 3. *The centrally extended $\mathbb{Z}_2^n$-Minkowski space-time is the $\mathbb{Z}_2^n$-manifold, which we will denote as $\mathcal{M}_4^{[n]}$ that comes equipped with privileged global coordinates*
$$(x^\mu, \theta^I_\alpha, z^{JK}),$$
where via symmetry we can insist on $J < K$, and the $\mathbb{Z}_2^n$-Lie group structure defined via (2). By construction, x^μ is a Lorentz vector, θ^I_α are Majorana spinors and z_{IJ} are Lorentz scalars. The supersymmetry transformations on centrally extended $\mathbb{Z}_2^n$-Minkowski space-time are given by
$$x^\mu \mapsto x^\mu + \frac{1}{4}\epsilon^I_\beta \theta^J_\alpha \delta_{JI}(C\gamma^\mu)^{\alpha\beta}, \tag{3}$$
$$\theta^I_\alpha \mapsto \theta^I_\alpha + \epsilon^I_\alpha, \tag{4}$$
$$z^{JK} \mapsto z^{JK} + \frac{1}{2}\epsilon^{(K}_{\beta}\,\theta^{J)}_{\alpha} C^{\alpha\beta}, \tag{5}$$
where ϵ^I_α is a Majorana spinor-valued parameter.

To be explicit, the degrees of the coordinates are as follows:

$$\begin{aligned} \deg(x^\mu) &= \underline{0}, && \text{(even)} \\ \deg(\theta^I_\alpha) &= \deg(I) \in \mathbb{Z}_2^n \setminus \{\underline{0}\}, && \text{(odd)} \\ \deg(z^{IJ}) &= \deg(I) + \deg(J) \in \mathbb{Z}_2^n \setminus \{\underline{0}\}. && \text{(even)} \end{aligned}$$

In particular, we have nilpotent/odd coordinates θ and non-nilpotent/even coordinates z yet, in general, we will have non-trivial commutation relations between these coordinates governed by the $\mathbb{Z}_2^n$-degrees. As we have global coordinates we can construct the sheaf theoretical definition of a $\mathbb{Z}_2^n$-manifold directly, in particular $\mathbb{Z}_2^n$-Minkowski space-time is isomorphic to a $\mathbb{Z}_2^n$-graded Cartesian space. That is,

$$\mathcal{M}_4^{[n]} \cong \left(\mathbb{R}^4,\ C^\infty_{\mathbb{R}^4}[[\theta, z]]\right).$$

Naturally, we see that the underlying smooth manifold is just standard Minkowski space-time.

Remark 7. *A $\mathbb{Z}_2^3$-graded generalisation of the supersymmetry algebra was given by Le Roy [56] who had applications to the strong interaction in mind. However, he did not present a geometric understanding thereof.*

3.4. Invariant Differential Forms

A direct calculation shows that we have a basis of the left-invariant differential forms given by

$$e^\mu = \mathrm{d}x^\mu + \frac{1}{4}\mathrm{d}\theta^I_\beta \theta^J_\alpha \delta_{JI} (C\gamma^\mu)^{\alpha\beta}, \qquad \psi^I_\alpha = \mathrm{d}\theta^I_\alpha, \qquad e^{JK} = \mathrm{d}z^{JK} - \theta_\beta^{(K}\mathrm{d}\theta_\alpha^{\ J)} C^{\alpha\beta}.$$

We consider differential forms on $\mathbb{Z}_2^n$-manifolds to be functions on the $\mathbb{Z}_2^{n+1}$-manifold built from the tangent bundle, and we then append an additional component to the degree which is zero for the base coordinates and one for the fibre coordinates (see [37]). In particular, we have

$$\deg(\mathrm{d}e^\mu) = (\underline{0}, 1), \qquad \deg(\mathrm{d}\psi^I_\alpha) = (\underline{I}, 1), \qquad \deg(\mathrm{d}z^{JK}) = (\underline{J} + \underline{K}, 1),$$

where for notational case we set $\underline{I} := \deg(I) \in \mathbb{Z}_2^n$, etc.

Note that these conventions are similar to the Deligne–Freed conventions for superdifferential forms [57]. The left-invariant differential forms we can, as a standard, consider as canonical vielbeins. Choosing these vielbeins is no more than a change of basis of the space of one-forms. In particular, the canonical vielbeiens are not all closed and so we have *torsion*:

$$\mathrm{d}e^\mu = \frac{1}{4}(C\gamma^\mu)^{\alpha\beta}\psi^I_\beta \psi^J_\beta \delta_{JI}, \qquad \mathrm{d}\psi^I_\alpha = 0, \qquad \mathrm{d}e^{JK} = 0.$$

Thus, $\mathbb{Z}_2^n$-Minkowski space-time—just like the standard case—is flat, but has a non-vanishing torsion.

3.5. Superfields

To fully and correctly understand *superfields*, one must employ the categorical notion of the internal Homs—loosely we need to consider maps between $\mathbb{Z}_2^n$-manifolds that are parametrised by arbitrary $\mathbb{Z}_2^n$-manifolds. However, formally we can proceed as in the standard super-case (see for example [13,14]).

Note that we have to understand maps between $\mathbb{Z}_2^n$-manifolds when written in terms of local coordinates as formal power series in coordinates of non-zero degree. However, for the case at hand, all the Majorana spinor coordinates are odd and so are nilpotent. The complication—as with the standard case of extended supersymmetry—is the fact that power series in z does not truncate. Thus, an unconstrained superfield will be a formal power series, i.e., the theory will have an infinite number of axillary fields.

Expanding any superfield in terms of the formal coordinates, we have

$$\Phi^{A}(x,\theta,z) = \phi^{A}(x) + \theta^{I}_{\alpha}\chi^{\alpha A}_{I}(x) + \text{higher terms},$$

where A is some extra index or collection of indices, that may be related to spin, isospin, etc. We will not be more specific here. The degree of a superfield can be arbitrary, though homogeneous (if not, then we can decompose the superfield into homogeneous pieces). To lowest order the $\mathbb{Z}_2^n$-supersymmetry transformations are

$$\begin{aligned} \delta\phi^{A}(x) &\simeq \epsilon^{I}_{\alpha}\chi^{\alpha A}_{I}(x), \\ \delta\chi^{\alpha A}_{I}(x) &\simeq -\frac{1}{4}\epsilon^{J}_{\beta}\delta_{JI}(C\gamma^{\mu})^{\beta\alpha}\frac{\partial\phi^{A}}{\partial x^{\mu}}(x). \end{aligned}$$

The form of the $\mathbb{Z}_2^n$-supersymmetry transformations is of course expected by comparison with the standard case of extended supersymmetry. Note that the transformations mix fields whose degrees are related by $\deg(I)$, which we have taken to be odd. Thus, we have a generalisation of standard (quasi-classical) supersymmetry. The similarity here with Green's parastatistics should be noted (see [26]). In particular, if we take the superfield to be degree zero, i.e., $\deg(\Phi) = \underline{0}$, then $\phi^{A}(x)$ is also degree zero and so could represent a physical boson. Similarly, $\chi^{\alpha A}_{I}(x)$ are then odd and thus interpreted to be nilpotent Majorana spinor fields. However, in general pairs of such Majorana spinor fields need *not* anticommute due to the $\mathbb{Z}_2^n$-grading. Thus, the 'physical content' of the theory resembles (to lowest order) a system consisting of bosons and parafermions (written in terms of their 'Green components') together with a 'parasupersymmetry'.

Remark 8. *To our knowledge, the first work on parasupersymmetry in the context of quantum mechanics was that of Rubakov & Spiridonov [58]. The earliest work that we are aware of that discusses a parafermion generalisation of supersymmetry in a field theoretical setting is Jarvis [53].*

Remark 9. *It is known that 'reasonable' systems of paraparticles in relativistic quantum field theory are equivalent to systems of particles with standard statistics that carry an extra quantum number, see for example Araki [59], Doplicher & Roberts [60], and Drühl, Haag & Roberts [61]. Generically, given an algebra of physical observables, there will always be some ambiguity in the non-observable field content. Very loosely, the ambiguity in field content allows a redefinition (a Klein transformation) of the fields such that the redefined fields obey standard statistics. Thus, the general belief is that the difference between parastatistics and normal statistics is a matter of choice with regards to the field content of the theory under study. In particular, via field redefinitions, we can keep the well known spin-statistics theorem. However, the known equivalence of paraparticles and standard particles only holds for theories that satisfy the DHR superselection constraints, which loosely say that all observable events are localised in space and time. These constraints are not satisfied long-range interactions such as those found in electromagnetic theory. Moreover, interacting field theories and their superselection rules are far less well understood than free theories—this is a current generic problem with algebraic field theory. A good discussion of the equivalence of paraparticles and particles, together with the limitations of what is mathematically established, can be found in [62]. Thus, there is still some possibility of, as of yet unseen (c.f. [63]), fundamental parabosons and parafermions. Indeed, it has even been suggested that paraparticles may be associated with dark matter, see [64,65]. Even more remarkable is the suggestion that parastatistics can be used to solve the problem of the cosmological constant, see [66].*

3.6. $\mathbb{Z}_2^2$-Minkowski Space-Time

As a specific and illustrative example let us consider $\mathcal{M}_4^{[2]}$. This simplifies some of the expressions and resembles $\mathcal{N} = 2$ extended superspace closely. However, one has to be careful with the signs involved with $\mathbb{Z}_2^2$-graded commutators. Because of the symmetry of Z_{IJ} for the $n = 2$ case, we can rewrite the non-trivial part of the $\mathbb{Z}_2^2$-super translation algebra as

$$[Q_I^\alpha, Q_J^\beta] = \frac{1}{2}\delta_{IJ}(C\gamma^\mu)^{\alpha\beta}P_\mu + \frac{1}{2}C^{\alpha\beta}(\sigma_1)_{IJ}Z,$$

where

$$\sigma_1 := \begin{pmatrix} 0 & 1 \\ 1 & 0 \end{pmatrix}$$

is the first Pauli matrix. Here $I, J \in \{1,2\}$. Note that when $I = J$ we have an anticommutator in the 'ungraded' language. This is of course in parallel with standard $\mathcal{N} = 2$ supersymmetry. When $I \neq J$ we have a commutator and this is in contrast to the standard case.

Remark 10. *Interestingly, there seems to be no direct analogue of the Bogomol'nyi-Prasad-Sommerfield (BPS) bound: there is no direct inequality like '$m \geq |z|$', where m is the mass of a particle. For one, such an inequality is hard to understand for formal variables: we have to understand z as being of degree $(1,1)$ and so cannot simply be a real or complex number. Secondly, there is a subtle sign change in the commutator $[Q_I^\alpha, Q_J^\beta]$ when $I \neq J$ as compared with the standard case. This implies that we cannot use the standard reasoning to deduce the BPS bound as we do not have a positive definite operator when $I \neq J$. BPS states in standard supersymmetric theories are generically important as they can be determined using non-perturbative methods. They are one of the 'gifts' of supersymmetry. However, the situation for $\mathbb{Z}_2^n$-graded theories is far less clear.*

Using constructions of Section 3.3, we see that $\mathbb{Z}_2^2$-Minkowski space-time comes equipped with global coordinates

$$(\underbrace{x^\mu}_{(0,0)}, \underbrace{\theta_\alpha^1}_{(0,1)}, \underbrace{\theta_\beta^2}_{(1,0)}, \underbrace{z}_{(1,1)}).$$

Explicitly the $\mathbb{Z}_2^2$-commutation rules are:

$$\theta_\alpha^1\theta_\beta^2 = +\theta_\beta^2\theta_\alpha^1, \qquad z\theta_\alpha^I = -\theta_\alpha^I z,$$

with all other commutation rules being the same as the standard case of $\mathcal{N} = 2$ supersymmetry. Naturally, we still have $(\theta_\alpha^I)^2 = 0$, and while z picks up a minus sign when commuted with any θ, it is not nilpotent (it is a formal coordinate).

The supersymmetry transformations (3)–(5) simplify to

$$\begin{aligned} x^\mu &\mapsto x^\mu + \frac{1}{4}\epsilon_\beta^I\theta_\alpha^J\delta_{JI}(C\gamma^\mu)^{\alpha\beta}, \\ \theta_\alpha^I &\mapsto \theta_\alpha^I + \epsilon_\alpha^I, \\ z &\mapsto z + \frac{1}{4}\epsilon_\beta^I\theta_\alpha^J(\sigma_1)_{JI}C^{\alpha\beta}. \end{aligned}$$

The left-invariant vector fields of the related group action are easily derived and are given by (using standard notation)

$$\begin{aligned} P_\mu &= \frac{\partial}{\partial x^\mu}, \\ Q_I^\alpha &= \frac{\partial}{\partial\theta_\alpha^I} + \frac{1}{4}\theta_\beta^J\delta_{JI}(C\gamma^\mu)^{\beta\alpha}\frac{\partial}{\partial x^\mu} + \frac{1}{4}\theta_\beta^J(\sigma_1)_{JI}C^{\beta\alpha}\frac{\partial}{\partial z}, \\ Z &= \frac{\partial}{\partial z}. \end{aligned}$$

As a standard, we will call the vector fields Q the SUSY generators We can also directly find the right-invariant vector fields, and doing so leads to the SUSY covariant derivatives

$$D_I^\alpha = \frac{\partial}{\partial \theta_\alpha^I} - \frac{1}{4}\theta_\beta^J \delta_{JI}(C\gamma^\mu)^{\beta\alpha}\frac{\partial}{\partial x^\mu} - \frac{1}{4}\theta_\beta^J(\sigma_1)_{JI}C^{\beta\alpha}\frac{\partial}{\partial z}.$$

The reader can directly check that the SUSY covariant derivatives graded commute with the SUSY generators, i.e., $[D, Q] = 0$. Moreover, a direct computation gives the expected result (as it should as we are using right-invariant vector fields)

$$[D_I^\alpha, D_J^\beta] = -\frac{1}{2}\delta_{IJ}(C\gamma^\mu)^{\alpha\beta}P_\mu - \frac{1}{2}C^{\alpha\beta}(\sigma_1)_{IJ}Z = -[Q_I^\alpha, Q_J^\beta]. \tag{6}$$

Note that

$$\mathcal{D} := \text{Span}\{D_I^\alpha\},$$

is a distribution, i.e., a local direct factor of $\mathsf{Vect}(\mathcal{M}_4^{[2]})$, (see [40] for details of distributions on $\mathbb{Z}_2^n$-supermanifolds). One can see that we have a distribution by verifying that $(\partial_x, D, \partial_z)$ form a local basis for the vector fields on $\mathcal{M}_4^{[2]}$. Following Manin [67], we refer to this distribution as the canonical $\mathbb{Z}_2^2$-SUSY structure on $\mathcal{M}_4^{[2]}$.

Given $\mathcal{D}$, (or any distribution in fact) we can construct the short exact sequence of modules

$$0 \longrightarrow \mathcal{D} \overset{\iota}{\hookrightarrow} \mathsf{Vect}(\mathcal{M}_4^{[2]}) \overset{\varphi}{\longrightarrow} \mathcal{E},$$

where $\mathcal{E} := \mathsf{Vect}(\mathcal{M}_4^{[2]}) \setminus \mathcal{D}$. Locally the map φ is just dropping the components of the vector field that lie in the distribution. The Frobenius curvature of $\mathcal{D}$ is defined as

$$\begin{aligned} R : \mathcal{D} \times \mathcal{D} &\longrightarrow \mathcal{E} \\ (X, Y) &\mapsto R(X, Y) := \varphi([X, Y]). \end{aligned}$$

The characteristic distribution of $\mathcal{D}$, denoted $\mathcal{C}$, consists of all $X \in \mathcal{D}$ such that $R(X, -) = 0$. As $\mathcal{C}$ consists of only the zero vector, $\mathcal{D}$ is said to be maximally non-integrable. To see this we just need to consider the commutator

$$[X, D_J^\beta] = X_\alpha^I[D_I^\alpha, D_J^\beta] \pm D_J^\beta(X_\alpha^I)D_I^\alpha, \tag{7}$$

where the coefficients X_α^I may depend on all the coordinates. The second term in the above is clearly an element of $\mathcal{D}$, and so it dropped when looking at the Frobenius curvature. The first term, due to (6), is in $\mathcal{E}$ and so to construct the characteristic distribution we need to examine when this vanishes. This condition reduces to a set of linear equations

$$(C\gamma^\mu)^{\alpha\beta}X_\beta^I = 0, \qquad C^{\alpha\beta}X_\beta^I = 0.$$

Due to the fact that the Dirac gamma matrices are non-singular, the only solution to the above is $X_\beta^I = 0$, as required.

4. Closing Remarks

In this paper, we have suggested one reasonable (if somewhat ad hoc) way to construct a $\mathbb{Z}_2^n$-manifold version of extended super-Minkowski space-time. We fixed the dimension of the underlying manifold to be four, though one can consider other dimensions. We have not attempted to build any models here. In the standard theory, one needs to use constraints to obtain a finite expansion of the superfield. This is usually done in terms of the SUSY covariant derivatives—mathematically these are the right invariant vector fields of the associated super Lie group. It is expected that one can mimic this for $\mathbb{Z}_2^n$-case (for low n anyway). Currently, the obstruction to model building is the lack of a workable integration theory on $\mathbb{Z}_2^n$-manifolds (see Poncin [68] for progress in this direction). However, even with all the mathematical tools in place, it is not clear what the physical relevance of such theories would be. Moreover, it is fair to say that $\mathbb{Z}_2^n$-graded objects are not so well-known throughout the theoretical physics community.

It is recognised that for standard $\mathcal{N}$-extended supersymmetry, the automorphism group of the algebra plays an important rôle. This is especially so in the context of $\mathcal{N}$-extended superconformal algebras. It is thus a natural question to seek to understand the automorphism group of the $\mathbb{Z}_2^n$-extended Poincaré algebra, which due to the grading could have a richer structure than the standard case. Currently this structure remains to be uncovered. Moreover, defining the $\mathbb{Z}_2^n$-conformal algebra could lead to further novel structures. For example, the notion of a $\mathbb{Z}_2^n$-Riemannian surface is, so far, completely unexplored.

We have not touched upon the notion of R-symmetry in this paper. In the standard context of extended supersymmetry, the R-symmetry transforms different supercharges into each other. The immediate problem for the $\mathbb{Z}_2^n$-case is the charges are assigned a different degree and so the corresponding R-symmetries would themselves need to carry $\mathbb{Z}_2^n$-degree. This highlights another subtle difference when extending the grading from $\mathbb{Z}_2$ to $\mathbb{Z}_2^n$. However, given the importance of R-symmetry in the standard case, the 'higher' analogue requires investigation.

Another related avenue for future exploration is the study of $\mathbb{Z}_2^n$-graded versions of extended de Sitter and anti-de Sitter superalgebras and their integration to $\mathbb{Z}_2^n$-Lie groups. The reader should note the absence of Majorana spinors on de Sitter space-time, though this only means that we need to accept spinor objects along with their charge-conjugates in the constructions. Importantly, there is increasing observational evidence that the expansion of the Universe is accelerating and that a positive cosmological constant fits the data well. Thus, having a handle on de Sitter $\mathbb{Z}_2^n$-supergravity could shed light on the problem of dark energy. Of course, we are being speculative, but this definitely motivates further work.

Acknowledgments: The author thanks Norbert Poncin for many discussions about $\mathbb{Z}_2^n$-geometry. A special thank you goes to Steven Duplij for many fruitful discussions. The author also thanks anonymous referees for their constructive comments. Part of this work was reported on at the miniworkshop "Supergeometry and Applications", University of Luxembourg, 14–15 December 2017.

References

1. Gervais, J.-L.; Sakita, B. Field theory interpretation of supergauges in dual models. *Nucl. Phys. B* **1971**, *34*, 632–639. [CrossRef]
2. Gol'fand, Y.; Lichtman, E. Extension of the Algebra of Poincare Group Generators and Violation of P Invariance. *JETP Lett.* **1971**, *13*, 323.
3. Volkov, D.; Akulov, V. Possible Universal Neutrino Interaction. *JETP Lett.* **1972**, *16*, 438.
4. Wess, J.; Zumino, B. Supergauge transformations in four dimensions. *Nucl. Phys. B* **1974**, *70*, 39–50. [CrossRef]
5. Autermann, C. Experimental status of supersymmetry after the LHC Run-I. *Prog. Part. Nucl. Phys.* **2016**, *90*, 125–155. [CrossRef]
6. Kaku, M. *Introduction to Superstrings and M-Theory*, 2nd ed.; Graduate Texts in Contemporary Physics; Springer: New York, NY, USA, 1999.
7. Berezin, F.A.; Leĭtes, D.A. Supermanifolds. *Soviet Math. Dokl.* **1976**, *16*, 1218–1222.
8. Mac Lane, S. *Homology*; Academic Press, Inc.: New York, NY, USA; Springer: Berlin/Heidelberg, Germany; Göttingen, Germany, 1963.
9. Salam, A.; Strathdee, J. Super-gauge transformations. *Nucl. Phys. B* **1974**, *76*, 477–482. [CrossRef]
10. Kuzenko, S.M. Lectures on nonlinear sigma-models in projective superspace. *J. Phys. A* **2010**, *43* 443001. [CrossRef]
11. Lindström, U.; Roček, M. Properties of hyperkahler manifolds and their twistor spaces. *Commun. Math. Phys.* **2010**, *293*, 257. [CrossRef]

12. Galperin, A.S.; Ivanov, E.A.; Ogievetsky, V.I.; Sokatchev, E.S. *Harmonic Superspace*; Corrected Reprint of the 2001 Original; Cambridge University Press: Cambridge, MA, USA, 2007.
13. West, P. *Introduction to Supersymmetry and Supergravity*, 2nd ed.; World Scientific Publishing Co., Inc.: Teaneck, NJ, USA, 1990.
14. Wess, J.; Bagger, J. *Supersymmetry and Supergravity*, 2nd ed.; Princeton University Press: Princeton, NJ, USA, 1992.
15. Gates, S.J.; Grisaru, M.T.; Roček, M.; Siegel, W. *Superspace. One Thousand and One Lessons in Supersymmetry*; Benjamin/Cummings Publishing Co., Inc., Advanced Book Program: Reading, MA, USA, 1983.
16. Duplij, S.; Siegel, W.; Bagger, J. *Concise Encyclopedia of Supersymmetry and Noncommutative Structures in Mathematics and Physics*; Kluwer Academic Publishers: Dordrecht, The Netherlands, 2004.
17. Connes, A.; Douglas, M.; Schwarz, A. Noncommutative geometry and matrix theory: Compactification on tori. *J. High Energy Phys.* **1998**, *35*. [CrossRef]
18. Ferrara, S.; Lledó M.A.; Macia, O. Supersymmetry in noncommutative superspaces. *J. High Energy Phys.* **2003**, 068. [CrossRef]
19. Schwarz, A. Noncommutative supergeometry, duality and deformations. *Nucl. Phys. B* **2003**, *650*, 475–496. [CrossRef]
20. Schwarz, J.H.; van Nieuwenhuizen, P. Speculation concerning a fermionic substructure of space-time. *Lett. Nuovo Cimento* **1982**, *34*, 21–25. [CrossRef]
21. Seiberg, N. Noncommutative superspace, $\mathcal{N}=1/2$ supersymmetry, field theory and string theory. *J. High Energy Phys.* **2003**, 10. [CrossRef]
22. Seiberg, N.; Witten, E. String theory and noncommutative geometry. *J. High Energy Phys.* **1999**, *9*. [CrossRef]
23. Frydryszak, A. Nilpotent classical mechanics. *Int. J. Mod. Phys. A* **2007**, *22*, 2513–2533. [CrossRef]
24. Frydryszak, A. Nilpotent quantum mechanics. *Int. J. Mod. Phys. A* **2010**, *25*, 951–983. [CrossRef]
25. Frydryszak, A. Nilpotent Quantum Mechanics and SUSY. *Part. Nucl. Lett.* **2011**, *8*, 279–281. [CrossRef]
26. Green, H.S. A generalized method of field quantization. *Phys. Rev.* **1953**, *90*, 270–273. [CrossRef]
27. Volkov, D.V. On the quantization of half-integer spin fields. *Soviet Phys. JETP* **1959**, *9*, 1107–1111.
28. Zheltukhin, A.A. On a para-Grassmann extension of the Neveu-Schwarz-Ramond algebra. *Teoret. Mat. Fiz.* **1987**, *71*, 218–225. [CrossRef]
29. Vasil'ev, M.A. de Sitter supergravity with positive cosmological constant and generalised Lie superalgebras. *Class. Quantum Gravity* **1985**, *2*, 645–659. [CrossRef]
30. Yang, W.; Jing, S. A new kind of graded Lie algebra and parastatistical supersymmetry. *Sci. China Ser. A* **2001**, *44*, 1167–1173. [CrossRef]
31. Yang, W.; Jing, S.; Ping, L. Graded Lie algebra generating of parastatistical algebraic relations. *Commun. Theor. Phys.* **2001**, *36*, 647–650.
32. Aizawa, N.; Kuznetsova, Z.; Tanaka, H.; Toppan, F. $\mathbb{Z}_2 \times \mathbb{Z}_2$-graded Lie Symmetries of the Lévy-Leblond Equations. *Prog. Theor. Exp. Phys.* **2016**, 123A01. [CrossRef]
33. Tolstoy, V.N. Super-de Sitter and alternative super-Poincaré symmetries. In *Lie Theory and Its Applications in Physics*; Springer: Berlin, Germany, 2014.
34. de Goursac, A. Noncommutative supergeometry and quantum supergroups. *J. Phys. Conf. Ser.* **2015**, *597*, 012028. [CrossRef]
35. Grosse, H.; Reiter, G. Simple fuzzy (super)spaces. *Int. J. Mod. Phys. B* **2000**, *14*, 2467–2470. [CrossRef]
36. Haag, R.; Łopuszański, J.T.; Sohnius, M. All possible generators of supersymmetries of the S-matrix. *Nucl. Phys. B* **1975**, *88*, 257–274. [CrossRef]
37. Covolo, T.; Grabowski, J.; Poncin, N. The category of $\mathbb{Z}_2^n$-supermanifolds. *J. Math. Phys.* **2016**, *57*, 073503. [CrossRef]
38. Covolo, T.; Grabowski, J.; Poncin, N. Splitting theorem for $\mathbb{Z}_2^n$-supermanifolds. *J. Geom. Phys.* **2016**, *110*, 393–401. [CrossRef]
39. Covolo, T.; Kwok, S.; Poncin, N. Differential calculus on $\mathbb{Z}_2^n$-supermanifolds. *arXiv* **2016**, arXiv:1608.00949.
40. Covolo, T.; Kwok, S.; Poncin, N. The Frobenius theorem for $\mathbb{Z}_2^n$-supermanifolds. *arXiv* **2016**, arXiv:1608.00961.
41. Molotkov, V. Infinite-dimensional and colored supermanifolds. *J. Nonlinear Math. Phys.* **2010**, *17*, 375–446. [CrossRef]
42. Marcinek, M. Colour superspaces and supermanifolds. *Rep. Math. Phys.* **1991**, *29*, 55–74. [CrossRef]

43. Grabowski, J.; Rotkiewicz, M. Higher vector bundles and multi-graded symplectic manifolds. *J. Geom. Phys.* **2009**, *59*, 1285–1305. [CrossRef]
44. Voronov, T. Graded manifolds and Drinfeld doubles for Lie bialgebroids. In *Quantization, Poisson Brackets and Beyond*; American Mathematical Society: Providence, RI, USA, 2002.
45. Covolo, T.; Ovsienko, V.; Poncin, N. Higher trace and Berezinian of matrices over a Clifford algebra. *J. Geom. Phys.* **2012**, *62*, 2294–2319. [CrossRef]
46. Rittenberg, V.; Wyler, D. Generalized superalgebras. *Nucl. Phys. B* **1978**, *139*, 189–202. [CrossRef]
47. Scheunert, M. Generalized Lie algebras. *J. Math. Phys.* **1979**, *20*, 712–720. [CrossRef]
48. Bruce, A.J.; Poncin, N. Functional analytic issues in $\mathbb{Z}_2^n$-Geometry. *arXiv* **2018**, arXiv:1807.11739v1.
49. Bruce, A.J.; Poncin, N. Products in the category of $\mathbb{Z}_2^n$-manifolds. *arXiv* **2018**, arXiv:1807.11740v1.
50. Bellucci, S.; Krivonos, S. Supersymmetric mechanics in superspace. *Lect. Notes Phys.* **2006**, *698*, 49–96.
51. Beckers, J.; Debergh, N. Poincaré invariance and quantum parasuperfields. *Int. J. Mod. Phys. A* **1993**, *8*, 5041–5061. [CrossRef]
52. Nikitin, A.G.; Galkin, V.A. Extended Poincaré parasuperalgebra with central charges and invariant wave equations. *J. Phys. A* **2000**, *33*, 8525–8547. [CrossRef]
53. Jarvis, P.D. A parafermion generalization of Poincaré supersymmetry. *Aust. J. Phys.* **1978**, *31*, 461–469. [CrossRef]
54. Witten, E.; Olive, D.I. Supersymmetry Algebras That Include Topological Charges. *Phys. Lett. B* **1978**, *78*, 97–101. [CrossRef]
55. Duplij, S.A.; Soroka, D.V.; Soroka, V.A.J. A special fermionic generalization of lineal gravity. *Zhejiang Univ. Sci. A* **2006**, *7*, 629. [CrossRef]
56. Le Roy, B. $\mathbb{Z}_2^3$-graded colored supersymmetry. *Czech. J. Phys.* **1997**, *47*, 47–54. [CrossRef]
57. Deligne, P.; Freed, D.S. Sign manifesto. In *Quantum Fields and Strings: A Course for Mathematicians*; American Mathematical Society: Providence, RI, USA, 1999.
58. Rubakov, V.A.; Spiridonov, V.P. Parasupersymmetric quantum mechanics. *Mod. Phys. Lett. A* **1988**, *3*, 1337–1347. [CrossRef]
59. Araki, H. On the connection of spin and commutation relations between different fields. *J. Math. Phys.* **1961**, *2*, 267–270. [CrossRef]
60. Doplicher, S.; Roberts, J.E. Why there is a field algebra with a compact gauge group describing the superselection structure in particle physics. *Commun. Math. Phys.* **1990**, *131*, 51–107. [CrossRef]
61. Drühl, K.; Haag, R.; Roberts, J.E. On parastatistics. *Commun. Math. Phys.* **1970**, *18*, 204–226. [CrossRef]
62. Baker, D.J.; Halvorson, H.; Swanson, N. The conventionality of parastatistics. *Br. J. Philos. Sci.* **2015**, *66*, 929–976. [CrossRef]
63. Greenberg, O.W.; Messiah, A.M.L. Selection rules for parafields and the absence of para particles in nature. *Phys. Rev.* **1965**, *138*, 1155–1167. [CrossRef]
64. Kitabayashi, T.; Yasuè M. Parafermionic dark matter. *Phys. Rev. D* **2018**, *98*, 043504. [CrossRef]
65. Nelson, C.A.; Kraynova, M.; Mera, C.S.; Shapiro, A.M. Diagrams and Parastatistical Factors for Cascade Emission of a Pair of Paraparticles. *Phys. Rev. D* **2016**, *93*, 034039. [CrossRef]
66. Moffat, J.W. Charge Conjugation Invariance of the Vacuum and the Cosmological Constant Problem. *Phys. Lett. B* **2005**, *627*, 9–17. [CrossRef]
67. Manin, Y.I. *Topics in Noncommutative Geometry*; Princeton University Press: Princeton, NJ, USA, 1991.
68. Poncin, N. Towards integration on colored supermanifolds. In *Geometry of Jets and Fields*; Banach Center Publications: Warsaw, Poland, 2016.

3

Do Fractals Confirm the General Theory of Relativity?

Irina Rozgacheva
All-Russian Institute for Scientific and Technical Information of Russian Academy of Sciences (VINITI RAS), Usievich 20, Moscow 125190, Russia; rozgacheva@yandex.ru

Abstract: The relatively high abundance of fractal properties of complex systems on Earth and in space is considered an argument in support of the general relativity of the geometric theory of gravity. The fractality may be called the fractal symmetry of physical interactions providing self-similarities of complex systems. Fractal symmetry is discrete. A class of geometric solutions of the general relativity equations for a complex scalar field is offered. This class allows analogy to spatial fractals in large-scale structures of the universe due to its invariance with respect to the discrete scale transformation of the interval $ds \leftrightarrow qd\overline{s}$. The method of constructing such solutions is described. As an application, the treatment of spatial variations of the Hubble constant H_0^{HST} (Riess et al., 2016) is considered. It is noted that the values H_0^{HST} form an almost fractal set. It has been shown that: a) the variation H_0^{HST} may be connected with the local gravitational perturbations of the space-time metrics in the vicinity of the galaxies containing Cepheids and supernovae selected for measurements; b) the value of the variation H_0^{HST} can be a consequence of variations in the space-time metric on the outskirts of the local supercluster, and their self-similarity indicates the fractal distribution of matter in this region.

Keywords: fractals; self-similarity; fractal symmetry; general relativity; geometric solutions; Hubble constant; gravitational perturbations

---

1. Introduction

For more than a hundred years, on 29 May 1919, the Eddington expedition made observations of a solar eclipse. The purpose of the observation was to test one of the three effects proposed by Einstein to confirm his general relativity (GR)—the curvature of light lines from a distant star in the gravitational field of the Sun. The results of Eddington's observations confirmed the predictions of GR [1]. It was the first experiment to test the main postulate of this theory as a geometrical model of gravitational interaction. Over the past century, many direct experiments have been made. The absolute majority of these experiments confirmed Einstein's theory [2].

However, at present, criticism of GR has renewed, in particular, because of the appearance of, the so-called "dark sector" in cosmology. Astrophysicists and cosmologists introduced the dark matter, and then the dark energy, to explain the characteristics of the rotation curves of galaxies, the stability of galactic disks, the growth of large-scale structures in the early universe, the Hubble parameter (accelerated expansion of space), and the anisotropy of the CMB radiation. Here these problems will not be discussed; but some relevant publications are given in [3]. So far, the "dark sector" methodology could not provide a satisfactory explanation of these problems because of the large variety of dark energy and dark matter models.

Modifications of the GR appeared in response to the "paradoxes of the dark sector". The main disadvantage of these modifications is that they introduce additional fundamental constants of gravitational interaction, which do not allow the use of the effects of the "dark sector". However, these parameters are defined from the same astrophysical observations of galaxies which need to be explained. It does not cause satisfaction. In GR, the fundamental constants measured in terrestrial experiments are used.

Modern information technologies allow the numerical solution of the non-linear equations of GR or its modifications. They are also used in observations, in processing observational data, and in the analysis of these data within a chosen cosmological model. The authors of various cosmological simulations fit their results to observations with good accuracy, which becomes better with an increase in the number of free parameters for baryonic matter (for example, different types of viscosity), for dark matter (from cold to hot), and for dark energy (different variants of scalar field potentials).

In work [3], facts indicating that the dark matter hypothesis is groundless from the point of view of the interpretation of the observations are given. Galaxies and clusters of galaxies are located inside the cosmic web. They are open systems that interact with the circumgalactic medium and the intergalactic medium. The description of the dynamics of such open systems within the framework of GR has not yet been done. Therefore, there is no indisputable evidence that would justify the rejection of GR.

In this article, the relatively high abundance of fractal properties of complex systems on Earth and in space is considered as an argument in support of the GR as a geometric theory of gravity. Geometrically, fractals are self-similar structures. Not always this self-similarity can be observed directly, especially in astrophysical objects and cosmological structures. Meanwhile, a sign of self-similar discrete sets is power dependencies between the physical characteristics of complex systems. For example, let us consider the Tully-Fisher relation $L \sim v^3$ in spirals for the luminosity L and the rotation rate of spiral galaxies v. Suppose that the values of the luminosities of galaxies and the rotation rates form geometric progressions (the most obvious example of self-similarity in the state space of the system): $L_{i+1} = aL_i$, $v_{i+1} = bv_i$, where a and b are constant parameters, $i = 1, 2, \ldots$. Number i can be written as $i = \frac{\ln \frac{v_i}{v_1}}{\ln b}$. Then for luminosity values we find $L_i = (v_i)^{\frac{\ln a}{\ln b}} \left(\frac{L_1}{(v_1)^{\frac{\ln a}{\ln b}}} \right) \sim (v_i)^{\frac{\ln a}{\ln b}}$. The Tully-Fisher relation follows from this formula if we assume $\frac{\ln a}{\ln b} = 3$. Spiral galaxies are similar physical objects. The power of the Tully-Fisher relation indicates that these objects form a fractal discrete set in the state space of these objects; we observe the projection of this set onto the $L - v$ plane as the Tully-Fisher relation.

The power of the relationship is often met in very different astrophysics objects. Let's give several examples:

- Main sequence stars obey a mass-luminosity relation, $L \sim m^\alpha$, $\alpha \approx 3.52$;
- Period-luminosity relation for pulsating variable stars, $L \sim P^\alpha$, for classical Cepheids $\alpha \approx 1.15$;
- Luminosity functions for stars, galaxies, and quasars, $\Phi \sim L^\alpha$, $-2.6 < \alpha < -0.6$;
- Faber-Jackson relation for luminosity and their central stellar velocity dispersion of stars of elliptical galaxies, $L \sim \sigma^4$;
- Tully-Fisher relation for the luminosity and rotation rate of spiral galaxies, $L \sim v^3$;
- Brightness distribution in the images of bright elliptical galaxies, $I \sim r^{-\alpha}\left(1 + \left(\frac{r}{r_0}\right)^\beta\right)^\gamma$, $-1.37 < \alpha < 0.85$, $0.28 < \beta < 1$, $-7.57 < \gamma < -0.18$;
- Power-law spectra of the radio emission of the jets from active galactic nuclei, $F \sim \nu^{-\alpha}$, $0.6 < \alpha < 0.7$ (jets are composed of plasma clouds with a magnetic field);
- Baldwin's ratio in the active galactic nucleus for the equivalent width of the emission lines and the luminosity of its galaxy, $EW(CaII) \sim L^{-\alpha}$, $\alpha > 1$;
- Red shift distribution of absorption lines in quasar spectra, $\frac{dN}{dz} \sim (1+z)^\alpha$, $1.67 < \alpha < 2.09$;
- Spatial correlation function of galaxies in the clusters of galaxies and clusters in superclusters, $\xi(r) \sim r^{-\alpha}$, $1.6 < \alpha < 2.2$.

The given power relationships are of course correlations. However, new observations only specify them.

The relative high abundance of fractals indicates that perhaps fractality is a fundamental property of physical interactions. The fractality may be called the fractal symmetry of physical interactions providing self-similarity of complex structures. Fractal symmetry is discrete, since a complex system with fractal symmetry consists of self-similar subsystems and the physical parameters of these subsystems form geometric progressions. Fractality can be compared with quantization (discreteness) of the states of the systems in the microcosm.

Can fractal symmetry be a consequence of the geometric model of gravity in GR? This question is logical if we take into account the geometric properties of the phase spaces of the systems, which we call fractals. It is also logical from a historical point of view in the development of science. For example, in the classical work of Minkowski (1909, [4]) it was proposed to back out of the hypothesis of ether, but to combine space and time into a four-dimensional homogeneous and isotropic set of points (the Minkowski world). Minkowski proved his offer relying on postulates of the special theory of relativity and the Lorentz invariance of the Maxwell's equations. According to Minkowski, we discover a homogeneous and isotropic space-time through the invariance of Maxwell's equations with respect to the Lorentz transformations. From the modern point of view, Minkowski considered the Lorentz's symmetry for an electromagnetic field as a consequence of the properties of four-dimensional space-time: the vector potential of the electromagnetic field is also transformed like any vector in four-dimensional space-time. Theoretical cosmology as a science about the structure and evolution of space-time of the universe began with this research by Minkowski. It should be noted that the fundamental hypothesis by Weyl, where he proposed (1918, 1929 [5,6]) to use a scalar field $\Phi(x^k)$ (now it is called dilation) with transformation $\Phi \leftrightarrow \sigma^{1/2}\widetilde{\Phi}$ and the continuous conformal symmetry of space-time $ds^2 \leftrightarrow \sigma\left(x^k\right)\widetilde{ds}^2$ for the electron model. This field should change the Newtonian gravitational interaction of all systems and its consequences would be self-similar properties of these systems. However, the existence of the dilaton field is not detected.

Now symmetries play an important role in the development of our ideas about space-time and the physics of matter. The description of the observed properties of matter by means of dynamic equations and symmetries of the interaction fields allows us to understand the nature of the universe at the micro and macro levels. This paradigm inspires cosmologists.

In this paper, a class of geometric solutions of GR is proposed for a complex scalar field. The solutions admit the possibility of spatial fractals in the large-scale structure of the universe [7–9]. Quasar groups are an example of such fractals. It was found in [8], that the number of groups N_i with size ϑ_i depends on the size $N_i \sim (\vartheta_i)^{-2.02}$.Let's consider the quasar group with number i as the so-called "island system", that has a space-time with an interval ds_i. If the intervals of different island systems form a geometric progression $ds_{i+1} = q ds_i$ (q is a number, discrete scale transformation), then a distant observer using the metric of the Minkowski's space will observe the geometric progression $\vartheta_{i+1} = q\vartheta_i$ for the angular sizes of the quasar groups. The introduction of a geometric progression for interval means, that we consider the space-time, consists of an ensemble of geometrically self-similar regions.

Below we consider an example of the solution of the GR equations, which allows for the discrete scale transformation of the space-time interval. The complex scalar field is used as a source in the GR equations. It is assumed that the field has a unitary symmetry U (1), which is given in the algebraic form:

$$(\psi_1)^2 + (\psi_2)^2 = \Psi^2 = const. \tag{1}$$

The complex field is used in the scalar electrodynamics to introduce charges, that are conserved due to symmetry (1). In the Standard Model of the physics of elementary particles SU(n) symmetries are introduced as a generalization of the circle (1) for multidimensional fields.

In the problem considered here, the discrete transform of the field phase $\varphi \leftrightarrow \widetilde{\varphi} + \alpha$ (α is a constant) without changing its amplitude Ψ (unitary symmetry) corresponds to a field transform $\widetilde{\psi} \leftrightarrow e^{-i\alpha}\psi$ and a discrete transform of metric tensor $g_{mn}(\varphi) \leftrightarrow \frac{\widetilde{U}_0}{U_0}\widetilde{g}_{mn}(\widetilde{\varphi})$, where U_0 and $\widetilde{U}_0$ are the parameters of the field potentials $U = U_0\psi\psi^*$, $\widetilde{U} = \widetilde{U}_0\widetilde{\psi}\widetilde{\psi}^*$, an asterisk means complex conjugation. In this case, the two regions of space-time are geometrically similar, their intervals are connected by a scale transformation $ds^2 \leftrightarrow \frac{\widetilde{U}_0}{U_0}\widetilde{ds^2}$ (an analogue of the conformal discrete transformation). An ensemble of such regions can be called a fractal, if $\left(ds^2\right)_{i+1} = \left(\frac{\widetilde{U}_0}{U_0}\right)\left(\widetilde{ds^2}\right)_{i'}, i = 1, 2, \ldots$.

Let's define a fractal cosmological model as a model in which fractal properties are characteristic of the spatial distribution of matter. The space-time of this model consists of a set of regions in space-time, the metric tensors of which are connected by discrete scale transformations.

Note, that in the fractal cosmological model, the simulation of the large-scale structure of the galaxy distribution should require less computational time, because our need is to model the self-similar parts of this structure and not their ensembles as a whole.

The rest of the article is organized as follows. Section 2 presents the method of constructing a geometrical solution of the Lagrange and Einstein equations for a complex scalar field. Section 3 discusses the contradiction in the measurements of the Hubble constant and its interpretation within the framework of the fractal cosmological model.

2. Exact Geometrical Solution of the Lagrange and Einstein Equations for a Complex Scalar Field with Symmetry U (1)

In the preprint [10], a method is presented for constructing the class of exact solutions discussed here for the GR equations of a complex scalar field. The following describes the revised and expanded version of this method. Let's consider a system with the Hilbert—Einstein's action

$$S = -\frac{c^3}{16\pi G}\int\left(R - \frac{8\pi G}{c^4}L\right)\sqrt{-g}d^4x, \tag{2}$$

where the curvature scalar is R, $g < 0$ is the determinant of the metric tensor g_{mn}, the space-time interval is $ds^2 = g_{mn}dx^m dx^n$, the indices run through the values 0, 1, 2, 3, the metric signature is $(+ - - -)$. Here, the physical system of units is maintained to simplify the comparison with the observations.

The Lagrangian of a complex scalar field has the form:

$$L = \frac{1}{hc}\left(g^{mn}\frac{\partial\psi}{\partial x^m}\frac{\partial\psi^*}{\partial x^n} - U(\psi\psi^*)\right), \tag{3}$$

where h is the Planck constant, $U(\psi\psi^*)$ is the field potential, which we choose in a simple form $U(\psi\psi^*) = U_0\psi\psi^* = U_0\Psi^2$ for a field $\psi = \Psi e^{i\varphi}$ with symmetry (1).

From the Lagrange equation

$$\left(\frac{\partial^2\psi}{\partial x^m\partial x^n} - \Gamma^l_{mn}\frac{\partial\psi}{\partial x^l}\right)g^{mn} = -\frac{\partial U}{\partial\psi^*}, \tag{4}$$

we derive an equation for the field phase $\varphi(x^m)$:

$$\left(-\psi\frac{\partial\varphi}{\partial x^m}\frac{\partial\varphi}{\partial x^n} + i\psi\frac{\partial^2\varphi}{\partial x^m\partial x^n} - i\psi\Gamma^l_{mn}\frac{\partial\varphi}{\partial x^l}\right)g^{mn} = -U_0\psi. \tag{5}$$

The goal is to choose the dependence of the Christoffel's symbols Γ^l_{mn} and the metric tensor g_{mn} on the derivatives $\frac{\partial\varphi}{\partial x^m}$ in such away so that Equation (5) for the phase $\varphi(x^m)$ is identically satisfied, and the Einstein equation for the metric tensor g_{mn} can be reduced to an algebraic form for the vector $e_n = \frac{\partial\varphi}{\partial x^n}$. In this case, the Einstein's equation is invariant relative to the discrete phase transformation $\varphi \leftrightarrow \widetilde{\varphi} + \alpha$. Achieving this goal is carried out in two stages.

2.1. Selection of the Christoffel's Symbols and Metric Tensor

By direct substitution, we can see that Equation (5) becomes an identity if the following formulas are used:

$$e_m e_n g^{mn} = U_0, \tag{6}$$

$$\Gamma^l_{mn} = \frac{1}{U_0}\frac{\partial^2\varphi}{\partial x^m \partial x^n}\left(e^l + b^l\right), \tag{7}$$

$$e_l b^l = 0, \tag{8}$$

where b^l is a vector.

The left-hand sides of formulas (6) and (8) are the tensors, therefore the formulas are covariance invariant. For the covariant conservation of formula (7), it is necessary that the covariant derivatives be equal to zero, $e_{n,m} = 0$, $b^l{}_{,m} = 0$ and

$$\left(\frac{\partial^2\varphi}{\partial x^m \partial x^n}\right)_{,l} = \left(\frac{\partial e_n}{\partial x^m}\right)_{,l} = 0. \tag{9}$$

We define the metric tensor by the following formula:

$$g_{mn} = \frac{4}{U_0} e_m e_n + a_m e_n + a_n e_m, \tag{10}$$

where there a_m is a vector. From the condition $g_{mn} g^{mn} = 4$, the equation follows, $a_m e^m = 0$.

The equation for the covariant derivative $g_{mn,k} = 0$ is identically satisfied, if $e_{n,m} = 0$ and $a_{m,n} = 0$, then

$$\frac{\partial a_m}{\partial x^l} = -\frac{3}{U_0}\frac{\partial e_m}{\partial x^l}, \tag{11}$$

$$\frac{\partial e_m}{\partial x^n} = \frac{\partial e_n}{\partial x^m}, \frac{\partial a_m}{\partial x^n} = \frac{\partial a_n}{\partial x^m}, \frac{\partial b_m}{\partial x^n} = \frac{\partial b_n}{\partial x^m}. \tag{12}$$

Calculation of the Cristoffel's symbols $\Gamma^n_{ml} = \frac{1}{2} g^{kn}\left(\frac{\partial g_{mk}}{\partial x^l} + \frac{\partial g_{lk}}{\partial x^m} - \frac{\partial g_{ml}}{\partial x^k}\right)$ and comparison with the definition (7) lead to the following equation:

$$\frac{1}{U_0}\frac{\partial e_m}{\partial x^l}(e^n + b^n) = \frac{4}{U_0} e^n \frac{\partial e_m}{\partial x^l} + a^n \frac{\partial e_m}{\partial x^l} + e^n \frac{\partial a_m}{\partial x^l}, \tag{13}$$

if the calculations take into account the equalities (12). Equation (13) turns into an identity, if we take into account Equation (12) and

$$b^n = U_0 a^n. \tag{14}$$

Formulas (6)–(14) determine the relationships between the vector e_m and g_{mn}, Γ^l_{mn}, a_m, and b^l.

2.2. Algebraic form of the Einstein's Equations

To determine the vector e_n and metric tensor (10), one must use the Einstein's equation $R_{km} = \frac{8\pi G}{c^4}\left(T_{km} - \frac{1}{2} g_{km} T\right)$ (here we use the notation of the classic textbook [11]). The energy-momentum tensor of the complex field is $T_{km} = \frac{2\Psi^2}{hc} U_0 e_k e_m$. The calculation of the Ricci tensor R_{mn} through the symbols Γ^l_{mn} and Equation (12) allows us to obtain the Einstein's equation in the following form:

$$\left(\frac{\partial^2 e_m}{\partial x^l \partial x^n} - \frac{\partial^2 e_m}{\partial x^n \partial x^l}\right)\left(e^l + U_0 a^l\right) = -\zeta U_0 (2 e_m e_n + U_0 a_m e_n + U_0 a_n e_m), \tag{15}$$

where $\zeta = \frac{8\pi G}{hc^5}\Psi^2$. The derivatives of the left side of Equation (15) are determined from Equation (9):

$$\frac{\partial^2 e_m}{\partial x^n \partial x^l} = \Gamma^k_{mn}\frac{\partial e_k}{\partial x^l} + \Gamma^k_{ln}\frac{\partial e_k}{\partial x^m} = \frac{1}{U_0}\left(e^k + U_0 a^k\right)\left(\frac{\partial e_m}{\partial x^n}\frac{\partial e_k}{\partial x^l} + \frac{\partial e_l}{\partial x^n}\frac{\partial e_k}{\partial x^m}e_n\right). \tag{16}$$

In this case, Equation (15) is rewritten as follows:

$$\left(\frac{\partial e_m}{\partial x^k}\frac{\partial e_n}{\partial x^l} - \frac{\partial e_m}{\partial x^n}\frac{\partial e_k}{\partial x^l}\right)\left(e^k + U_0 a^k\right)\left(e^l + U_0 a^l\right) = -\zeta(U_0)^2(2e_m e_n + U_0 a_m e_n + U_0 a_n e_m). \tag{17}$$

The Equation (17) has an algebraic equation for a vector e_m if we introduce another definition for the tensor $\frac{\partial e_m}{\partial x^n}$:

$$\frac{\partial e_m}{\partial x^n} = e_m\gamma_n + e_n\gamma_m, \tag{18}$$

where γ_m is a vector. We also use the general form of the solution of Equation (11):

$$U_0 a_m = -3e_m + \xi_m, \tag{19}$$

where the vector ξ_m does not depend on coordinates x^k, i.e., $\frac{\partial \xi_m}{\partial x^n}$ and $\xi_m e^m = 3U_0$. As a result, Equation (17) is written in algebraic form:

$$e_m e_n(y-2z)^2 - U_0(y-2z)(e_n\gamma_m + e_m\gamma_n) + (U_0)^2\gamma_m\gamma_n = \zeta(U_0)^2(4e_m e_n - \xi_m e_n - \xi_n e_m), \tag{20}$$

where $y = \gamma_k\xi^k$ and $z = \gamma_k e^k$. The vector e_m can be found by convolving Equation (20) with the vector e^n:

$$e_m = U_0\frac{(y-3z)\gamma_m - \zeta(U_0)^2\xi_m}{(y-2z)(y-3z) - \zeta(U_0)^2}. \tag{21}$$

Equations (18) and (21) define the class of geometrical solutions for problem (1)–(3) for the metric tensor

$$g_{mn} = \frac{1}{U_0}(-2e_m e_n + e_n\xi_m + e_m\xi_n), \tag{22}$$

and the Cristoffel's symbols

$$\Gamma^l_{mn} = \frac{1}{U_0}(e_n\gamma_m + e_m\gamma_n)\left(-2e^l + \xi^l\right), \tag{23}$$

if the vectors γ_k and ξ_k are chosen. Options, that lead to contradictions, should be excluded from this choice: $y = 3z$, $y = 0$, $z = 0$, $y = 2z$, and $\gamma_m\gamma^m = 0$.

So, the Einstein's equation is reduced to the algebraic form (20). Using it and Equation (18), we find the function $e_m(x^n)$.

Convolution of Equation (20) with $e^m e^n$, $\gamma^m\gamma^n$, and $\xi^m\xi^n$ gives three corresponding equations:

$$(y-3z)^2 = -2\zeta(U_0)^2, \tag{24}$$

$$(z(y-2z) - U_0\gamma_m\gamma^m)^2 = -2\zeta(U_0)^2 z(y-2z), \tag{25}$$

$$2(y-z)^2 = 3\zeta U_0(6U_0 - \xi_n\xi^n). \tag{26}$$

Equation (24) defines the relationship between y and z, Equations (25) and (26) define the functions $\gamma_m\gamma^m$ and $\xi_n\xi^n$.

Using formula (24) in formula (21) we get, $e_m = U_0 \frac{\gamma_m + \beta U_0 \xi_m}{z + 3\beta U_0}$, where $\beta = \left(-\frac{\zeta}{2}\right)^{1/2}$ is an imaginary number. Then the Equation (18) is reduced to the form (the condition $\frac{\partial \xi_m}{\partial x^n} = 0$ is used):

$$\frac{\partial \gamma_m}{\partial x^n} - \frac{\gamma_m + \beta U_0 \xi_m}{z + 3\beta U_0} \frac{\partial z}{\partial x^n} = 2\gamma_m \gamma_n + \beta U_0 (\xi_m \gamma_n + \xi_n \gamma_m). \tag{27}$$

The Equation (27) includes three unknowns—two functions $\gamma_m(x^k)$, $z(x^k)$, and a constant vector ξ_m. Their choice determines the vector $e_m(x^n)$. As an example, we present the following choice of equations for functions $\gamma_m(x^k)$, $z(x^k)$:

$$\frac{\partial \gamma_m}{\partial x^n} = 2\gamma_m \gamma_n + \beta U_0 (\gamma_n + \beta \xi_n) \xi_m, \tag{28}$$

$$\frac{\partial z}{\partial x^n} = \beta (z + 3\beta U_0) \xi_n. \tag{29}$$

They turn Equation (27) into an identity. The complex solution of Equation (29) has the form:

$$z = e^{\beta \xi_n x^n} - 3\beta U_0 = \cos\left(\sqrt{\frac{\zeta}{2}} \xi_n x^n\right) + i\left(\sin\left(\sqrt{\frac{\zeta}{2}} \xi_n x^n\right) - 3U_0 \sqrt{\frac{\zeta}{2}}\right). \tag{30}$$

The solution of Equation (28) is more bulky:

$$\gamma_m = \beta U_0 \xi_m e^{2\int \frac{y}{\xi_m \xi^m} d\eta} \int \frac{\frac{y}{(\xi_m \xi^m)} + \beta}{e^{2\int \frac{y}{\xi_m \xi^m} d\eta}} d\eta, \tag{31}$$

where $d\eta = \xi_n dx^n$, the functions $\xi_n \xi^n$ and y are determined from Equations (25) and (26) using solution (30). To determine the vector e_m and metric tensor (22), it is necessary to use the real parts of the functions (30) and (31). The given solution of problem (1)–(3) has a pulsating space-time.

Despite the relative bulkiness of the described method of constructing a solution to problem (1)–(3), it is simple and easily verified by analytical calculations, and also allows numerical modeling of all solutions.

Note that for the problem in question, Lagrangian (3) is zero, but the field energy density is not zero:

$$E = \frac{1}{hc}\left(g^{mn} \frac{\partial \psi}{\partial x^m} \frac{\partial \psi^*}{\partial x^n} + U(\psi \psi^*)\right) = \frac{2}{hc} U_0 \Psi^2. \tag{32}$$

The interval of the phase trajectory for the circle (1) is equal to

$$dF^2 = (d\psi_1)^2 + (d\psi_2)^2 = d\psi d\psi^* = \Psi^2 e_m e_n dx^m dx^n. \tag{33}$$

Using formulas (22), one can transform formula (33) to the following form:

$$dF^2 = \frac{1}{2} \Psi^2 U_0 \left(-ds^2 + \frac{1}{U_0} (e_m \xi_n + e_n \xi_m) dx^m dx^n\right). \tag{34}$$

Formula (34) shows how the local geometry of the phase space (1) is related to the local space-time geometry.

When the phase transformation is $\varphi \leftrightarrow \widetilde{\varphi} + \alpha$ and the field amplitude is $\Psi = \widetilde{\Psi}$, but the potential parameter is changed $U_0 \leftrightarrow \widetilde{U}_0$, then the Lagrange's equation, the Einstein's equations, covariant vectors e_m, ξ_m, γ_m, the Christoffel's symbols do not change. However, the metric tensor is multiplied by a constant coefficient:

$$g_{mn}(\varphi) \leftrightarrow \frac{\widetilde{U}_0}{U_0} \widetilde{g}_{mn}(\widetilde{\varphi}). \tag{35}$$

As a result, the transformation of the phase of the complex field leads to the compression or stretching of the space-time interval $ds^2 \leftrightarrow \frac{\widetilde{U}_0}{U_0}\widetilde{ds^2}$. This means that the volumes of space-time with fields ψ and $\widetilde{\psi}$ are similar geometrically, therefore, the directly observed Newtonian law of law does not change. On the contrary, continuous conformal transformations distort the geometry and are accompanied by a change in the Newtonian law of gravitation.

Thus, if there are two regions of space-time in which complex scalar fields are connected by a phase transformation $\varphi \leftrightarrow \widetilde{\varphi} + \alpha$, then the spatial scales in these regions are connected by a transformation $dl \leftrightarrow \left(\frac{\widetilde{U}_0}{U_0}\right)^{1/2} \widetilde{dl}$. Let there be a set of regions in which the phases of the fields and the spatial scales are connected by the equations $\varphi_i = \varphi + \alpha_i$ and $(dl)_{i+1} \leftrightarrow \left(\frac{\widetilde{U}_0}{U_0}\right)^{1/2} (dl)_i$, respectively. Then a sample of physically similar structures from these regions will form a fractal.

3. Spatial Variations of the Hubble Constant and Local Gravitational Perturbations

Hubble's article on "relation between distance and radial velocity among extra-galactic nebular" was published 90 years ago [12]. This discovery was the first law of the observational cosmology. The Hubble constant is the most important cosmological parameter. It is used to estimate cosmological distances to galaxies and their clusters. It is used in the theoretical cosmology through the Hubble parameter, $H = \frac{\dot{a}}{a}$, where a is the scale factor of the cosmological model and a dot above the letter means the differentiation according to the cosmological time of the model t. The Hubble parameter characterizes the speed of the expansion of the universe space. In a homogeneous and isotropic cosmological model, the Hubble parameter depends only on cosmological time t (at each moment of time, the expansion rate at all points of the 3D-space is identical). Then in the modern era, the Hubble constant $H_0 = H(t_0)$ has to be identical in all directions.

The results of the latest local measurements H_0 were very carefully performed in [13,14] using data on Cepheids in 23 galaxies, supernovae SN Ia were observed in 19 galaxies (redshifts $0.01 < z < 0.15$, observations by the Hubble Space Telescope). The average value is equal to $H_0^{HST} = (73.48 \pm 1.66) km/s/Mpc$ for the standard cosmological model.

Global estimates H_0 are obtained by fitting cosmological parameters of the model to observations of the CMB anisotropy and galaxy clustering. According to the latest data from the Planck collaboration for a spatially flat model, we have $H_0^{Pl} = (67.4 \pm 0.5) km/s/Mpc$ [15].

In recent publications, an unexpectedly large difference between H_0^{HST} and H_0^{Pl} is explained by the possibility of the presence of still unclear errors of measurements or hypotheses about the dark energy model (for example, [16]) or about the inhomogeneity of dark matter distribution (for example, [17,18]).

Here we propose a simpler hypothesis to explain the inequality $H_0^{HST} \neq H_0^{Pl}$ remaining within the GR framework and without the use of exotic forms of matter. The hypothesis itself appeared due to a careful look at the measurement data H_0^{HST} in various galaxies [13]. A sample of these data is shown in Table 1. It appears that the measured values H_0^{HST} form a manifold that is approximately described by the power of low characterizing fractals. It has the following form $H_0^{HST} \approx q_i H^*$, and $q = 1.023$. Here $H^* = 65 km/s/Mpc$ and the values of the exponent i are shown in Table 1. The proximity of value $q = 1.023$ to 1 may indicate a very small deviation from the isotropic expansion of space.

Let's suppose that this form reflects the real local properties of the space expansion in the vicinity of the supernovae (or galaxies) that were used for the measurement. Variations of the expansion speed relative to the value H_0^{Pl} indicate the presence of local gravitational perturbations in the vicinity of each source (local changes of the space-time metric relative to the metric of the background reference system). The fractality indicates that the processes generating these gravitational perturbations have an identical nature. Such processes can be gravitational waves generated by the asymmetric collapse of a supernova core or cosmological gravitational perturbations. The fractality of the Hubble constant should also be observed in the fractal cosmological model.

Table 1. Hubble constant (data [13]).

Galaxy	H_0^{HST} (km/s/Mpc)	i
M101	68.39	2
N1015	80.09	9
N1309	70.24	3
N1365	68.39	2
N1448	77.77	8
N2442	73.42	5
N3021	63.94	−1
N3370	76.00	7
N3447	74.37	6
N3972	77.98	8
N3982	64.80	−1
N4038	79.69	9
N4258	72.25	5
N4424	63.97	−1
N4536	71.48	4
N4639	77.98	8
N5584	78.67	9
N5917	72.75	5
N7250	74.75	6
U9391	66.53	1

Let's consider local gravitational perturbations in the vicinity of the galaxy which hosts Cepheids and supernovae. These sources are used to measure the Hubble constant in [13]. We use the synchronous space-time metric

$$ds^2 = g_{ik}dx^i dx^k = a^2 d\eta^2 + \left(a^2\delta_{\alpha\beta} + h_{\alpha\beta}\right)dx^\alpha dx^\beta, \tag{36}$$

where Latin indices run through values 0, 1, 2, 3, and Greek 1, 2, 3. The coordinates of the background reference system are η and x^α, the cosmological time t is related to the conformal time η by the Equation $ad\eta = cdt$, and $h_{\alpha\beta}$ is a local gravitational perturbation.

The commoving distance of galaxy r_g is measured along the isotropic geodesic of the space-time with metric (1). For the isotropic geodesic $ds = 0$, then $c^2dt^2 = -\left(g_{\alpha\beta} + h_{\alpha\beta}\right)k^\alpha k^\beta dr^2$. Here the isotropic wave vector is $k^i = \frac{dx^i}{dr}$, $k_i k^i = 0$, and $k^\alpha = \frac{q^\alpha}{a^2}$, $k_\alpha = q_\alpha$, q^α and q_α don't depend on t, $k_\alpha k^\alpha = \frac{q^2}{a^2}$.

The commoving distance of galaxy r_g is equal to

$$r_g = \int_0^{z_g} k_\alpha dx^\alpha = c\int_0^{z_g} \frac{dt}{dz}\frac{k_\alpha k^\alpha}{\sqrt{-\left(g_{\alpha\beta} + h_{\alpha\beta}\right)k^\alpha k^\beta}}dz = cq^2\int_0^{z_g}\frac{dt}{dz}\frac{1}{a}\frac{dz}{\sqrt{\left(\delta_{\alpha\beta} - \frac{h_{\alpha\beta}}{a^2}\right)q^\alpha q^\beta}} \tag{37}$$

Let's consider that $\frac{dt}{dz} = \frac{1}{H_0}\frac{1}{\sqrt{\Omega_m(1+z)^3+\Omega_\Lambda}}$, where the cosmological parameters for the flat ΛCDM model are $\Omega_m = 0.28$, $\Omega_\Lambda = 0.72$. The luminosity distance of galaxy is $d_L = \left(1 + z_g\right)r_g$. It is used in the distance modulus equation of source, $(m - M)_{PL} = 5\log\frac{d_L}{Mpc} + 25$, where $(m - M)_{PL}$ is a distance module for Cepheids. The authors of [11] found values H_0^{HST} for dozens of Cepheids in each galaxy. The most probable value H_0^{HST} (included in Table 1) was determined from the condition of equality of the distance modules for Cepheids and a supernova in the same galaxy.

Using Equation (37), we found the expression for the Hubble constant taking into account the gravitational perturbation:

$$H_0 = cq\left(1+z_g\right)10^{5-\frac{(m-M)_{PL}}{5}} \int_0^{z_g} \frac{1+z}{\sqrt{\Omega_m(1+z)^3+\Omega_\Lambda}} \frac{dz}{\sqrt{\left(1-\frac{h_{\alpha\beta}q^\alpha q^\beta}{a^2q^2}\right)}}. \quad (38)$$

The Hubble constant for the background frame H_0^{Pl} is obtained from Equation (38) with $\frac{h_{\alpha\beta}q^\alpha q^\beta}{a^2q^2} = 0$. Then the relative variation of the Hubble constant is:

$$\frac{H_0 - H_0^{Pl}}{H_0^{Pl}} \approx \frac{\frac{1}{2}\int_0^{z_g} \frac{1+z}{\sqrt{\Omega_m(1+z)^3+\Omega_\Lambda}} \frac{h_{\alpha\beta}q^\alpha q^\beta}{a^2q^2} dz}{\int_0^{z_g} \frac{1+z}{\sqrt{\Omega_m(1+z)^3+\Omega_\Lambda}} dz}, \quad (39)$$

where $\frac{h_{\alpha\beta}q^\alpha q^\beta}{a^2q^2} \ll 1$.

For a local perturbation of the metric we can assume that the main contribution to the integral of the term of a fraction (38) is a perturbation in the vicinity of the galaxy at an epoch $z = z_g$. Then, to estimate the variation of the value of the Hubble constant, we used two conditions $\frac{h_{\alpha\beta}q^\alpha q^\beta}{a^2q^2} \approx \frac{q^\alpha q^\beta}{q^2}\left(\frac{h_{\alpha\beta}}{a^2}\right)_{z_g}$, and $h_{\alpha\beta}(z < z_g) \approx 0$. In this case we obtained:

$$\frac{H_0 - H_0^{Pl}}{H_0^{Pl}} \approx \frac{1}{2}\frac{q^\alpha q^\beta}{q^2}\left(\frac{h_{\alpha\beta}}{a^2}\right)_{z_g}. \quad (40)$$

Let's consider the case, when the galaxies used in [13] belong to the outskirts of the local supercluster. The medium density contrast in the superclusters is of the order $\frac{\delta\varepsilon}{\varepsilon} \approx 0.1$ (see, for example, [19]). This density contrast corresponds to the metric perturbation $h_{\alpha\beta}$ and $\frac{\delta\varepsilon}{\varepsilon} \approx \frac{q^\alpha q^\beta}{q^2}\left(\frac{h_{\alpha\beta}}{a^2}\right)_{z_g}$. As a result, the variation of the Hubble constant is $\frac{H_0 - H_0^{Pl}}{H_0^{Pl}} \approx \frac{1}{2}\frac{\delta\varepsilon}{\varepsilon} \approx 0.05$. This value is comparable to the average variation of the Hubble constant in measurements [11]. The above example shows that variations of the Hubble constant can be used to study local gravitational perturbations.

In the fractal cosmological model, the space-time is considered as a set of self-similar regions of space-time. This self-similarity is due, in particular, to the fact that the transition from one region to another is reduced to stretching or shrinking the interval. Metric tensors of two regions are related by the relation: $\widetilde{g}_{\alpha\beta} = \frac{U_0}{\widetilde{U}_0}g_{\alpha\beta}$, where $\frac{U_0}{\widetilde{U}_0}$ is a constant number, it can be different for different pairs of regions. Consider the metric $\widetilde{g}_{\alpha\beta}$ in the vicinity of the galaxy as weakly perturbed relative to the metric of a homogeneous and isotropic space-time $g_{\alpha\beta}$ for the region in which the observer is located. Then $\widetilde{g}_{\alpha\beta} = g_{\alpha\beta} + h_{\alpha\beta}$, and $h_{\alpha\beta} = \left(\frac{U_0}{\widetilde{U}_0} - 1\right)g_{\alpha\beta}$. The direction of the isotropic geodetic upon transition from one region to another does not change. Using Equation (39), we found that in such a model, the observer will find a variation of the Hubble constant, $\frac{H_0 - H_0^{Pl}}{H_0^{Pl}} \approx \frac{1}{2}\left(\frac{\frac{U_0}{\widetilde{U}_0}-1}{a^2}\right)_{z_g}$. The values of the Hubble constant for a set of self-similar regions will form a fractal set, $H_{0n} \approx \delta^i H_0^{Pl}$, where $\delta = 1 + \frac{1}{2}\left(\frac{\frac{U_0}{\widetilde{U}_0}-1}{a^2}\right)_{z_g}$.

4. Discussion

Above, it is proposed to consider the puzzle of fractals as a consequence of the properties of space-time and the symmetry of the interaction fields of matter. In this case, it is necessary to use the geometric theory of gravity, which is similar to GR. We have described the method of constructing the problems solution (1)–(3) and shown that it had simple and easily verified analytical solutions with geometrical properties. It is important that these solutions for complex scalar fields be generalized to multidimensional fields.

This class of geometric solutions of the general relativity equation allows analogy to spatial fractals in large-scale structures of the universe due to its invariance with respect to discrete scale transformation of interval, $ds \leftrightarrow q d\widetilde{s}$. As an application, the treatment of spatial variations of the Hubble constant H_0^{HST} [13] is considered. It is noted that the values H_0^{HST} form almost fractal sets. It has been shown that the variations of H_0^{HST} may be connected with local gravitational perturbations of the space-time metrics in the vicinity of the galaxies containing Cepheids and supernovae selected for measurement. The value of variations H_0^{HST} can indicate the presence of local variations in the space-time metric (local gravitational perturbations) on the outskirts of the local supercluster, and their self-similarity indicates the fractal distribution of matter in this region. This example shows that the variations of the Hubble constant can be used to study local gravitational perturbations.

Acknowledgments: The author thanks the All-Russian Institute for Scientific and Technical Information of Russian Academy of Sciences for the opportunity to use the polythematic database VINITI.

References

1. Dyson, F.W.; Eddington, A.S.; Davidson, C.R. A Determination of the Deflection of Light by the Sun's Gravitational Field, from Observations Made at the Total Eclipse of May 29, 1919. *Mem. R. Astron. Soc.* **1920**, *220*, 291–333. [CrossRef]
2. Iorio, L. Editorial for the Special Issue 100 Years of Chronogeometrodynamics: The Status of the Einstein's Theory of Gravitation in Its Centennial Year. *Universe* **2015**, *1*, 38–81. [CrossRef]
3. Rozgacheva, I.K.; Agapov, A.A. Is Dark Matter Needed in Galaxies? *Astron. Rep.* **2018**, *62*, 551–556. [CrossRef]
4. Minkowski, H. Space and Time. *Phys. ZS.* **1909**, *10*, 104–111. [CrossRef]
5. Weyl, H. Gravitation und Elektrizität. *Sitz. Preuss. Akad. Wiss.* **1918**, 465–480.
6. Weyl, H. Elektron und Gravitation. *Zeitschrift für Physik* **1929**, *56*, 330–352. [CrossRef]
7. Baryshev, Y.; Teerikorpi, P. Fundamental Questions of Practical Cosmology. In *Astrophysics and Space Science Library*; Springer Science: Dordrecht, The Netherlands, 2012. [CrossRef]
8. Rozgacheva, I.K.; Borisov, A.A.; Agapov, A.A.; Pozdneev, I.A.; Shchetinina, O.A. Fractal properties of the large-scale structure. *arXiv* **2012**, arXiv:1201.5554.
9. Nottale, L. *Scale Relativity and Fractal Space Time*; Imperial College Press: London, UK, 2011.
10. Rozgacheva, I.K.; Agapov, A.A. The fractal cosmological model. *arXiv* **2011**, arXiv:1103.0552.
11. Landau, L.D.; Lifshitz, E.M. *Theory of Field*; Nauka: Moscow, Russia, 1988; Volume II.
12. Hubble, E. A relation between distance and radial velocity among extra-galactic nebulae. *Proc. Natl. Acad. Sci. USA* **1929**, *15*, 168–173. [CrossRef] [PubMed]
13. Riess, A.G.; Macri, L.M.; Hoffmann, S.L.; Scolnic, D.; Casertano, S.; Filippenko, A.V.; Chornock, R. A 2.4% Determination of the Local Value of the Hubble Constant. *Astrophys. J.* **2016**, *826*, 56. [CrossRef]
14. Riess, A.G.; Casertano, S.; Yuan, W.; Macri, L.; Anderson, J.; Mackenty, J.W.; Bowers, J.B.; Clubb, K.I.; Filippenko, A.V.; Jones, D.O.; et al. Parallaxes of Galactic Cepheids from Spatially Scanning the Hubble Space Telescope: Implications for the Hubble Constant. *Astrophys. J.* **2018**, *855*, 136. [CrossRef]
15. Aghanim, N.; Akrami, Y.; Ashdown, M.; Aumont, J.; Baccigalupi, C.; Ballardini, M.; Banday, A.J.; Barreiro, R.B.; Bartolo, N.; Basak, S.; et al. Planck 2018 results. VI. Cosmological parameters. *arXiv* **2018**, arXiv:1807.06209.

16. Ryan, J.; Chen, Y.; Ratra, B. Baryon acoustic oscillation, Hubble parameter, and angular size measurement constraints on the Hubble constant, dark energy dynamics, and spatial curvature. *arXiv* **2019**, arXiv:1902.03196.
17. Cusin, G.; Pitrou, C.; Uzan, J.P. Are we living near the center of a local void? *J. Cosmol. Astropart. Phys.* **2017**, *2017*, 038. [CrossRef]
18. Dumin, Y.V. Can the Dark-Matter Deficit in the High-Redshift Galaxies Explain the Persistent Discrepancy in Hubble Constants? *arXiv* **2018**, arXiv:1804.00562.
19. Tansella, V.; Bonvin, C.; Durrer, R.; Gosh, B.; Sellentin, E. The full-sky relativistic correlation function and power spectrum of galaxy number counts. Part I: theoretical aspects. *J. Cosmol. Astropart. Phys.* **2018**, *1803*, 19. [CrossRef]

4

A Continuous Coordinate System for the Plane by Triangular Symmetry

Benedek Nagy * and Khaled Abuhmaidan

Department of Mathematics, Faculty of Arts and Sciences, Eastern Mediterranean University, via Mersin 10, Turkey, Famagusta 99450, North Cyprus; kabuhumaidan@yahoo.com
* Correspondence: nbenedek.inf@gmail.com

Abstract: The concept of the grid is broadly used in digital geometry and other fields of computer science. It consists of discrete points with integer coordinates. Coordinate systems are essential for making grids easy to use. Up to now, for the triangular grid, only discrete coordinate systems have been investigated. These have limited capabilities for some image-processing applications, including transformations like rotations or interpolation. In this paper, we introduce the continuous triangular coordinate system as an extension of the discrete triangular and hexagonal coordinate systems. The new system addresses each point of the plane with a coordinate triplet. Conversion between the Cartesian coordinate system and the new system is described. The sum of three coordinate values lies in the closed interval $[-1, 1]$, which gives many other vital properties of this coordinate system.

Keywords: barycentric coordinate system; coordinate system; hexagonal grid; triangular grid; tri-hexagonal grid; transformations

---

1. Introduction

The concept of the grid is essential and heavily used in digital geometry and in digital image processing. A grid is comprised of discrete points addressed with integer vectors. There are three regular tessellations of the plane, which define the square, hexagonal, and triangular grids (named after the form of the pixels used as tiles) [1]. Most of the applications use the square grid because its orthogonal coordinate system, known as the Cartesian coordinate system (CCS), which fits very well to it. This addresses each square pixel of the grid by a pair of independent integers. The dual of the square grid (the grid formed by the nodes, which are the crossing points of the gridlines) is again a square grid. Therefore, essentially, the same CCS is used as well. Working with real images, we may need to perform operations that do not map the grid to itself, e.g., zooming or rotations. The Cartesian coordinate system allows real numbers to be used in such cases. Moreover, the digitization operation can easily be defined by a rounding operation.

The hexagonal grid, tiling the plane by the same size regular hexagons (hexagonal pixels), has been used for decades in image processing applications [2], in cartography [3,4], in biological simulations [5], and in other fields, since the digital geometry of the hexagonal grid provides better results than the square grid in various cases. In addition, it is used in various table and computer games based on its compactness. It is the simplest 2D grid, since the only usual neighborhood using the nearest neighbors is simpler and less confusing than the two types of neighbors in the square grid [6]. The neighborhood of a pixel contains six other hexagons (see Figure 1a). In contrast to the square grid, the dual of the hexagonal grid is not the hexagonal, but the triangular grid (see Figure 1a,b). Adequate and elegant coordinate systems for these kinds of grids are required for their use in both theory and applications, e.g., in image processing or engineering applications. In Reference [7], a three-coordinate-valued system of zero-sum triplets is used to describe the hexagonal grid capturing the triangular symmetry of the grid. In Figure 1a, the first coordinate value is ascending right-upwardly, the second values are ascending into the right-downward direction, and the third one is ascending into the left-upward direction [7,8]. We should mention that this system could be seen as the extension of the oblique coordinate system using two independent integer values [9] by concerning the third value to obtain zero-sum for every triplet. The digital distance based on the neighborhood relation is computed in Reference [9]. Since the vectors describing the grid are not orthogonal, some geometric descriptions based on Cartesian coordinates are not very clear. However, to simplify the expressions of the constrained three-dimensional coordinate system is recommended. We should also mention that 0-sum triplets allowing real numbers were used in Reference [8] to describe rotations (that may not map the hexagonal grid to itself). In this way, a useful digitization operator is found. Her's system was mentioned and used in References [10,11] for various imaging-related disciplines.

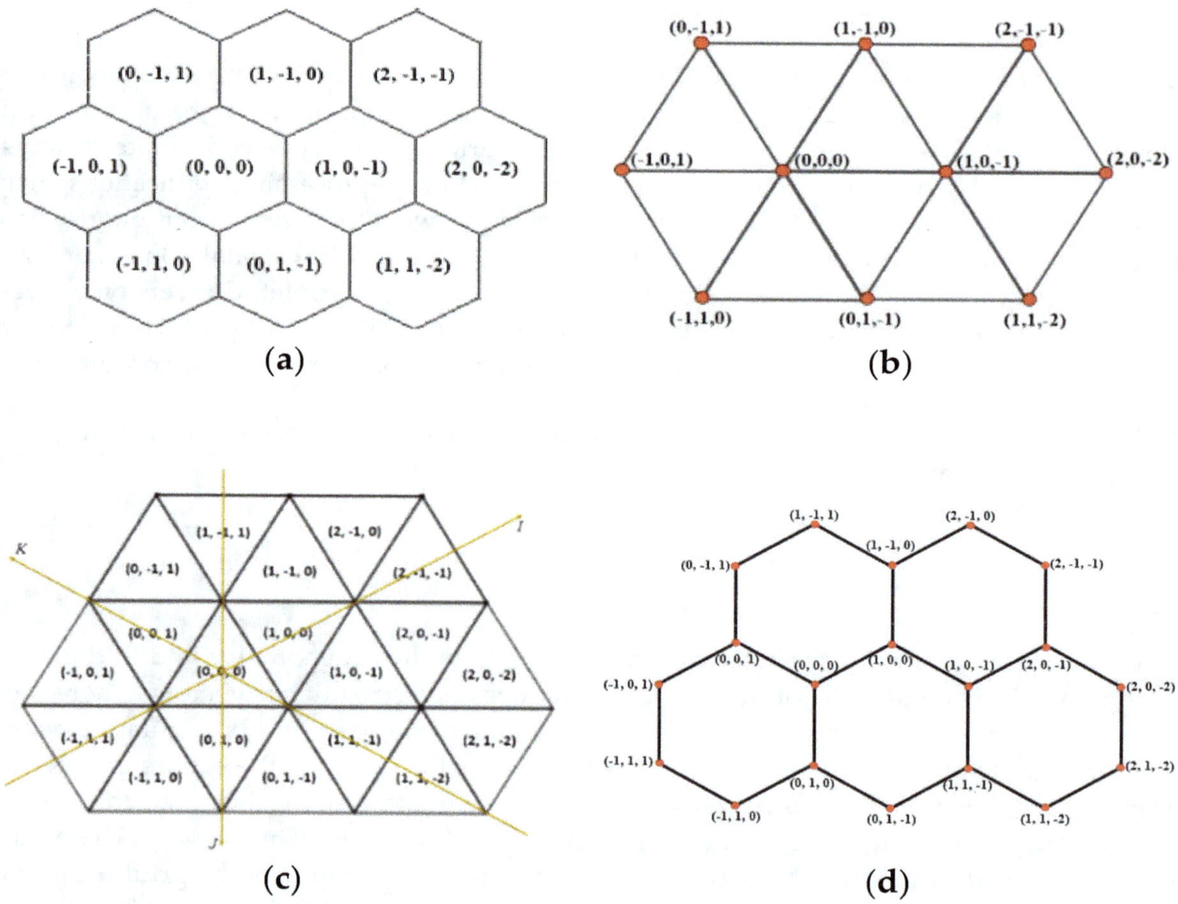

Figure 1. The coordinate system for the hexagonal grid (**a**) and its dual (**b**). The coordinate system for the triangular grid (**c**) and its dual (**d**).

The triangular grid is the third regular grid. It is generated by tiling the plane regularly with equilateral triangles. Although it is the most complex among the three regular tessellations (it has the largest number and types of neighbors), it has various advantages in applications, e.g., by the flexibility of the used neighborhood. The triangular grid is built by triangles in two different orientations. Moreover, it is not a point lattice since some of the grid vectors, i.e., the vectors connecting the midpoints of triangles with different orientations, do not translate the grid into itself [12]. However, the triangular grid gives a valid alternative for applications in image processing. In some cases, it gives better results than the usual square grid due to its better symmetric properties and larger, natural neighborhood structure. Each pixel has 12 neighbors sharing at least a corner. They are categorized into three types of neighbor relations [13,14]. The triangular grid has similar symmetry to the hexagonal grid. Therefore, in Reference [15], a coordinate system with zero-sum and one-sum triplets are used to describe this grid (the three values are not independent, since this is also a 2D grid, see Figure 1c,d). The angle between any two of the three coordinate axes is 120° as for the hexagonal grid. In this description, two pixels are neighbors if their coordinate values differ by at most one. Moreover, at the closest neighbors, only one coordinate differs [15,16]. Observe that exactly the same coordinate triplets are used on the left and on the right-hand side in Figure 1c,d, respectively, which shows the duality of the triangular and hexagonal grids. In Reference [17], a combinatorial coordinate system was shown that addresses the pixels, the corners, and the edges between them. That coordinate system gives an efficient tool to work with digital cell complexes on the triangular grid. Various image processing algorithms have been defined and implemented for the triangular grid recently such as discrete tomography [18,19], thinning [20], and mathematical morphology [21]. We should also mention that triangulation is a frequently used technique in imaging and computer graphics. However, the obtained grid is usually not the regular triangular grid. Lastly, we should note that several non-regular grids have various applications in 2D and in higher dimensions.

The coordination is vital because it is a leading tool in making a simple, easily usable, and effectively programmable system with integer numbers (coordinates). The isometric transformations are described in Reference [12]. However, up to now, there was no such extension of this coordinate system that is able to address the entire plane. For various applications, including, e.g., arbitrary angled rotations, an extension of the coordinate system is needed. We note here that Her's zero-sum triplets could match only up to half of the grid points in the triangular grid (Figure 1c,d), and, therefore, the coordinate system addressing the whole plain cannot be used for the triangular grid.

In this paper, we introduce a continuous coordinate system for the plain based on the symmetry of the triangular grid, where every point of the 2D plane has its unique coordinate triplet. We use three coordinate values to describe the triangular grid as in Reference [15] but also to address the points of the plane "between" and "around" the nodes of the dual grid. Our new coordinate system is shown to be an extension of the hexagonal and of the triangular coordinate systems. Moreover, our system builds upon the coordinate system for the so-called, tri-hexagonal grid, also called the three-plane triangular grid in Reference [16].

For further applications, we also provide a mapping between our continuous coordinate system for the triangular grid and the Cartesian coordinate system of the 2D plane.

The rest of this paper is as follows. The next, preliminary section, describes some important discrete coordinate systems and the barycentric coordinate system. The continuous coordinate system for the triangular grid is then introduced in Section 3. Conversions to and from Cartesian coordinates are also presented. Some important properties of the new coordinate system are presented in Section 4. Lastly, conclusions close the paper.

2. Preliminaries

In this paper, as usual, Z^3 denotes the cubic grid, whose points are addressed by integer triplets, according to the three coordinates x, y, z.

In order to create a continuous coordinate system for the triangular grid that enables us to uniquely address any point of the triangular grid, we combine discrete triangular coordinate systems from Reference [22] (see also Figure 1 for some examples) with the barycentric coordinate system (BCS), discovered by Möbius (see References [1,23]).

In Figure 2, a coordinate system for the tri-hexagonal grid (the three-plane triangular grid of regions [16]) and its dual is given. This grid resembles a mix of the triangular and hexagonal grids since it is a combination of the one-plane and two-plane grids [7,22]. The new coordinate system will be an extension of the discrete triangular and hexagonal coordinate systems.

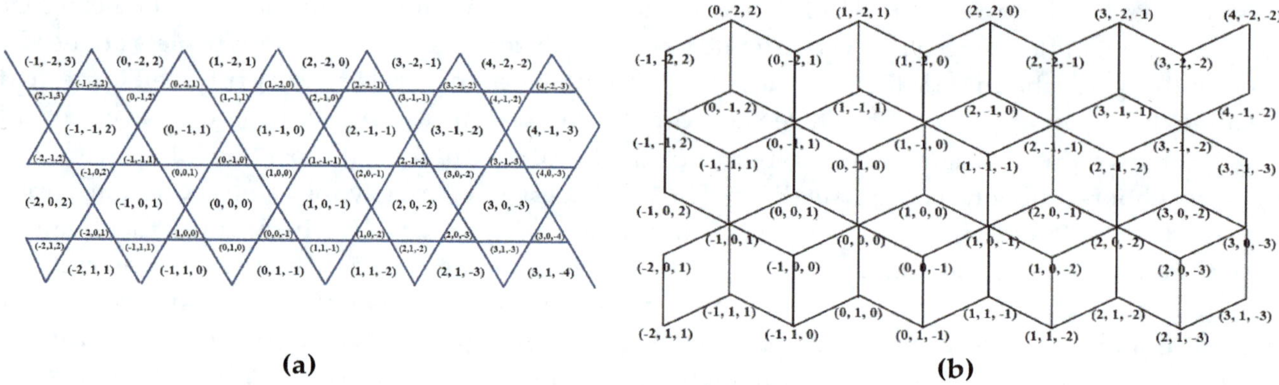

Figure 2. Representation of the tri-hexagonal coordinate system (**a**) and its dual (**b**). The same coordinate system is used to address the pixels (**a**) and the nodes of the dual grid (**b**).

Next, a brief description is given for the discrete triangular coordinate system and BCS.

2.1. Discrete Triangular Coordinate System

The discrete Hexagonal Coordinate System uses 0-sum triplets (Figure 1a,b). The discrete Triangular Coordinate System [22] is a symmetric coordinate system that addresses each pixel by an integer triplet. The three coordinate axes have angles of 120° as in the hexagonal grid. The sum of the triplets is equal to 0 or 1, which refers to the two types of orientations of triangles (△, ▽). The triangles with zero-sum are the "even" triangles (△), and the triangles with one-sum triplets are the "odd" triangles (▽) (see Figure 1).

For finding an appropriate extension to this system that is able to address all points of the 2D plane, we start by addressing the midpoints of triangles with integer triplets of +1 and −1 sum. Therefore, we call them "positive △" and "negative ▽" triangles, respectively. According to Figure 3, the coordinate system of Figure 2b is used to address midpoints of triangles of the triangular grid. Observe that each triplet assigned to a midpoint (see the blue triplets in Figure 3) builds up from the coordinate values shared by two of the corners of the given pixel (see the three red triplets around each blue triplet). There is already an important difference between our proposed and Her's zero-sum coordinate system, which includes the use of zero-sum triplets to address these midpoints as well (actually, three fractional values for each midpoint), which was a very good and efficient choice to extend the coordinate system of the hexagonal grid. However, it does not meet our requirements. Therefore, we have fixed these coordinate values in another way. We should also mention that Her's system inside the regular triangles can be seen as an application of the barycentric coordinate system (see next subsection) based on the values assigned to the corners of the triangle.

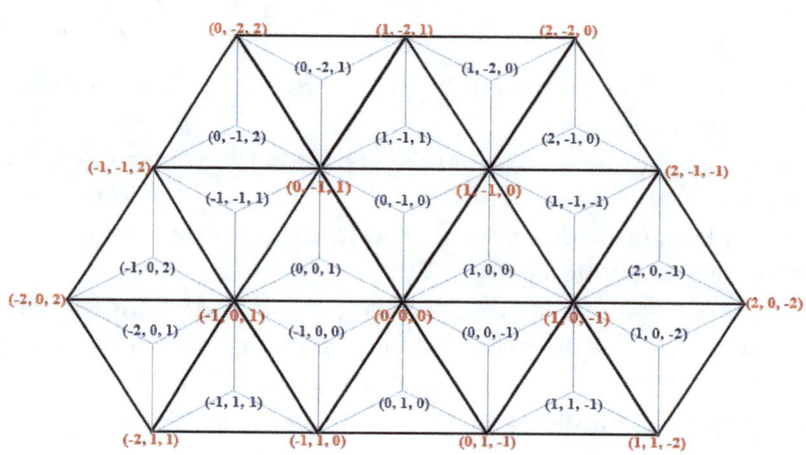

Figure 3. The coordinate system for the tri-hexagonal grid is used for the triangular grid (and for its dual at the same time).

2.2. The Barycentric Coordinate System (BCS)

One of the main motivations of the barycentric technique is to use a coordinate system only for a finite (bordered) segment of the plane, which is also more "balanced" inside this region than the values of the Cartesian frame.

The barycentric technique uses coordinate triplets to address any points inside (and on the border) of a given triangle. We put three one-sum weights, (w, v, u), to the corners known as a, b, and c of the triangle and the mass center p inside the triangle is assigned to the triplet of weights. It is also known that, if the area of the triangle abc is one unit, then the areas of bcp, acp, and abp are exactly w, v, and u, respectively. The coordinates of p can be computed from the coordinates of the corners of the triangle by weighted average, i.e., $p = wa + vb + uc$ (where p, a, b, and c are the vectors representing the positions of these points).

This formula can easily be transformed to the following formula using the fact that $w = 1 - v - u$:

$$p = a + v\,(b - a) + u\,(c - a) \tag{1}$$

Actually, since the three barycentric coordinate values (w, v, and u) of point p are not independent (sum of 1), we may use only two of them, v and u, to address point p similarly as in an oblique coordinate system. One may understand Equation (1) as stating that the starting point is a, and we can go to the direction of b and the direction of c by some distance, which is indicated by v and u, respectively. However, in our approach, the starting point a is the midpoint of the regular triangle and, according to the values of v and u, the coordinate triplet for point p is calculated (see Figure 4 and Example 1).

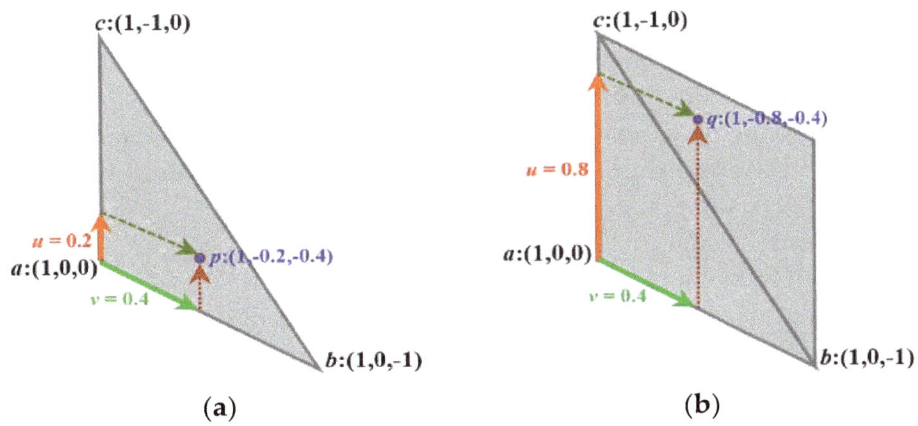

Figure 4. A composition of the barycentric technique and discrete coordinate system to address points p and q in the triangular plane by coordinate triplets in (**a**) and (**b**), respectively.

In the classical barycentric technique, the point is considered to be inside a triangle, as long as the sum of (u and v) is between 0 and 1: $0 < u + v < 1$. If the sum is equal to 1, then the point will be on the edge (cb), while it will be out of the triangle if the sum is less than 0 or greater than 1 (see Figure 4a). Now, we relax the condition of the barycentric technique and allow the sum of u and v to be any real number between 0 and 2 besides the conditions $0 \le u \le 1$ and $0 \le v \le 1$ hold (see Figure 4b). In this way, we may also address some points outside of the triangle. In Figure 4 and Example 1, a composition of the barycentric technique and the discrete coordinate system (assigned to the corners of an isosceles triangle in the regular triangle) is given to address other points of the plane.

Example 1. *Consider the triangle defined by corners* $(1, 0, 0)$, $(1, 0, -1)$, *and* $(1, -1, 0)$, *which represent the three vertices* a, b, *and* c, *respectively. Let* $u = 0.2$ *and* $v = 0.4$, *where* $0 < u + v < 1$. *Then, based on Equation (1), the coordinate triplet of point* p *is* $(1, -0.2, -0.4)$. *If* $u = 0.8$ *and* $v = 0.4$ $(0 < u + v < 2)$, *then the coordinate triplet of point* q *is* $(1, -0.8, -0.4)$ *(see Figure 4).*

3. Continuous Coordinate System for Reflecting the Triangular Symmetry

In order to create a continuous triangular coordinate system that works efficiently, we combine the discrete coordinate system for the triangle grid with BCS. In the discrete triangular coordinate system, integer coordinate triplets with various sums were used. In BCS, coordinate triplets with fractional values address the points inside a triangle. We develop a new system, which uses triplets on the entire plane. We start by dividing each equilateral triangle of the triangular grid into three isosceles obtuse-angled triangles, which will possess areas A, B, and C, as shown in Figure 5. In this case, the midpoint m between areas will be the start point (the red point), which is represented by the letter a in Equation (1). This point will be used to calculate the coordinates of the points in the three areas A, B, and C.

(a) (b)

Figure 5. Dividing positive (**a**) and negative (**b**) triangles to three areas A, B, and C. The letters assigned to the isosceles triangles are based on the orientation of the sides.

As we have already mentioned above, we use coordinate triplets with sum +1 and −1 for these midpoints, depending on the orientation of the original triangle. The sum of 1 represents the midpoint of the positive triangles △ and the sum of −1 represents the midpoint of negative triangles ▽. Therefore, using the barycentric Equation (1) based on these midpoints, we obtain a unique triplet for each point in each area of the plane, which we will describe below.

Based on the barycentric Equation (1), we know that the values u and v are limited by $0 \leq u + v \leq 1$ (inside or on the border of the given triangle), which gives the ability to address the points inside areas A, B, and C of each type of triangle (△, ▽), separately. However, let us consider the case in which the sum of u and v satisfies $0 \leq u + v \leq 2$, such that the conditions $0 \leq u \leq 1$ and $0 \leq v \leq 1$ hold. Then, consequently, each midpoint can be used to address not only the points in the area located in this original triangle but also the points in the neighboring area denoted with the same letters. To illustrate this, the green area in Figure 6a can be completely addressed by using either midpoint $a^{(+)}$ or $a^{(-)}$ as the starting point in Equation (1).

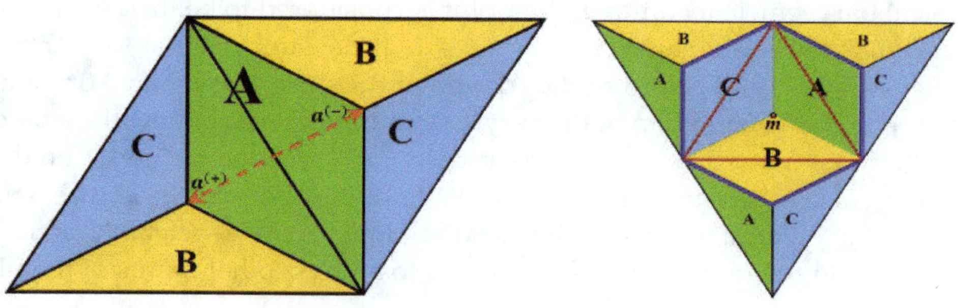

(a) (b)

Figure 6. (**a**) By using either $a^{(+)}$ or $a^{(-)}$, the whole green area A could be addressed. (**b**) The hexagon surrounded by the thick dark blue line shows the entire area that can be addressed by using a positive midpoint m.

Proposition 1. *To address the points inside a rhombus A, B, or C, the coordinate triplet of a point does not depend on the choice of whether the midpoint of the positive or the negative triangle is used for addressing.*

Proof. Assume point (p) in area A of the negative triangle. Let $a^{(+)} = (a_1, a_2, a_3)$, $a^{(-)} = (a_1, a_2 - 1, a_3 - 1)$, $p = (p_1, p_2, p_3)$, $b = (b_1, b_2, b_3)$, and $c = (c_1, c_2, c_3)$ where $a^{(+)}$ is the midpoint of the positive triangle and $a^{(-)}$ is the midpoint of the negative one (see Figure 7).

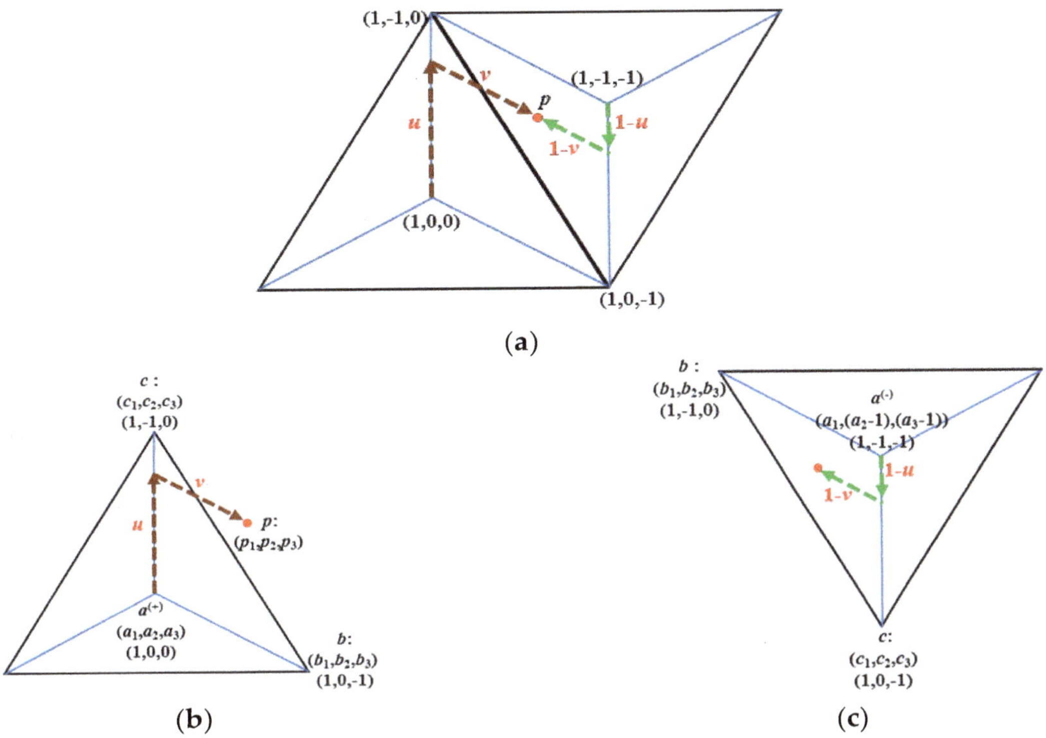

Figure 7. Proving how point p can be calculated by either the positive or negative midpoint ($a^{(+)}$ or $a^{(-)}$). (**a**) Shows the position of point p with respect to both a positive and a negative triangle, while (**b**) and (**c**) represent the calculation of the coordinates of point p based on the positive and negative triangles, respectively.

As we mentioned earlier, each triplet assigned to a midpoint builds up from the coordinate values shared by two of the corners of the given pixel (see Section 2.1). Thus, we have some equalities.

In the positive triangle, we have:

$$a_1 = b_1 = c_1,\ b_2 = a_2,\ c_3 = a_3 \text{ and } c_2 = a_2 - 1,$$

In the negative triangle, we have:

$$a_1 = b_1 = c_1,\ b_2 = a_2,\ c_3 = a_3 \text{ and } b_2 = a_2 - 1.$$

Now, to compute the first coordinate value, a_1, from the positive triangle, we have the following.

$$p_{1(+)} = a_1 + v\,(b_1 - a_1) + u\,(c_1 - a_1) \text{ since } a_1 = b_1 = c_1 \text{ then } p_{1(+)} = a_1.$$

From the negative side, we have the following.

$$p_{1(-)} = a_1 + (1 - v)\,(b_1 - a_1) + (1 - u)\,(c_1 - a_1) \text{ then also } p_{1(-)} = a_1.$$

For the second coordinate value, a_2, from the positive triangle, we have the following.

$$p_{2(+)} = a_2 + v\,(b_2 - a_2) + u\,(c_2 - a_2),$$

since $b_2 = a_2$, then:

$$p_{2(+)} = a_2 + u\,(c_2 - a_2).$$

Using $c_2 = a_2 - 1$, we get $p_{2(+)} = a_2 - u$.
From the negative side, we have the following.

$$p_{2(-)} = (a_2 - 1) + (1 - v)\,(b_2 - (a_2 - 1)) + (1 - u)\,(c_2 - (a_2 - 1)),$$

since $b_2 = (a_2 - 1)$, it is:

$$p_{2(-)} = (a_2 - 1) + (1 - u)\,(c_2 - (a_2 - 1)),$$

since $c_2 = a_2$, then we have $p_{2(-)} = a_2 - u$.
Lastly, for the third coordinate value, a_3, from the positive triangle, we have:

$$p_{3(+)} = a_3 + v\,(b_3 - a_3) + u\,(c_3 - a_3).$$

Since $c_3 = a_3$,

$$p_{3(+)} = a_3 + v\,(b_3 - a_3).$$

Furthermore, $b_3 = a_3 - 1$, which yields to $p_{3(+)} = a_3 - v$.
From the negative side, we have the following.

$$p_{3(-)} = (a_3 - 1) + (1 - v)\,(b_3 - (a_3 - 1)) + (1 - u)\,(c_3 - (a_3 - 1)).$$

Since $c_3 = a_3 - 1$ and $b_3 = a_3$, it can be written as:

$$p_{3(-)} = (a_3 - 1) + (1 - v)\,(b_3 - (a_3 - 1)) = a_3 - v.$$

Having the point inside other regions, the proof goes in a similar manner. □

As we have already mentioned, a popular way to understand BCS for a point (inside a triangle) goes by the ratio of the areas defined by the triangles determined by the point and two of the triangle corners. In fact, our system uses a similar technique to address the points inside a triangle since it is stated in the next corollary based on the previous proposition.

Corollary 1. *Let p be any point inside or on the border of an obtuse-angled triangle determined by a midpoint* $a = (a_1, a_2, a_3)$, *and corners* $b = (b_1, b_2, b_3)$, $c = (c_1, c_2, c_3)$. *Let the barycentric coordinates of p be* (w, v, u), *with* $w + v + u = 1$, *i.e., by assigning these weights to a and b and c, respectively, the weighted midpoint is at p. Then, the coordinates of* $p = (p_1, p_2, p_3)$ *are exactly* $p_i = w\,a_i + v\,b_i + u\,c_i$ *for* $i = 1, 2, 3$.

Notice that the three points a, b, and c above must have a fixed coordinate value (depending on the type of the triangle). The weighted average of this coordinate value will be the same for any point inside or on the border of this obtuse-angled triangle.

As we have seen a given triplet of corners, including a midpoint can be used to address points not only on the inside but also on the border of the triangle determined by them. The type of these regions is important in this issue. In Figure 6b, the thick, dark blue line shows the entire area that the positive midpoint can address. The key issue is to use triplets of the discrete coordinate system and to use only two barycentric fractional values inside, by using the directions of the sides of the appropriate rhombus in which the point is located. The sides of a rhombus are actually parallel to two of the coordinate axes.

Hereafter, for simplicity, we will use only the positive midpoints for further calculations, while ignoring the negative ones. The triangular plane can be seen in Figure 8a.

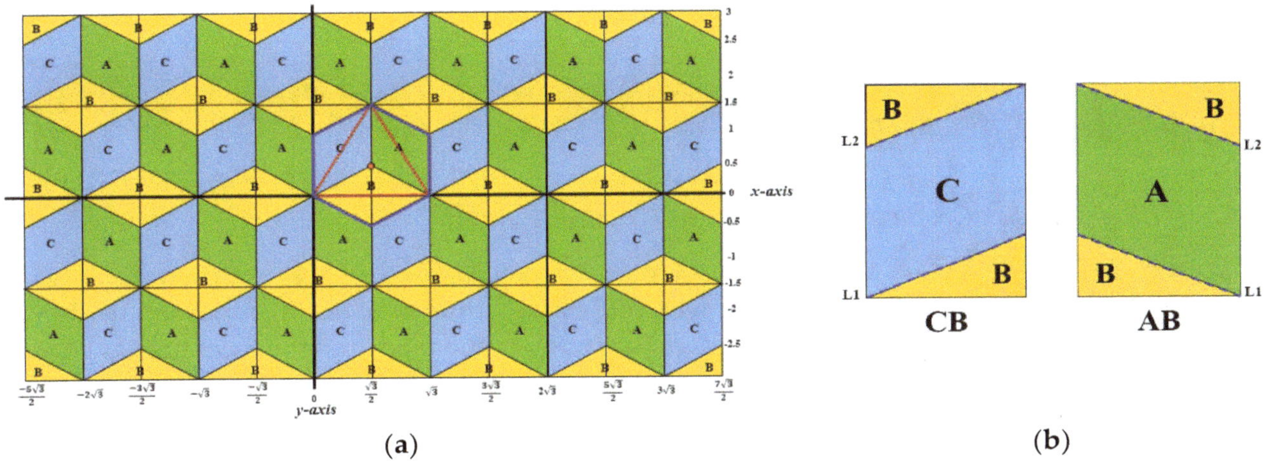

Figure 8. (**a**) Re-structuring the triangular plane to fit the Cartesian plane. (**b**) The two distinguished rectangles of the plane.

In the next two subsections, we will illustrate the conversion between the continuous coordinate system to/from the Cartesian coordinate system. Namely, we can convert the coordinate triplet of a certain point in our new coordinate system to its corresponding Cartesian coordinates and vice versa.

3.1. *Converting Triplets to Cartesian Coordinates*

Assume that we use (i, j, k) as a coordinate triplet of a point, where I, J, and K are the axes of the triangular plane (see Figure 1), and suppose (x, y) is used to indicate the same point in the Cartesian plane where X and Y are the axes.

For the conversion, we fix the side-length of the triangle of the triangular grid to $\sqrt{3}$. Consequently, its height is 1.5 (see the dashed blue lines in Figure 9). Then, the following matrix equation computes the corresponding coordinate values x and y for the given triplet (i, j, k):

$$\begin{pmatrix} \frac{\sqrt{3}}{2} & 0 & -\frac{\sqrt{3}}{2} \\ \frac{1}{2} & -1 & \frac{1}{2} \end{pmatrix} \cdot \begin{pmatrix} i \\ j \\ k \end{pmatrix} = \frac{1}{2} \cdot \begin{pmatrix} \sqrt{3}(i-k) \\ i-2j+k \end{pmatrix} = \begin{pmatrix} x \\ y \end{pmatrix} \tag{2}$$

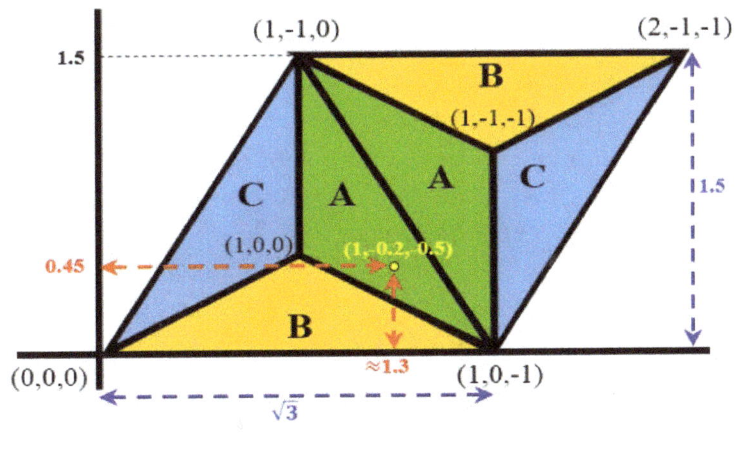

Figure 9. The dashed red lines indicate the Cartesian coordinates of the point.

Example 2. *Let (1, −0.2, −0.5) be a point in the triangular plane. Then, based on (2):*

$$\begin{pmatrix} \frac{\sqrt{3}}{2} & 0 & -\frac{\sqrt{3}}{2} \\ \frac{1}{2} & -1 & \frac{1}{2} \end{pmatrix} \cdot \begin{pmatrix} 1 \\ -0.2 \\ -0.5 \end{pmatrix} \approx \begin{pmatrix} 1.3 \\ 0.45 \end{pmatrix}$$

Thus, $(x, y) \approx (1.3, 0.45)$, *as shown in Figure 9.*

3.2. Converting Cartesian Coordinates to Equivalent Triplets

One of the simplest ways to do such a conversion is to determine the midpoint and the two corner points, which defines the triangle in which the given point locates (inside or on the border). Then, by computing the barycentric coordinates (weights) of the point with respect to these triangle corners, by Corollary 1, the continuous coordinate triplet is computed. In this subsection, we present a slightly different method with little more details to convert Cartesian coordinates to continuous triangular coordinates.

As we already mentioned earlier, the midpoints of positive triangles are used to address all points in the triangular plane. Therefore, every positive midpoint will address areas A, B, and C in neighbor (negative) triangles, as already seen in Figure 6b. Consequently, the triangular plane will be re-structured, which can be seen in Figure 8a. However, two kinds of rectangles can be clearly distinguished in this plane, called CB and AB (see Figure 8b).

We may start by specifying the area (i.e., rhombus) A, B, or C that a Cartesian point (x, y) belongs to. Then, we can use appropriate formulae that are assigned to each type of area, which is a process we will describe later in this section (see Table 1). Hence, the following three steps are used to specify the area.

Step 1 Which quarter of the Cartesian plane is involved? Note that the 1st and 3rd quarters have the same structure, while the 2nd and 4th quarters have another one.
Step 2 Which rectangle is involved (AB or CB)?
Step 3 Which area is involved (A, B, or C)?

Table 1. The coordinate triplets formulae, based on area type, where $\langle \ldots \rangle$ is a rounding operation *.

Coordinate Triplet	Area A	Area B	Area C
i	$\langle \frac{x}{\sqrt{3}} \rangle + \langle \frac{y}{3} \rangle$	$\frac{x\sqrt{3}}{3} + y + j$	$\frac{2x}{\sqrt{3}} + k$
j	$\frac{i+k}{2} - y$	$\langle \frac{-2y}{3} \rangle$	$\frac{i+k}{2} - y$
k	$i - \frac{2x}{\sqrt{3}}$	$i - \frac{2x}{\sqrt{3}}$	$\langle \frac{y}{3} \rangle - \langle \frac{x}{\sqrt{3}} \rangle$

* rounding operation returns the nearest integer to the real number, such that numbers exactly the same distance from two integers are rounded to the larger absolute valued one, e.g., $\langle 1.5 \rangle = 2$, $\langle -1.5 \rangle = -2$ and $\langle -0.4 \rangle = 0$.

We show how these steps can be computed by pseudo codes. The first step is the easiest one since we can inquire whether the values of (x, y) are greater or equal to zero or not. Then, this task is completed and the involved quarter is specified (see Code 1).

Code 1.

```
IF ((x ≥ 0) AND (y ≥ 0)) OR ((x < 0) AND (y < 0))
THEN "1st or 3rd quarter"
ELSE "2nd or 4th quarter"
```

To complete the second step, note that the basic measurements of every rectangle are known, with a height and width equal to 1.5 and $\left(\sqrt{3}/2\right)$, respectively. Then, the CB rectangle is involved whenever the integer part of x and y, are both even or both odd. Otherwise, the AB rectangle is involved. See Code 2 to clarify this step.

Code 2.

```
IF ((int (2x/√3) mod 2 = 0) AND (int (y/1.5) mod 2) = 0) OR
(int (2x/√3) mod 2 = 1) AND (int (y/1.5) mod 2) = 1))
THEN "CB Rectangle is involved"
ELSE "AB Rectangle is involved"
```

where:

- int takes the integer part of the (decimal) number,
- mod is the modulus (or remainder, here after division by two).

Now, the involved rectangle is specified and the third step follows. Since we have two types of rectangles AB and CB, and they are symmetric, we will discuss only one of them known as type AB rectangles.

In rectangle AB, a point (x, y) will belong to either part A or part B. To decide which one is involved, we consider Line 1 (L1) and Line 2 (L2) in Figure 8b, where they divide the rectangle into parts A and B. If the point is between L1 and L2, then part A is involved. Otherwise, it is in part B. However, L1 and L2 are considered to be within the area A in cases when the point is on the lines. Equations (3) and (4) of a straight line are used for Line 1 and Line 2, respectively.

$$m \cdot x + r_1 - y = 0, \tag{3}$$

$$m \cdot x + r_2 - y = 0, \tag{4}$$

where:

- m is the slope of L1 and L2, which is a constant here, equal to $\left(-\sqrt{3}/3\right)$,
- r_1 and r_2 are the *y-axis* intercept with L1 and L2, respectively, where $r_2 = r_1 + 1$

This step is started by substituting the point (x, y) in Equations (3) and (4). Then, it is determined that part A is involved if the left side of Equation (3) is not greater than 0 and the left side of Equation (4) is not less than 0. Otherwise, part B is involved. Code 3 is used to clarify this step.

Code 3.

```
IF ((r1 − x√3/3 − y) ≤ 0) AND ((r2 − x√3/3 − y) ≥ 0)
THEN "Area A is involved"
ELSE "Area B is involved"
```

The *y-axis* intercept with Line 1, r_1, in area AB, is computed by computing any point (x, y) on Line 1. Therefore, the point at the bottom-right corner of rectangle AB is computed for this purpose (see the red point in Figure 10). Hence, Equation (5) is used to find the value of r_1. In Equation (5), we used the modulus (mod) as a function naturally extended to real numbers, i.e., it gives the remainder after the division by a real number (and that is between 0 and the divisor). Moreover, by adding 1 to r_1, we get the value of the *y-axis* intercept with Line 2, r_2 (see Code 3).

$$r_1 = y - \left(y \bmod \frac{3}{2}\right) + \frac{\sqrt{3}}{3} \cdot \left(x + \frac{\sqrt{3}}{2} - \left(x \bmod \frac{\sqrt{3}}{2}\right)\right) \tag{5}$$

Similar strategies are used when the CB rectangle is involved, taking care that the slopes of Line 1 and Line 2 will be equal to $\left(\sqrt{3}/3\right)$.

Lastly, when the involved area A, B, or C is specified, particular formulae are used that are specified in Table 1. Example 3 is used for a further explanation.

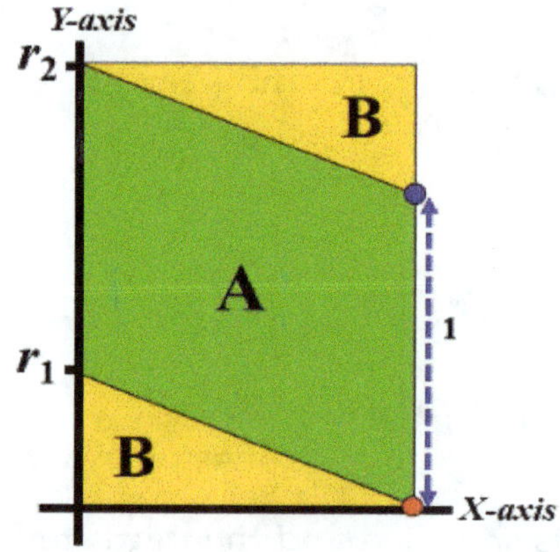

Figure 10. The red point is used to compute the value of r_1, which is the Y-axis intercept with Line 1.

Example 3. *Consider a point with Cartesian coordinates: $(x, y) = (1.299, 0.45)$. In order to convert this Cartesian coordinate pair to its corresponding triangular triplet, the three steps below will be followed.*

Step 1 The 1st or 3rd quarter is involved.
Step 2 It's odd and even. Therefore, rectangle AB is involved.
Step 3 Area A is matched.

Thus, formulae of area A (see Table 1) should be applied in this order, so:

(1) $i = \langle 0.75 \rangle + \langle 0.15 \rangle = 1 + 0 = 1$
(2) $k = 1 - \frac{2x}{\sqrt{3}} \approx -0.500.$
(3) $j \approx -y + 0.5 \cdot 0.5 = -0.2$

The corresponding triplet is $(i, j, k) \approx (1, -0.2, -0.5)$, which is approximately the same as in Example 2, as it should be.

To show the conversion of points on other areas (e.g., Area B or Area C), the following examples are given:

Example 4. *(Point in Area B)*

(a) *Converting from Continuous Coordinate System to CCS.*

Let $(i, j, k) = (0.683, 0, -0.183)$ be a point in the triangular plane. To convert to CCS, Equation (2) is used:

$$\begin{pmatrix} \frac{\sqrt{3}}{2} & 0 & -\frac{\sqrt{3}}{2} \\ \frac{1}{2} & -1 & \frac{1}{2} \end{pmatrix} \cdot \begin{pmatrix} 0.683 \\ 0 \\ -0.183 \end{pmatrix} \approx \begin{pmatrix} 0.75 \\ 0.25 \end{pmatrix}$$

(b) *Converting from CCS to a Continuous Coordinate System.*

Let $(x, y) = (0.75, 0.25)$. The corresponding Continuous Coordinate System triplet can be calculated based on the three steps above as follows.

Step 1 It belongs to the 1st quarter.
Step 2 It belongs to rectangle CB.
Step 3 Area B is matched.

Thus, formulae of area B from Table 1 are applied in the following order:

(1) $j = \langle \frac{-2y}{3} \rangle = 0$

(2) $i = \frac{x\sqrt{3}}{3} + y + j \approx 0.433 + 0.25 + 0 = 0.683$

(3) $k = i - \frac{2x}{\sqrt{3}} \approx 0.683 - 0.866 = -0.183$

The corresponding triplet is (i, j, k) = (0.683, 0, −0.183), which is exactly the original value.

Example 5. *(Point in Area C)*

(a) *Converting from the Continuous Coordinate System to CCS.*

Let (i, j, k) = (0.346, −0.626, 0) be a point in the triangular plane. To convert to CCS, we use Equation (2) below.

$$\begin{pmatrix} \frac{\sqrt{3}}{2} & 0 & -\frac{\sqrt{3}}{2} \\ \frac{1}{2} & -1 & \frac{1}{2} \end{pmatrix} \cdot \begin{pmatrix} 0.346 \\ -0.626 \\ 0 \end{pmatrix} \approx \begin{pmatrix} 0.299 \\ 0.799 \end{pmatrix}$$

(b) *Converting from CCS to a Continuous Coordinate System*

Let (x, y) = (0.299, 0.799). The corresponding Continuous Coordinate System triplet can be calculated based on the three steps below.

Step 1 *It belongs to the first quarter.*
Step 2 *It belongs to rectangle CB.*
Step 3 *Area C is matched.*

Thus, formulae of area C from Table 1 are applied in the following order.

(1) $k = \langle \frac{y}{3} \rangle - \langle \frac{x}{\sqrt{3}} \rangle = 0 - 0 = 0$

(2) $i = \frac{2x}{\sqrt{3}} + k \approx 0.346 + 0 = 0.346$

(3) $j = \frac{i+k}{2} - y \approx 0.173 - 0.799 = -0.626$

The corresponding triplet is (i, j, k) = (0.346, −0.626, 0), which is exactly the original triplet.

Example 6. *(A mid-point)*

(a) *Converting from Continuous Coordinate System to CCS.*

Let (i, j, k) = (1, 0, 0) be a point in the triangular plane. To convert to CCS, Equation (2) is used below.

$$\begin{pmatrix} \frac{\sqrt{3}}{2} & 0 & -\frac{\sqrt{3}}{2} \\ \frac{1}{2} & -1 & \frac{1}{2} \end{pmatrix} \cdot \begin{pmatrix} 1 \\ 0 \\ 0 \end{pmatrix} = \begin{pmatrix} \frac{\sqrt{3}}{2} \\ \frac{1}{2} \end{pmatrix}$$

(b) *Converting from CCS to a Continuous Coordinate System.*

Let (x, y) = ($\sqrt{3}/2$, 0.5). The corresponding Continuous Coordinate System triplet can be calculated based on the three steps below.

Step 1 *It belongs to the first quarter.*
Step 2 *Based on Code 2, it belongs to rectangle CB (but, since it is a mid-point, then either rectangle AB or CB may be used).*
Step 3 *Area B is matched.*

Thus, formulae of area B from Table 1 are applied in the following order.

(1) $j = \langle \frac{-2y}{3} \rangle = 0$

(2) $i = \frac{x\sqrt{3}}{3} + y + j = 0.5 + 0.5 + 0 = 1$

(3) $k = i - \frac{2x}{\sqrt{3}} = 1 - 1 = 0$

The corresponding triplet is (i, j, k) = (1, 0, 0), which is exactly the original value.

4. Properties of the Continuous Triangular Coordinate System

In this section, we will focus on the most important properties of this continuous coordinate system.

4.1. On the Triplets of a General Point

In Figure 6a, consider the red straight line between $a^{(+)}$ and $a^{(-)}$, in the green area. Then, the sum of the coordinate values of the points on this line would change continuously from 1 until −1. Depending on the sum, we can classify the points as follows:

If a point with 3 CV's sum is:

- equal to 1, then it indicates the positive midpoint (i.e., $a^{(+)}$);
- equal to −1, then it indicates the negative midpoint (i.e., $a^{(-)}$);
- equal to 0, then it indicates the point on the triangle's edge;
- positive, then the point belongs to the positive triangle;
- negative, then the point belongs to the negative triangle.

Theorem 1. *The sum of the coordinate triplet of any point in the plane is in the range of the closed interval* $[-1, 1]$.

Proof. Consider an area A of a positive triangle with the corners $a = (a_1, a_2, a_3)$, $b = (b_1, b_2, b_3)$, and $c = (c_1, c_2, c_3)$, where b and c are vertices (corners of an equilateral triangle) of the grid, while a is the midpoint of a positive triangle and $p = (p_1, p_2, p_3)$ is a randomly chosen point belonging to this area (inside or on the border of A). Now based on the barycentric Equation (1), we have the following.

$$\sum_{i=1}^{3} p_i = \sum_{i=1}^{3} a_i + v \cdot \left(\sum_{i=1}^{3} (c_i - a_i)\right) + u \cdot \left(\sum_{i=1}^{3} (b_i - a_i)\right)$$

It is clear that $\sum_{i=1}^{3} a_i = 1$, whereas

$$\begin{aligned} \sum_{i=1}^{3} (c_i - a_i) &= (c_1 - a_1) + (c_2 - a_2) + (c_3 - a_3) \\ &= (c_1 + c_2 + c_3) - (a_1 + a_2 + a_3) \\ &= 0 - 1 = -1 \end{aligned}$$

Similarly, $\sum_{i=1}^{3} (b_i - a_i) = -1$.

Then, by substitution, we have the following.

$$\sum_{i=1}^{3} p_i = 1 - u - v = 1 - (u + v). \quad (6)$$

Since $0 \leq u + v \leq 2$, the maximal and minimal value of the sum of any coordinate triplet is equal to 1 (when $u + v = 0$) and to −1 (when $u + v = 2$), respectively. □

Theorem 2. *The sum of the coordinates of a triplet is non-negative in a positive triangle and non-positive in a negative triangle.*

Proof. Consider a point p that belongs to a positive triangle. Clearly, the coordinates of p are based on u and v such that $0 \leq u + v \leq 1$. Now, by substituting $u + v$ in Formula (6) (at the proof of Theorem 1, the summation will always be non-negative (moreover, it is positive inside the triangle).

Similarly, let p belong to a negative triangle, then $1 < u + v \leq 2$. Thus, by substituting into Formula (6), the sum will always be a non-positive value. □

Every point in the triangular plane has at least one integer value in its triplet. Moreover, the place of the integer value indicates its area (A, B, or C).

Theorem 3. *The* first *coordinate value of every point in area* A *is the same as the* first *coordinate value of the midpoint. The* second *coordinate value of every point in area* B *equals the* second *coordinate value of the midpoint. Similarly, the third coordinate value of any point in area* C *equals the third coordinate value of the midpoint* (Figure 11).

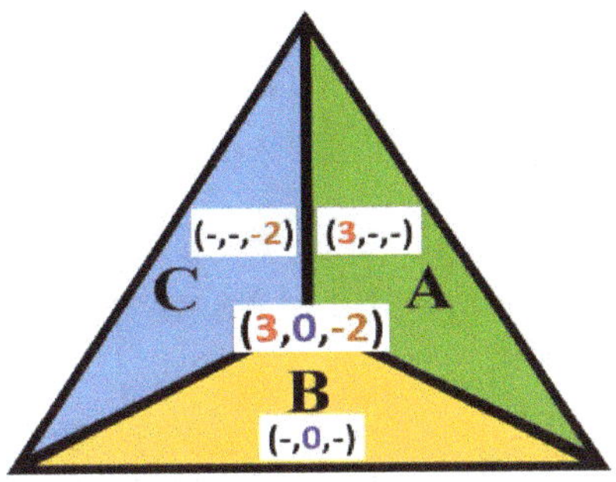

Figure 11. The corresponding constant coordinate value for each area.

Proof. Consider an area A of a positive triangle with the corners $a = (a_1, a_2, a_3)$, $b = (b_1, b_2, b_3)$, and $c = (c_1, c_2, c_3)$, where b and c are vertices (corners of an equilateral triangle) of the grid, while a is the midpoint and $p = (p_1, p_2, p_3)$ is a randomly chosen point belonging to this area (i.e., inside or on the border of the triangle abc). Since it is area A, we have $a_1 = b_1 = c_1$. Substituting this into Equation (1), $p_1 = a_1$ follows for any point p in this area. A similar proof can be considered for areas B and C. □

If a triplet contains two integer values, then the point is located on the line bordering the areas. For example, a triplet of the form $(1, 0, k)$ addresses a point on the line (side of the obtuse-angled triangle) between area A and B $(0 \leq k \leq -1)$. However, if three integers are a triplet, then this triplet addresses either a midpoint or a vertex (corner) of a triangle.

4.2. Relation to Discrete Coordinate Systems

In Reference [16], the hexagonal grid is called a one-plane triangular grid since it is a sub-plane of Z^3 and because of its symmetry. The triangular grid (nodes of the hexagonal grid) is called a two-plane triangular grid. Combining one-plane and two-plane grids produces the so-called three-plane triangular grid, which is known as the tri-hexagonal grid (Reference [24], Figure 2). In this subsection, their coordinate systems are compared to the new coordinate system.

Theorem 4. *The triplets containing only integers such that their sum equals* zero *represent exactly the hexagonal grid (one-plane triangular grid).*

Proof. See Figure 1b for the points of this grid. One may check that exactly those points are addressed with zero-sum integer triplets for which the Cartesian coordinate pair is described below.

$$\mathrm{H} = \left\{(x, y)\middle| x = \left(m\sqrt{3}\right)/2,\ y = 1.5n,\ \text{where } m, n \text{ are integers such that } m+n \text{ is even}\right\}.$$

□

Theorem 5. *The triplets containing only integers such that their sum is either* 0 *or* 1 *represent exactly the triangular grid (two-plane triangular grid).*

Proof. See Figure 1d for the points of this grid. The locations of the points with zero-sum coordinate triplets are already known by Theorem 4. Now, we give the locations of the points addressed with 1-sum (integer) triplets.

$$\mathrm{T} = \left\{(x,\ y)\middle| x = \left(m\sqrt{3}\right)/2,\ y = 1.5n - 1,\ \text{where } m, n \text{ are integers such that } m + n \text{ is even}\right\}.$$

One can easily see that the union of these two sets (H and T) of points gives back exactly the vertices of the hexagons of the figure, i.e., the coordinate system of the dual triangular grid. □

Theorem 6. *The triplets containing only integers such that their sum is either* 0 *or* ± 1 *represent exactly the tri-hexagonal grid (that is the three-plane triangular grid).*

Proof. See Figure 3 for the points of this grid. According to Theorem 5, the locations of the zero-sum and one-sum integer coordinate triplets are already shown. We need to show the locations of the points addressed with (integer) triplets that have -1-sum. They are:

$$\mathrm{M} = \left\{(x,\ y)\middle| x = \left(m\sqrt{3}\right)/2,\ y = 1.5n + 1,\ \text{where } m, n \text{ are integers such that } m + n \text{ is even}\right\}.$$

Actually, the points of this grid, T ∪ H ∪ M, are exactly those that were the base of the method of creating the coordinate system. □

5. Conclusions

The presented continuous coordinate system is an extension of some previously known discrete coordinate systems, e.g., of the symmetric coordinate frame for the triangular grid. This extension is needed and helpful for various applications, where the grid points are not necessarily mapped to grid points, e.g., arbitrary angled rotations, zooming or interpolation of images. We should also mention translations of images [25] since the triangular grid is not a point lattice. Mathematical morphology operators are also based on local translations [21]. Thus, our coordinate system provides a new tool for the research direction as well. The proposed system addresses each point of the 2D (triangular) plane. Conversion to and from the Cartesian coordinate system is provided. These mappings are inverses of each other. Thus, the new coordinate system is ready to use in various applications including those operations that do not necessarily map the grid to itself.

Author Contributions: B.N. and K.A. both designed research and wrote the paper. Both authors have read and approved the final manuscript.

Acknowledgments: Comments of the anonymous reviewers are gratefully acknowledged.

References

1. Coxeter, H.S.M. *Introduction to Geometry*, 2nd ed.; Wiley: New York, NY, USA, 1969.
2. Middleton, L.; Sivaswamy, J. *Hexagonal Image Processing—A Practical Approach*; Springer: London, UK, 2005.
3. Carr, D.B.; Olsen, A.R.; White, D. Hexagon mosaic maps for display of univariate and bivariate geographical data. *Cartogr. Geograph. Inform. Syst.* **1992**, *19*, 228–236. [CrossRef]
4. Sahr, K. Hexagonal discrete global grid systems for geospatial computing. *Arch. Photogramm. Cartogr. Remote Sens.* **2011**, *22*, 363–376.
5. Sakai, K.I. Studies on the competition in plants. VII. Effect on competition of a varying number of competing and non-competing individuals. *J. Genet.* **1957**, *55*, 227–234. [CrossRef]

6. Birch, C.P.; Oom, S.P.; Beecham, J.A. Rectangular and hexagonal grids used for observation, experiment and simulation in ecology. *Ecol. Model.* **2007**, *206*, 347–359. [CrossRef]
7. Her, I. A symmetrical coordinate frame on the hexagonal grid for computer graphics and vision. *J. Mech. ASME* **1993**, *115*, 447–449. [CrossRef]
8. Her, I. Geometric Transformations on the Hexagonal Grid. *IEEE Trans. Image Proc.* **1995**, *4*, 1213–1222. [CrossRef] [PubMed]
9. Luczak, E.; Rosenfeld, A. Distance on a hexagonal grid. *IEEE Trans. Comput.* **1976**, 532–533. [CrossRef]
10. Almansa, A. Sampling, Interpolation and Detection. Applications in Satellite Imaging. Ph.D. Thesis, École Normale Supérieure de Cachan-ENS Cachan, Cachan, France, 2002.
11. Pluta, K.; Romon, P.; Kenmochi, Y.; Passat, N. Honeycomb geometry: Rigid motions on the hexagonal grid. In Proceedings of the International Conference on Discrete Geometry for Computer Imagery, Vienna, Austria, 19 September 2017; LNCS; Springer: Cham, Switzerland, 2017; Volume 10502, pp. 33–45.
12. Nagy, B. Isometric transformations of the dual of the hexagonal lattice. In Proceedings of the 6th International Symposium on Image and Signal Processing and Analysis, Salzburg, Austria, 16–18 September 2009; pp. 432–437.
13. Deutsch, E.S. Thinning algorithms on rectangular, hexagonal and triangular arrays. *Commun. ACM* **1972**, *15*, 827–837. [CrossRef]
14. Bays, C. Cellular automata in the triangular tessellation. *Complex Syst.* **1994**, *8*, 127.
15. Nagy, B. Finding shortest path with neighbourhood sequences in triangular grids. In Proceedings of the 2nd International Symposium, in Image and Signal Processing and Analysis, Pula, Croatia, 19–21 June 2001; pp. 55–60.
16. Nagy, B. A symmetric coordinate frame for hexagonal networks. In Proceedings of the ACM Theoretical Computer Science-Information Society, Ljubljana, Slovenia, 11–15 October 2004; Volume 4, pp. 193–196.
17. Nagy, B. Cellular topology and topological coordinate systems on the hexagonal and on the triangular grids. *Ann. Math. Artif. Intel.* **2015**, *75*, 117–134. [CrossRef]
18. Nagy, B.; Lukic, T. Dense Projection Tomography on the Triangular Tiling. *Fundam. Inform.* **2016**, *145*, 125–141. [CrossRef]
19. Nagy, B.; Valentina, E. Memetic algorithms for reconstruction of binary images on triangular grids with 3 and 6 projections. *Appl. Soft Comput.* **2017**, *52*, 549–565. [CrossRef]
20. Kardos, P.; Palágyi, K. Topology preservation on the triangular grid. *Ann. Math. Artif. Intell.* **2015**, *75*, 53–68. [CrossRef]
21. Abdalla, M.; Nagy, B. Dilation and Erosion on the Triangular Tessellation: An Independent Approach. *IEEE Access* **2018**, *6*, 23108–23119. [CrossRef]
22. Nagy, B. Generalized triangular grids in digital geometry. *Acta Math. Acad. Paedagog. Nyházi* **2004**, *20*, 63–78.
23. Skala, V. Barycentric coordinates computation in homogeneous coordinates. *Comput. Graph.* **2008**, *32*, 120–127. [CrossRef]
24. Kovács, G.; Nagy, B.; Vizvári, B. Weighted Distances on the Trihexagonal Grid. In Proceedings of the International Conference on the Discrete Geometry for Computer Imagery, Vienna, Austria, 19 September 2017; LNCS; Springer: Cham, Switzerland, 2017; Volume 10502, pp. 82–93.
25. Abuhmaidan, K.; Nagy, B. Non-bijective translations on the triangular plane. In Proceedings of the 2018 IEEE 16th World Symposium on Applied Machine Intelligence and Informatics (SAMI), Kosice, Slovakia, 7–10 February 2018; pp. 183–188.

5

Algorithm for Neutrosophic Soft Sets in Stochastic Multi-Criteria Group Decision-Making Based on Prospect Theory

Yuanxiang Dong [1,*], Chenjing Hou [1], Yuchen Pan [2] and Ke Gong [3]

1 School of Management Science and Engineering, Shanxi University of Finance and Economics, Taiyuan 030006, China
2 Business School, Southwest University of Political Science & Law, Chongqing 401120, China
3 School of Economics and Management, Chongqing Jiaotong University, Chongqing 400074, China
* Correspondence: dongyx@sxufe.edu.cn.

Abstract: To address issues involving inconsistencies, this paper proposes a stochastic multi-criteria group decision making algorithm based on neutrosophic soft sets, which includes a pair of asymmetric functions: Truth-membership and false-membership, and an indeterminacy-membership function. For integrating an inherent stochastic, the algorithm expresses the weights of decision makers and parameter subjective weights by neutrosophic numbers instead of determinate values. Additionally, the algorithm is guided by the prospect theory, which incorporates psychological expectations of decision makers into decision making. To construct the prospect decision matrix, this research establishes a conflict degree measure of neutrosophic numbers and improves it to accommodate the stochastic multi-criteria group decision making. Moreover, we introduce the weighted average aggregation rule and weighted geometric aggregation rule of neutrosophic soft sets. Later, this study presents an algorithm for neutrosophic soft sets in the stochastic multi-criteria group decision making based on the prospect theory. Finally, we perform an illustrative example and a comparative analysis to prove the effectiveness and feasibility of the proposed algorithm.

Keywords: neutrosophic soft sets; inconsistent information; prospect theory; stochastic multi-criteria group decision making

---

1. Introduction

Many complex issues in engineering, economics, environmental science and medical science involve uncertainties. In order to address these issues, the theory of possibility, fuzzy set [1], rough set [2], and interval mathematic [3] have been developed successively. However, the above theories have their inherent defects, which are mainly reflected in the inadequacy of parameterization tools [4]. In 1999, Molodtsov [4] initiated the soft set theory for modeling uncertainties from the parameterized point of view.

After Molodtsov, the research interests in the soft set theory have been growing rapidly, such as the algebraic structure [5,6], topology [7,8], normal parameter reduction [9], medical diagnosis [10], game theory [4], and decision making under uncertainties [11,12]. In addition, the study of hybrid models that are developed by combining the soft set theory with other mathematical tools, such as rough sets [13], fuzzy sets [14], and intuitionistic fuzzy sets [15], has also been an important research topic.

Under uncertain environments, a mass of inconsistent information appears due to diversities of source platforms and the differences in the acquisition time. To address issues involving inconsistencies, Smarandache [16] initiated neutrosophic sets from the perspective of philosophy. Subsequently, Maji [17] integrated neutrosophic sets into soft sets to propose neutrosophic soft sets, which retain the characteristics of neutrosophic sets and have adequate parameterization tools. Neutrosophic soft sets are characteristic by three independent functions, including a pair of asymmetric functions: Truth-membership and false-membership, and an indeterminacy-membership function. Among them, the truth-membership and false-membership represent the degree of belongingness and non-belongingness of an element with respect to parameters. The indeterminacy-membership shows the neutrality degree of an element related to parameters.

In recent years, the theory extensions of neutrosophic soft sets have made a rapid progress. Sahin and Küçük [18] constructed generalised neutrosophic soft sets. Deli and Broumi [19] refined the concept and operations of Maji's neutrosophic soft sets. In addition, they also studied the neutrosophic soft matrix and their operators. Considering that the approximate range is usually used to describe complex situations when there is no sufficient information, Deli [20] expanded the values of the truth-membership, indeterminacy-membership, and false-membership to the form of interval values to construct interval-valued neutrosiphic soft sets. Karaaslan [21] introduced the possibility of neutrosophic soft sets by assigning probability to the three function values and defined related properties and operations. In addition, the concepts of single-valued neutrosophic refined soft sets [22], generalized neutrosophic soft expert sets [23], and neutrosophic soft rough sets [24] were presented successively.

Meanwhile, neutrosophic soft sets are also employed in the fields of clustering, prediction and decision making under uncertainties, among which decision making under uncertainties is the most widely applied. Deli [20] proposed a decision making method of interval-valued neutrosophic soft sets by level soft sets, and illustrated it by an example. Peng and Liu [25] constructed three decision making algorithms of neutrosophic soft sets by evaluation based on the distance from average solution (EDAS), similarity measure, and level soft sets, respectively. Abu Qamar and Hassan [26] presented the similarity, distance, and fuzzy degree measures of Q-neutrosophic soft sets, and put forward the corresponding decision rule. Karaaslan [21] constructed a decision making method for the possibility of neutrosophic soft sets based on the and-product.

However, the existing studies mainly focus on decision making methods under a single decision maker, few scholars have studied group decision making problems by neutrsphic soft sets. At the same time, we also noticed that the existing methods have the following defects. On one hand, the above methods are mainly based on the expected utility theory, which assumes that decision makers are completelyrational. Actually, in decision making processes, decision makers do not make decisions in a complete rational manner, mainly showing that psychological expectations will greatly affect the actual decision making behavior. On the other hand, the parameter subjective weights are directly given determinate values [25], which do not fully reflect the hesitancies of decision makers' judgments under uncertain environments.

To make up for the gaps of existing researches, this study constructs an algorithm for the stochastic multi-criteria group decision making based on neutrosophic soft sets. Stochastic means that the weights of decision makers and parameters are uncertain or completely unknown under uncertainties. In this paper, neutrosophic numbers rather than determinate values are adopted to express the stochastic of the weights of decision makers and parameters. This method employs the prospect theory [27] rather than the expected utility theory to integrate the hesitancies of alternatives by decision makers' judgements. The prospect theory, a new theory of bounded rationality, is proposed from the point of view of cognitive psychology. In addition, it integrates the influence of psychological expectations on actual decision making behaviors into the decision making model. Therefore, the prospect theory is more in line with actual decision making behaviors under uncertainties [28]. Then, to establish the prospect decision matrix, we put forward the conflict degree measure of neutrosophic numbers and modify it to adapt group decision making. Moreover, on the purpose of aggregating in group decision making processes, this study proposes the weighted average aggregation rule and weighted geometric aggregation rule of neutrosophic soft sets.

To promote our discussion, some fundamental concepts of neutrosophic sets, soft sets, neutrosophic soft sets, and prospect theory are reviewed in Section 2. In Section 3, we establish the measures of determinacy degree and conflict degree, and construct the weighted average aggregation rule and weighted geometric aggregation rule of a neutrosophic soft set. In Section 4, this paper presents an algorithm for neutrosophic soft sets in the stochastic multi-criteria group decision making based on the prospect theory. In Section 5, to demonstrate the feasibility and effectiveness of the proposed algorithm, we perform an illustrative example and a comparative analysis.

2. Preliminaries

In this section, we briefly recall some basic concepts of neutrosophic sets, soft sets, neutrosophic soft sets, and prospect theory. More detailed conceptual basics can be found in references [4,16,17,27] (pp. 1–2).

2.1. Neutrosophic Soft Sets

Definition 1 [16] (p. 1). *Let U be the initial universal set, a neutrosophic set $A = \{< u : T_{A(u)}, I_{A(u)}, F_{A(u)} >, u \in U\}$ consists of the truth-membership $T_{A(u)}$, the indeterminacy-membership $I_{A(u)}$, and false-membership $F_{A(u)}$ of element $u \in U$ to set A, where $T, I, F : U \to]^-0, 1^+[$.* $]^-0, 1^+[$ is a non-standard interval, and the left and right borders of it are imprecise. Between them, $(^-0) = \{0 - x : x \in R^*, x \text{ is infinitesimal}\}$, *and* $(1^+) = \{1 + x : x \in R^*, x \text{ is infinitesimal}\}$.

For convenience, we employ $u =< T, I, F >$ to represent the element u in the neutrosophic set A, and it can be called a neutrosophic number.

Considering that neutrosophic sets are proposed from the philosophical point of view, it is difficult to apply to practical problems, such as management and engineering problems. Then, Haibin et al. [29] developed single valued neutrosophic sets.

Definition 2 [29]. *Let U be the universal set, a single valued neutrosophic set A over U can be defined as $A = \{< u : T_{A(u)}, I_{A(u)}, F_{A(u)} >, u \in U\}$, where $T, I, F : U \to [0, 1]$. Similarly, the values of $T_{A(u)}$, $I_{A(u)}$* and $F_{A(u)}$ stand for the truth-membership, indeterminacy-membership, and false-membership of *u to A*, respectively.

Definition 3 [30]. *Let $u =< T, I, F >$ be a neutrosophic number, then the score function, accuracy function and certainty function are defined as follows, respectively.*

$$s(u) = \frac{2 + T - I - F}{3}, \tag{1}$$

$$a(u) = T - F, \tag{2}$$

$$c(u) = T, \tag{3}$$

The score function is an important index for evaluating neutrosophic numbers. For a neutrosophic number $R =< T_r, I_r, F_r >$, the truth-membership T_r is positively correlated with the score function, and the indeterminacy-membership I_r and false-membership F_r are negatively correlated with the score function. In terms of the accuracy function, the greater the difference between the truth-membership T_r and false-membership F_r is, the more affirmative the statement is. Additionally, in regard to the certainty function, it positively depends on the truth-membership T_r.

On the basis of Definition 3, the comparison method between two neutrosophic numbers is represented as follows.

Definition 4 [30]. *Let* $u_1 =< T_1, I_1, F_1 >, u_2 =< T_2, I_2, F_2 >$ *be two neutrosophic numbers, the comparison relationships between* u_1 *and* u_2 are as follows:

1. *If* $s(u_1) > s(u_2)$, u_1 *is superior to* u_2 *and it can be denoted by* $u_1 \succ u_2$;
2. *If* $s(u_1) = s(u_2), a(u_1) > a(u_2)$, u_1 *is superior to* u_2 *and is denoted by* $u_1 \succ u_2$;
3. *If* $s(u_1) = s(u_2), a(u_1) = a(u_2)$ *and* $c(u_1) > c(u_2)$, u_1 *is superior to* u_2 *and is denoted by* $u_1 \succ u_2$;
4. *If* $s(u_1) = s(u_2), a(u_1) = a(u_2)$ *and* $c(u_1) = c(u_2)$, u_1 *is equal to* u_2, *denoted by* $u_1 \succ u_2$.

Example 1. *For two neutrosophic numbers* $u_1 =< 0.8, 0.2, 0.4 >$ *and* $u_2 =< 0.7, 0.4, 0.1 >$, *we can obtain that* $s(u_1) = 2.2/3$, $s(u_2) = 2.2/3$, $a(u_1) = 1.2/3$, $a(u_2) = 1.8/3$, $c(u_1) = 2.4/3$ *and* $c(u_2) = 2.1/3$ *based on Definition 3. Considering Definition 4, we can infer that* u_2 *is superior to* u_1, *as denoted by* $u_2 \succ u_1$.

Definition 5 [31]. *Let* $u_1 =< T_1, I_1, F_1 >, u_2 =< T_2, I_2, F_2 >$ *be two neutrosophic numbers, then the normalized Hamming distance between* u_1 *and* u_2 is defined as follows:

$$D^{\triangle}(u_1, u_2) = \frac{(|T_1 - T_2| + |I_1 - I_2| + |F_1 - F_2|)}{3}. \tag{4}$$

Definition 6 [4] (p. 1). *Let U be the set of initial universe, E be the parameter set, and* $P(U)$ *be the power set of U. Then a pair (F, E)is called a soft set over U where F is a mapping given by* $F : E \to P(U)$.

Remark 1 [32]. *On account of the single valued neutrosophic set is an instance of the neutrosophic set, it is natural to infer that a single valued neutrosophic soft set is an instance of the neutrosophic soft set. However, Maji only considers neutrosophic soft sets, which take value from the standard subset of* $[0, 1]$ *rather than* $]^{-}0, 1^{+}[$, *so the definition of the single valued neutrosophic soft set is exactly the same as the concept of the neutrosophic soft set defined by Maji.*

Definition 7 [17] (p. 1). *Let U be the initial universal set, E be a set of parameters, and* $P(U)$ *be the set of all neutrosophic subsets of U. The collection* (F, E) *is regarded as a neutrosophic soft set over U, where F refers to the mapping* $F : E \to P(U)$.

Example 2. *Assume* $U = \{u_1, u_2, u_3\}$ *is a set of three cars under consideration, and* $E = \{e_1 = \text{cheap}, e_2 = \text{equipment}, e_3 = \text{fuel consumption}\}$ *be the set of parameters for describing the three. In this case, we can define a function* $F : E \to P(U)$ *as a neutrosophic soft set* (F, E), *and it is represented as follows:*

$$(F, E) = \left\{ \begin{array}{l} F(e_1) = \{< u_1, 0.8, 0.4, 0.3 >, < u_2, 0.5, 0.7, 0.3 >, < u_3, 0.2, 0.5, 0.8 >\} \\ F(e_2) = \{< u_1, 0.5, 0.7, 0, 4 >, < u_2, 0.7, 0.3, 0.2 >, < u_3, 0.5, 0.8, 0.5 >\} \\ F(e_3) = \{< u_1, 0.4, 0.6, 0.3 >, < u_2, 0.9, 0.3, 0.1 >, < u_3, 0.4, 0.7, 0.5 >\} \end{array} \right\}.$$

2.2. Prospect Theory

The prospect theory [27] (p. 2), proposed by Tversky and Kahneman, is a mainstream theory of behavioral science, and it studies human judgments or decision making behaviors under uncertain environments. The prospect theory mainly considers the value function and decision weight function. It implies three characteristics: Reference dependence, diminishing sensitivity and lose aversion. Reference dependence refers to the change of people's perception depending on the change of the relative value. Diminishing sensitivity means that utility decreases as income increases. Additionally, loss aversion signifies that people value losses more than gains.

The prospect theory states that decision makers choose the optimal alternative based on the prospect value, which is determined by the value function and decision weight function. The prospect value can be obtained as follows:

$$V = \sum v(x-r)\omega(p_t). \tag{5}$$

$v(x-r)$ is the value function as defined follows:

$$v(x-r) = \begin{cases} (x-r)^{\alpha}, & x \geq r \\ -\lambda(x-r)^{\beta}, & x < r \end{cases}, \tag{6}$$

where x is the evaluation value of an object, r is the reference point, then $(x-r)$ represents losses or gains. $x \geq r$ means gains, and the value function is concave; $x < r$ means losses, and the value function is convex. So α, β stand for the concave degree and convexity degree of the value function, respectively. λ is the risk aversion coefficient, and $\lambda > 1$ indicates that decision makers value risk more. By experimental verification, Tversky and Kahneman took the value of parameters as follows: $\alpha = \beta = 0.88$, $\lambda = 2.25$.

$\omega(P_t)$ is the decision weight function as defined follows:

$$\omega(p_t) = \frac{p_t{}^{\gamma}}{((p_t{}^{\gamma}) + ((1-p_t)^{\gamma}))^{\frac{1}{\gamma}}}, \tag{7}$$

where p_t is the objective possibility, and Tversky and Kahneman took the value of parameter γ as 0.61.

3. The Measures of Determinacy Degree and Conflict Degree and Neutrosophic Soft Set Aggregation Rules

In this section, we initiate the determinacy degree measure and conflict degree measure of neutrosophic numbers, and then develop two kinds of aggregation rules of a neutrosophic soft set.

3.1. The Measures of Determinacy Degree and Conflict Degree

This paper employs the Hamming distance of information theory, which is a well-known measure designed to provide insights into the similarity of information [33,34] and has been widely employed in distance measures [26,35], to measure the determinacy degree and conflict degree. Before this, we present the concept of a minimum conflict neutrosophic number and maximum conflict neutrosophic number.

Definition 8. *Let Minc* $=< 1,0,0 >$ *be the minimum conflict neutrosophic number, which means that the belongingness degree of an object is 1, and the non-belongingness degree and the neutrality degree of an object be zero, respectively. That is, the conflict degree of information is the smallest.*

Additionally, let Maxc $=< 0.5,1,0.5 >$ *be the maximum conflict neutrosophic number. That is, the neutrosophic number, whose neutrality degree is one, and the belongingness degree and non-belongingness degree is 0.5. In order words, the conflict degree of information is the greatest.*

Definition 9. *Let* $u =< T,I,F >$ *be a neutrosophic number, the determinacy degree of* u *based on Equation (4) can be defined as follows:*

$$d^{\Delta}(u) = \frac{(|T-1| + I + F)}{3}, \tag{8}$$

which measures the normalized Hamming distance between u *and the minimum conflict neutrosophic number.*

Similarly, the conflict degree of u *is determined by the normalized Hamming distance between* u *and the maximum conflict neutrosophic number, and defined as follows:*

$$c^{\Delta}(u) = \frac{(|T-0.5| + |I-1| + |F-0.5|)}{3} \tag{9}$$

Example 3. *Considering Example 1, the determinacy degree and conflict degree of u_1 can be computed as follows:* $d^{\Delta}(u_1) = 0.8/3$, $c^{\Delta}(u_1) = 1.2/3$.

3.2. *Aggregation Rules of a Neutrosophic Soft Set*

In this subsection, we define two kinds of aggregation rules of a neutrosophic soft set, namely the weighted average aggregation rule and weighted geometric aggregation rule.

Definition 10. *Weighted average aggregation rule. Let U be the initial universal set, E be the set of parameters, (F,E) be a neutrosophic soft set over U, as represented by $F(e_j)(x_i) =< F_T(e_j)(x_i), F_I(e_j)(x_i), F_F(e_j)(x_i) >$ $(i = 1,2,\ldots,m; j = 1,2,\ldots,n)$. Then, the weighted average aggregation rule of (F,E) can be denoted by $(F,E)^{\Gamma} = \{F^{\Gamma}(x_1), F^{\Gamma}(x_2),\ldots,F^{\Gamma}(x_m)\}$, and defined as*

$$F^{\Gamma}(x_i) = \prod_{j=1}^{n} F(e_j)(x_i)\omega_j =< 1 - \prod_{j=1}^{n}(1 - F_T(e_j)(x_i))^{\omega_j}, \prod_{j=1}^{n}(F_I(e_j)(x_i))^{\omega_j}, \prod_{j=1}^{n}(F_F(e_j)(x_i))^{\omega_j} > \quad (10)$$

where the vector $\omega = \{\omega_1, \omega_2, \ldots, \omega_n\}$ stands for the weights of parameters, and $\sum_{j=1}^{n} \omega_j = 1$.

Based on Definition 10, the weighted geometric aggregation rule of a neutrosophic soft set is constructed.

Definition 11. *Weighted geometric aggregation rule. Considering the neutrosophic soft set (F,E) in Definition 10, we define the weighted geometric aggregation rule as $(F,E)^{\Theta} = \{F^{\Theta}(x_1), F^{\Theta}(x_2),\ldots,F^{\Theta}(x_m)\}$, and*

$$F^{\Theta}(x_i) = \prod_{j=1}^{n}(F(e_j)(x_i))^{\omega_j} =< \prod_{j=1}^{n}(F_T(e_j)(x_i))^{\omega_j}, 1 - \prod_{j=1}^{n}(1 - (F_I(e_j)(x_i)))^{\omega_j}, 1 - \prod_{j=1}^{n}(1 - (F_F(e_j)(x_i)))^{\omega_j} > \quad (11)$$

where the vector $\omega = \{\omega_1, \omega_2, \ldots, \omega_n\}$stands for the weights of parameters, and $\sum_{j=1}^{n} \omega_j = 1$.

Example 4. *Consider Example 2. Assume that the weight vector of parameters is $\omega = \{0.4, 0.2, 0.3\}$, then we can obtain the results of the weighted average aggregation and weighted geometric aggregation as follows, respectively.*

$$(F,E)^{\Gamma} = \{< u_1, 0.6077, 0.5537, 0.3584 >, < u_2, 0.7015, 0.4749, 0.2244 >, < u_3, 0.3169, 0.6512, 0.6467 >\}.$$
$$(F,E)^{\Theta} = \{< u_1, 0.6049, 0.9905, 0.9987 >, < u_2, 0.6837, 0.9973, 0.9998 >, < u_3, 0.3474, 0.9798, 0.9885 >\}$$

4. Algorithm for Neutrosophic Soft Sets in Stochastic Multi-Criteria Group Decision Making Based on Prospect Theory

4.1. *Problem Description*

In this section, we give a concise description of a stochastic multi-criteria group decision making problem under neutrosophic soft sets. Let $U = \{x_1, x_2, \ldots, x_m\}$ be a set of m alternatives, $E = \{e_1, e_2, \ldots, e_n\}$ be a set of n parameters and $DM = \{Z_1, Z_2, \ldots Z_p\}$ be a set of p decision makers. Assume that $\omega^{(t)} =< \omega_T^{(t)}, \omega_I^{(t)}, \omega_F^{(t)} >$ $(t = 1,2,\ldots,p)$ is the neutrosophic weight of decision maker Z_t, $\delta_j^{(t)} =< \delta_{Tj}^{(t)}, \delta_{Ij}^{(t)}, \delta_{Fj}^{(t)} >$ is the neutrosophic subjective weight assigned for parameter e_j by decision maker Z_t, and the evaluation value of alternative x_i related to parameter e_j by decision maker Z_t is expressed as $F^{(t)}(e_j)(x_i) =< F_T^{(t)}(e_j)(x_i), F_I^{(t)}(e_j)(x_i), F_F^{(t)}(e_j)(x_i) >$. Given p neutrosophic soft sets $(F^{(t)}, E)$ $(t = 1,2,\ldots,p)$ of alternatives evaluated by decision makers, and the tabular representation of $(F^{(t)}, E)$ $(t = 1,2,\ldots,p)$ is shown in Table 1.

4.2. Determining the Determinacy Degree of Decision Makers

In stochastic multi-criteria group decision making problems, the weights of decision makers are stochastic and indeterminate. Therefore, how to obtain the weights as determinate values has become an important research topic. In this paper, we express the weights of decision makers as a neutrosophic number, and then compute the determinacy degree of decision makers to replace traditional weights.

Considering Definition 9, let $\omega_t = < \omega_t^T, \omega_t^I, \omega_t^F > (t = 1, 2, \ldots, p)$ be the neutrosophic weight of decision maker Z_t, then the determinacy degree of Z_t can be computed as follows by Equation (8):

$$d^{\Delta}(t) = \frac{1 - \frac{1}{3}(|\omega_t^T - 1| + \omega_t^I + \omega_t^F)}{\sum\limits_{t=1}^{p} 1 - \frac{1}{3}\left(|\omega_t^T - 1| + \omega_t^I + \omega_t^F\right)} (t = 1, 2, \ldots, p), \tag{12}$$

Table 1. Tabular representation of neutrosophic soft sets $(F^{(t)}, E)$ of alternatives.

		$(\mathbf{F}^{(1)}, \mathbf{E})$		
	e_1	e_2	$\ldots$	e_n
x_1	$F^{(1)}(e_1)(x_1)$	$F^{(1)}(e_2)(x_1)$	$\ldots$	$F^{(1)}(e_n)(x_1)$
x_2	$F^{(1)}(e_1)(x_2)$	$F^{(1)}(e_2)(x_2)$	$\ldots$	$F^{(1)}(e_n)(x_2)$
	$\vdots$	$\vdots$	$\ddots$	$\vdots$
x_m	$F^{(1)}(e_1)(x_m)$	$F^{(1)}(e_2)(x_m)$	$\ldots$	$F^{(1)}(e_n)(x_m)$
		$(\mathbf{F}^{(2)}, \mathbf{E})$		
	e_1	e_2	$\ldots$	e_n
x_1	$F^{(2)}(e_1)(x_1)$	$F^{(2)}(e_2)(x_1)$	$\ldots$	$F^{(2)}(e_n)(x_1)$
x_2	$F^{(2)}(e_1)(x_2)$	$F^{(2)}(e_2)(x_2)$	$\ldots$	$F^{(2)}(e_n)(x_2)$
$\vdots$	$\vdots$	$\vdots$	$\ddots$	$\vdots$
x_m	$F^{(2)}(e_1)(x_m)$	$F^{(2)}(e_2)(x_m)$	$\ldots$	$F^{(2)}(e_n)(x_m)$
$\vdots$	$\vdots$	$\vdots$	$\vdots$	$\vdots$
		$(\mathbf{F}^{(\mathbf{P})}, \mathbf{E})$		
	e_1	e_2	$\ldots$	e_n
x_1	$F^{(p)}(e_1)(x_1)$	$F^{(p)}(e_2)(x_1)$	$\ldots$	$F^{(p)}(e_n)(x_1)$
x_2	$F^{(p)}(e_1)(x_2)$	$F^{(p)}(e_2)(x_2)$	$\ldots$	$F^{(p)}(e_n)(x_2)$
$\vdots$	$\vdots$	$\vdots$	$\ddots$	$\vdots$
x_m	$F^{(p)}(e_1)(x_m)$	$F^{(p)}(e_2)(x_m)$	$\ldots$	$F^{(p)}(e_n)(x_m)$

4.3. Calculating the Comprehensive Weights of Parameters

In this paper, the parameter weights are determined by combining subjective weights with objective weights. Among them, subjective weights are obtained by aggregating neutrosophic subjective weights provided by decision makers, which is more accurate than the way directly given by determinate values [25] (p. 2). The objective weights are calculated by the information entropy method [35]. Then, the principle of minimum information entropy [36] is employed to obtain comprehensive weights of parameters by integrating subjective weights and objective weights. The system framework is presented in Figure 1.

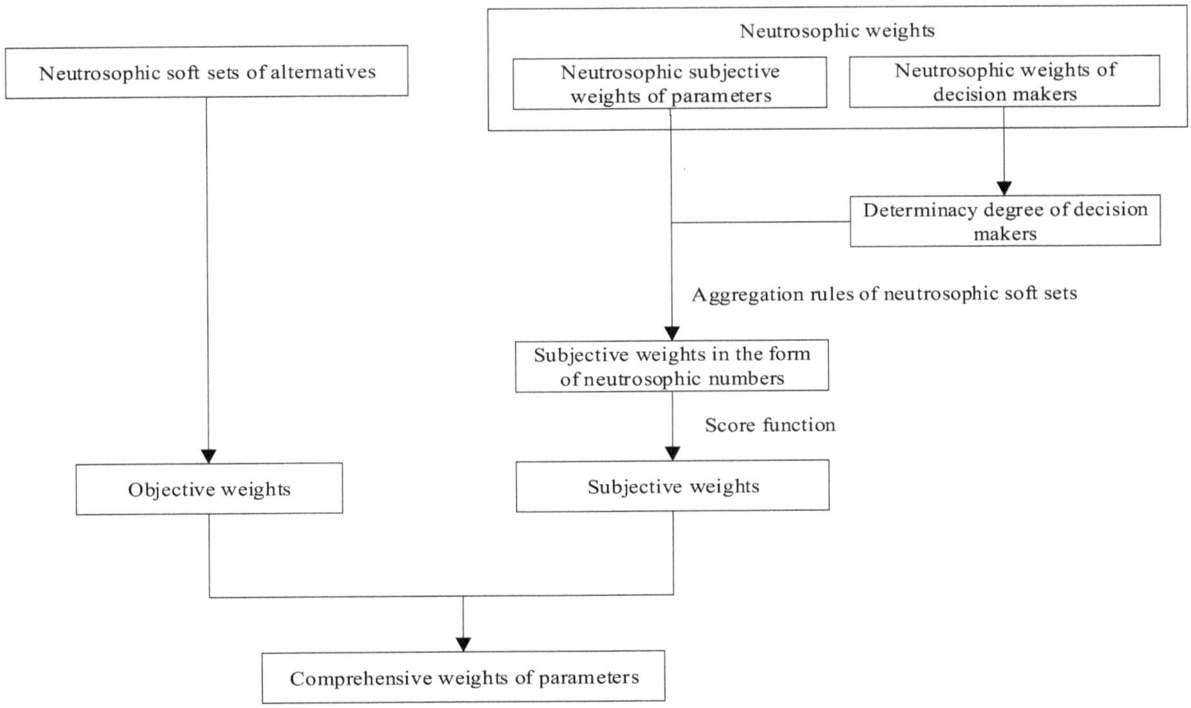

Figure 1. The system framework of the computing comprehensive weights of parameters.

4.3.1. Computing the Subjective Weights

Under the stochastic environment, the judgements of decision makers are full of hesitancies. Considering this situation, instead of giving determinate values, this paper firstly aggregates neutrosophic subjective weights of parameters to obtain subjective weights in the form of neutrosophic numbers. Based on this, subjective weights are computed by the score function as Equation (1).

Assume parameter set $E = \{e_1, e_2, \ldots, e_j\}$ is the initial universal set, the set of decision makers $Z = \{z_1, z_2, \ldots, z_t\}$ is the parameter set, and $P(Z)$ is the set of all neutrosophic subsets of E. The neutrosophic soft set (F, Z) over E can be integrated by the weighted geometric aggregation rule as $(F, Z)^{\Theta} = \{F^{\Theta}(e_1), F^{\Theta}(e_2), \ldots, F^{\Theta}(e_m)\}$, and

$$F^{\Theta}(e_j) = \prod_{t=1}^{p} \delta_j^{(t)\,\psi_t} =< \prod_{t=1}^{p} \delta_{Tj}^{(t)\,\psi_t}, 1 - \prod_{t=1}^{p} (1 - \delta_{Ij}^{(t)})^{\psi_t}, 1 - \prod_{t=1}^{p} (1 - \delta_{Fj}^{(t)})^{\psi_t} >, \tag{13}$$

where $\delta_j^{(t)} =< \delta_{jT}^{(t)}, \delta_{Ij}^{(t)}, \delta_{Fj}^{(t)} > (j = 1, 2, \ldots, n)$ is the neutrosophic subjective weight assigned for parameter e_j by Z_t, and ψ_t is the determinacy degree of Z_t.

Then, the subjective weights can be computed by the score function as shown below:

$$SW_j = \frac{2 + \prod_{t=1}^{p} \delta_{Tj}^{(t)\,\psi_t} - (1 - \prod_{t=1}^{p} (1 - \delta_{Ij}^{(t)})^{\psi_t}) - (1 - \prod_{t=1}^{p} (1 - \delta_{Fj}^{(t)})^{\psi_t})}{3}. \tag{14}$$

4.3.2. Obtaining the Objective Weights: Information Entropy Method

Considering that the computation of objective weights is not the focus of this paper, we obtain objective weights by the information entropy method. The information entropy is used to measure the uncertainty of events. The greater the information entropy is, the greater the uncertainty degree. That is, the smaller the amount of information it carries, the smaller the weight is. Note that the uncertainty of neutrosophic numbers consists of two factors, one is the truth-membership and false-membership, and the other is the indeterminacy-membership.

Based on the information entropy method, we can obtain that the information entropy of parameter e_j given by decision maker Z_t is defined as:

$$E_j^t = 1 - \frac{1}{m}\sum\nolimits_{i=1}^{m}\left(F_T^{(t)}\left(e_j\right)(x_i) + F_F^{(t)}\left(e_j\right)(x_i)\right)|F_I^{(t)}\left(e_j\right)(x_i) - F_I^{(t)c}\left(e_j\right)(x_i)|(j = 1, 2, \ldots, n). \tag{15}$$

Then, the comprehensive information entropy of parameter e_j is defined as follows:

$$E_j = \sum_{t=1}^{p} \varphi_t E_j^t (j = 1, 2, \ldots, n) \tag{16}$$

where φ_t is the determinacy degree of decision maker Z_t computed by Equation (8).

So, the objective weights are obtained as:

$$OW_j = \frac{1 - E_j}{\sum\limits_{j=1}^{n} 1 - E_j} (j = 1, 2, \ldots, n). \tag{17}$$

4.3.3. Calculating the Comprehensive Weights

Based on the principle of the minimum information entropy, the comprehensive weight of parameter ϖ_j can be calculated as follows:

$$\varpi_j = \frac{\sqrt{OW_j \cdot SW_j}}{\sum\limits_{j=1}^{n} \sqrt{OW_j \cdot SW_j}}, \tag{18}$$

where SW_j and OW_j represent the subjective weight and objective weight of parameter e_j, respectively.

4.4. Computing the Comprehensive Prospect Values

The comprehensive prospect values of alternatives are determined by the prospect decision matrix and the comprehensive weights of parameters. Next, we expound how to generate the prospect decision matrix and obtain comprehensive values of alternatives, respectively.

4.4.1. Constructing the Prospect Decision Matrix

The core of constructing the prospect decision matrix is to compute the value function and decision weight function. In terms of the value function, we need to analyze the distance between the reference point and the actual value. This paper regards the maximum conflict neutrosophic number as the reference point, then the distance can be treated as the conflict degree of the actual value. Additionally, actual values refer to the alternative evaluation values with respect to the parameters. As for the decision weight function, the objective possibility is seen as the determinacy degree of the decision makers. The system framework of constructing the prospect decision matrix is shown in Figure 2.

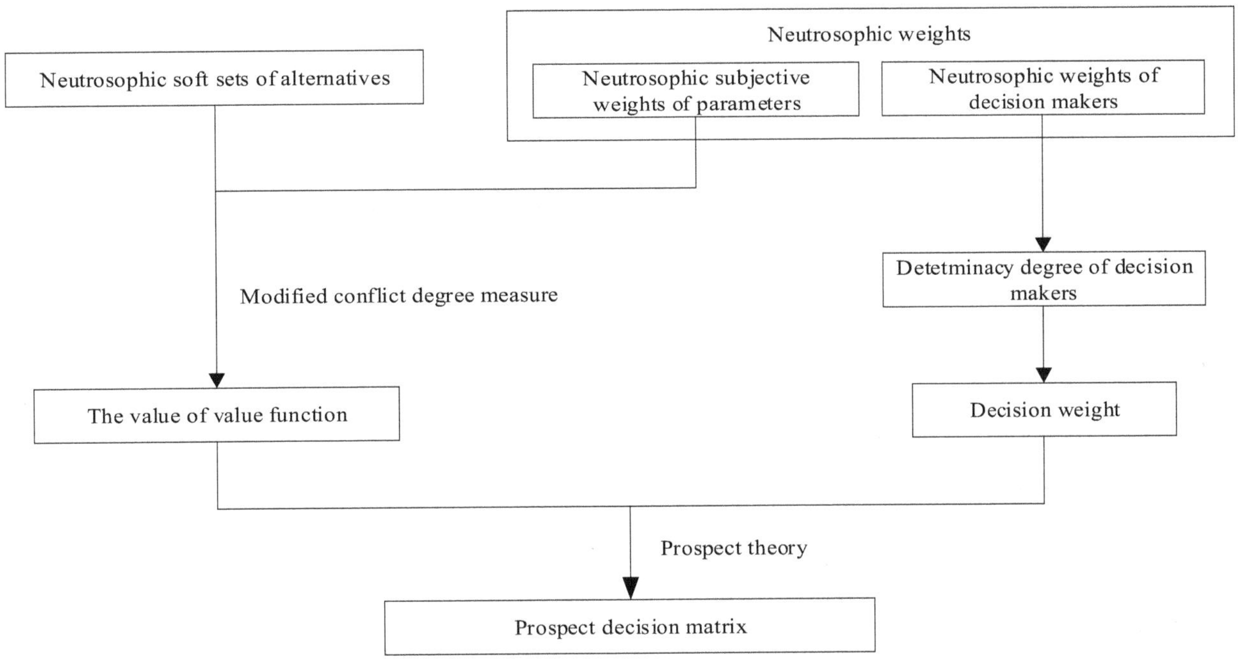

Figure 2. The system framework of constructing the prospect decision matrix.

We assume that the neutrosophic soft sets of alternatives and neutrosophic subjective weights of parameters are both provided by decision makers. So, the conflict degree of the alternative evaluation values with respect to the parameters should take the neutrosophic subjective weights of parameters into account. Based on the conflict degree measure given by Definition 9, we develop a modified conflict degree measure by introducing the neutrosophic subjective weights of parameters.

Assume $F(e_j)(x_i) =< F_T(e_j)(x_i), F_I(e_j)(x_i), F_F(e_j)(x_i) >$ is a neutrosophic number, which represents the value of alternative x_i related to parameter e_j, and $\alpha_j =< \alpha_{jT}, \alpha_{jI}, \alpha_{jF} >$ is the neutrosophic subjective weight of parameter e_j. Considering the sum of α_{jT}, α_{jI} and α_{jF} may not be one, this paper normalizes them to be more consistent with the reality. Therefore, the measure of the modified conflict degree of $F(e_j)(x_i)$ is defined as follows:

$$mc^{\Delta}(F(e_j)(x_i)) = \frac{\alpha_{jT} \cdot \left|F_T(e_j)(x_i) - 0.5\right|}{\alpha_{jT} + \alpha_{jI} + \alpha_{jF}} + \frac{\alpha_{jI} \cdot \left|F_I(e_j)(x_i) - 1\right|}{\alpha_{jT} + \alpha_{jI} + \alpha_{jF}} + \frac{\alpha_{jF} \cdot \left|F_F(e_j)(x_i) - 0.5\right|}{\alpha_{jT} + \alpha_{jI} + \alpha_{jF}}. \tag{19}$$

Subsequently, calculate the prospect value of each alternative with respect to the parameters as follows:

$$V_{ij} = \sum_{t=1}^{p} w(z_t) v(F^{(t)}(e_j)(x_i) - x_0), \tag{20}$$

where

$$v(F^{(t)}(e_j)(x_i) - x_0) = \begin{cases} (mc^{\Delta}(F^{(t)}(e_j)(x_i), x_0))^{0.88}, & F^{(t)}(e_j)(x_i) \geq x_0 \\ -2.25(mc^{\Delta}(F^{(t)}(e_j)(x_i), x_0))^{0.88}, & F^{(t)}(e_j)(x_i) < x_0 \end{cases}, \tag{21}$$

$$\omega(Z_t) = \frac{(\psi_t)^{0.61}}{((\psi_t)^{0.61} + (1 - \psi_t)^{0.61})^{\frac{1}{0.61}}}. \tag{22}$$

Then, we can obtain the prospect decision matrix.

4.4.2. Computing the Comprehensive Prospect Values

Based on comprehensive weights of parameters and the prospect decision matrix, we can compute the comprehensive prospect values for alternatives as follows:

$$V_i = \sum_{j=1}^{n} \varpi_j V_{ij}. \tag{23}$$

The system framework of computing the comprehensive prospect values is shown in Figure 3.

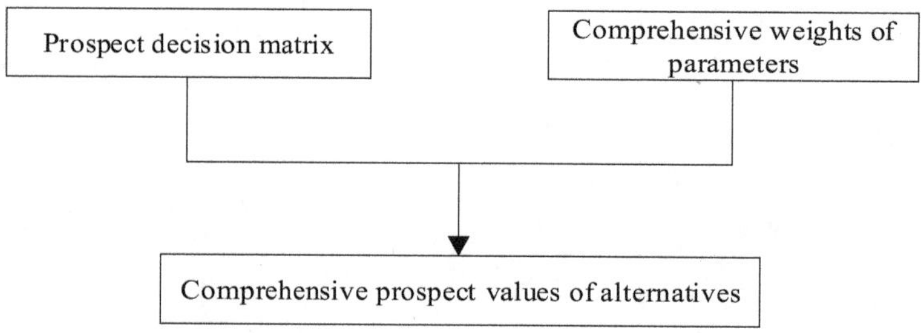

Figure 3. The system framework of computing the comprehensive prospect values of alternatives.

4.5. Algorithm for Neutrosophic Soft Sets in Stochastic Multi-Criteria Group Decision Making Based on Prospect Theory

In this section, a novel algorithm for neutrosophic soft sets in stochastic multi-criteria group decision making based on the prospect theory is proposed. The detailed operation steps of Algorithm 1 are presented below.

Algorithm 1: Neutrosophic soft sets in stochastic multi-criteria group decision making based on the prospect theory

---

Step 1: Input a neutrosophic set, which represents neutrosophic weights of decision makers and two neutrosophic soft sets, including alternatives description as shown in Table 1 and neutrosophic subjective weights of parameters evaluated by decision makers.
Step 2: Normalize the neutrosophic soft sets of alternatives as follows:

$$(\overset{\triangle}{F^{(t)}}, E) = \begin{cases} (F_T^{(t)}(e_j)(x_i), F_I^{(t)}(e_j)(x_i), F_F^{(t)}(e_j)(x_i)), & e_j \text{ is a benefit parameter} \\ (F_F^{(t)}(e_j)(x_i), 1 - F_I^{(t)}(e_j)(x_i), F_T^{(t)}(e_j)(x_i)), & e_j \text{ is a cost parameter} \end{cases} \tag{24}$$

Step 3: Compute the determinacy degree vector $\psi_t = (\psi_1, \psi_2, \ldots, \psi_p)$ of decision makers by Equation (8);
Step 4: Construct the prospect decision matrix based on Equation (20).
Step 5: Obtain the comprehensive weight vector $\varpi_j = (\varpi_1, \varpi_2, \ldots, \varpi_n)$ by Equation (18);
Step 6: Calculate the comprehensive prospect value V_i for each alternative through Equation (23).
Step 7: Make a decision by ranking alternatives based on comprehensive prospect values.

---

5. An Application of the Proposed Algorithm

In order to verify the feasibility of the proposed algorithm, we discuss the investment decision of a finance institution. Meanwhile, the existing five methods [17,25,37] (pp. 1–2) are employed for a comparative analysis to prove the feasibility and superiority of the proposed algorithm.

5.1. Example Analysis

Credit scoring can help financial institutions reduce financial risks and non-performing loans. Generally, financial institutions assess the credit score of borrowers based on basic information, such as age, profession, education, income, capital gains, residence and borrowing frequency. Recently, a financial institution wants to invest an amount of money in borrowers. The institution initially selects five borrowers as candidates. In addition, the institution makes a decision by analyzing the following four parameters: Highly educated, higher borrowing frequency, higher income and higher capital gains. Subsequently, the institution assembles a team composed of three decision makers to make the investment decision. Suppose that $U = \{u_1, u_2, u_3, u_4, u_5\}$ is the set of candidates, $E = \{e_1, e_2, e_3, e_4\}$ is the parameter set, and $DM = \{Z_1, Z_2, Z_3\}$ is the set of decision makers. Let the neutrosophic soft sets $(F^{(t)}, E)$ $(t = 1, 2, 3)$ be the alternative evaluation values with respect to the parameters given by decision makers as follows.

$$(F^{(1)},E) = \left\{ \begin{array}{l} F^{(1)}(e_1) = \left\{< \frac{u_1}{0.60,0.35,0.80} >,< \frac{u_2}{0.70,0.50,0.60} >,< \frac{u_3}{0.80,0.40,0.70} >,< \frac{u_4}{0.65,0.50,0.50} >,< \frac{u_5}{0.75,0.30,0.60} >\right\} \\ F^{(1)}(e_2) = \left\{< \frac{u_1}{0.50,0.80,0.20} >,< \frac{u_2}{0.60,0.30,0.70} >,< \frac{u_3}{0.70,0.35,0.80} >,< \frac{u_4}{0.80,0.30,0.70} >,< \frac{u_5}{0.80,0.20,0.55} >\right\} \\ F^{(1)}{}_1(e_3) = \left\{< \frac{u_1}{0.60,0.50,0.80} >,< \frac{u_2}{0.70,0.50,0.20} >,< \frac{u_3}{0.80,0.60,0.30} >,< \frac{u_4}{0.70,0.40,0.70} >,< \frac{u_5}{0.85,0.30,0.60} >\right\} \\ F^{(1)}(e_4) = \left\{< \frac{u_1}{0.50,0.80,0.60} >,< \frac{u_2}{0.40,0.70,0.30} >,< \frac{u_3}{0.60,0.40,0.70} >,< \frac{u_4}{0.60,0.35,0.80} >,< \frac{u_5}{0.70,0.30,0.40} >\right\} \end{array} \right\}$$

$$(F^{(2)},E) = \left\{ \begin{array}{l} F^{(2)}(e_1) = \left\{< \frac{u_1}{0.60,0.35,0.80} >,< \frac{u_2}{0.70,0.50,0.60} >,< \frac{u_3}{0.80,0.40,0.70} >,< \frac{u_4}{0.65,0.50,0.50} >,< \frac{u_5}{0.75,0.30,0.60} >\right\} \\ F^{(2)}(e_2) = \left\{< \frac{u_1}{0.50,0.80,0.20} >,< \frac{u_2}{0.60,0.30,0.70} >,< \frac{u_3}{0.70,0.35,0.80} >,< \frac{u_4}{0.80,0.30,0.70} >,< \frac{u_5}{0.80,0.20,0.55} >\right\} \\ F^{(2)}(e_3) = \left\{< \frac{u_1}{0.60,0.50,0.80} >,< \frac{u_2}{0.70,0.50,0.20} >,< \frac{u_3}{0.80,0.60,0.30} >,< \frac{u_4}{0.70,0.40,0.70} >,< \frac{u_5}{0.85,0.30,0.60} >\right\} \\ F^{(2)}(e_4) = \left\{< \frac{u_1}{0.50,0.80,0.60} >,< \frac{u_2}{0.40,0.70,0.30} >,< \frac{u_3}{0.60,0.40,0.70} >,< \frac{u_4}{0.60,0.35,0.80} >,< \frac{u_5}{0.70,0.30,0.40} >\right\} \end{array} \right\}$$

$$(F^{(3)},E) = \left\{ \begin{array}{l} F^{(3)}(e_1) = \left\{< \frac{u_1}{0.60,0.35,0.80} >,< \frac{u_2}{0.70,0.50,0.60} >,< \frac{u_3}{0.80,0.40,0.70} >,< \frac{u_4}{0.65,0.50,0.50} >,< \frac{u_5}{0.75,0.30,0.60} >\right\} \\ F^{(3)}(e_2) = \left\{< \frac{u_1}{0.50,0.80,0.20} >,< \frac{u_2}{0.60,0.30,0.70} >,< \frac{u_3}{0.70,0.35,0.80} >,< \frac{u_4}{0.80,0.30,0.70} >,< \frac{u_5}{0.80,0.20,0.55} >\right\} \\ F^{(3)}(e_3) = \left\{< \frac{u_1}{0.60,0.50,0.80} >,< \frac{u_2}{0.70,0.50,0.20} >,< \frac{u_3}{0.80,0.60,0.30} >,< \frac{u_4}{0.70,0.40,0.70} >,< \frac{u_5}{0.85,0.30,0.60} >\right\} \\ F^{(3)}(e_4) = \left\{< \frac{u_1}{0.50,0.80,0.60} >,< \frac{u_2}{0.40,0.70,0.30} >,< \frac{u_3}{0.60,0.40,0.70} >,< \frac{u_4}{0.60,0.35,0.80} >,< \frac{u_5}{0.70,0.30,0.40} >\right\} \end{array} \right\}$$

The neutrisophic set D represents the neutrosophic weights of decision makers, and the neutrisophic soft set (F, Z) stands for neutrosophic subjective weights of parameters. They are valued as follows:

$$D = \{< Z_1, 0.3, 0.5, 0.7 >, < Z_2, 0.1, 0.4, 0.6 >, < Z_3, 0.6, 0.5, 0.2 >\}$$

$$(F,Z) = \left\{ \begin{array}{l} F(Z_1) = \left\{< \frac{e_1}{0.40,0.60,0.50} >,< \frac{e_2}{0.35,0.70,0.60} >,< \frac{e_3}{0.40,0.60,0.55} >,< \frac{e_4}{0.40,0.60,0.75} >\right\} \\ F(Z_2) = \left\{< \frac{e_1}{0.70,0.45,0.30} >,< \frac{e_2}{0.50,0.80,0.60} >,< \frac{e_3}{0.70,0.55,0.40} >,< \frac{e_4}{0.70,0.40,0.65} >\right\} \\ F(Z_3) = \left\{< \frac{e_1}{0.65,0.70,0.40} >,< \frac{e_2}{0.60,0.35,0.75} >,< \frac{e_3}{0.40,0.65,0.70} >,< \frac{e_4}{0.35,0.60,0.50} >\right\} \end{array} \right\}$$

Step 1: Input the neutrosophic soft sets $(F^{(t)}, E)(t = 1, 2, 3)$, (F, Z) and the neutrosophic set D.

Step2: There is no need to normalize the neutrosophic soft sets $(F^{(t)}, E)(t = 1, 2, 3)$ of alternatives, because the parameters adopted in this study are benefit parameters.

Step 3: Compute the determinacy degree vector of decision makers based on Equation (8) as follows:

$$\psi_t = \{0.3478, 0.4130, 0.2391\}$$

Step 4: Construct the prospect decision matrix based on Equation (20).

$$V_{\text{ij}} = \begin{pmatrix} 0.3878 & 0.2846 & 0.3574 & 0.2274 \\ 0.3035 & 0.3751 & 0.3571 & 0.2712 \\ 0.4536 & 0.3834 & 0.3226 & 0.3180 \\ 0.3345 & 0.3294 & 0.3120 & 0.3776 \\ 0.3482 & 0.4482 & 0.4055 & 0.3481 \end{pmatrix}.$$

Step 5: Determine the comprehensive weight vector $\varpi_j = (\varpi_1, \varpi_2, \ldots, \varpi_n)$ for the parameters as Equation (18), and the neutrosophic subjective weights are aggregated by the weighted geometric aggregation rule as Equation (11).

$$\varpi_j = (0.2991, 0.2260, 0.2898, 0.1851)$$

Step 6: Obtain the comprehensive prospect value V_i by Equation (23).

$$V_1 = 0.3269, V_2 = 0.3292, V_3 = 0.3746, V_4 = 0.3348, V_5 = 0.3874.$$

Step 7: Make a decision by ranking the comprehensive prospect value of the five candidates.

$$x_5 \succ x_3 \succ x_4 \succ x_2 \succ x_1$$

Therefore, we can see that the optimal candidate is x_5. x_3, x_4 are suboptimal, and x_2, x_1 are the worst.

Furthermore, we also utilize the weighted average aggregation rule to compute the subjective weights of parameters. In addition, the computational procedure is shown as follows.

Step 1–4: Be consistent with the above steps 1–4.

Step 5: Determine the comprehensive weight vector $\varpi_j = (\varpi_1, \varpi_2, \ldots, \varpi_n)$ for the parameters as Equation (18), and the neutrosophic subjective weights are aggregated by the weighted average aggregation rule.

$$\varpi_j = (0.2903, 0.2127, 0.2523, 0.2447).$$

Step 6: Obtain the comprehensive prospect value V_i by Equation (23).

$$V_1 = 0.3254, V_2 = 0.3295, V_3 = 0.3744, V_4 = 0.3348, V_5 = 0.3876.$$

Step 7: Make a decision by ranking the five candidates.

$$x_5 \succ x_3 \succ x_4 \succ x_2 \succ x_1.$$

So the best optimal is still x_5, the following are x_3, x_4, and the worst are x_2, x_1.

Obviously, we can see that the ranking orders obtained by two aggregation rules of the neutrosophic soft set are the same.

5.2. Comparative Analysis

A comparative analysis with existing methods is performed to justify the feasibility and superiority of the proposed method. The existing methods include the method proposed by Maji [17] (p. 1), the three methods carried out by Peng and Liu [25] (p. 2) and the aggregated neutrosophic set method [37] (p. 11).

In the decision making method outlined by Maji [17] (p. 1), the final ranking is obtained based on the comparison matrix through briefly comparing with three membership function values. The three neutrosophic soft decision making methods [25] (p. 2) include the non-linear weighted comprehensive method to determine parameter comprehensive weights by combining objective weights and subjective weights. Objective weights are computed by the grey system method, and subjective weights are directly given determinate values. Then, three neutrosophic soft decision making methods are constructed based on EDAS, similarity measure, and the level soft set to rank alternatives in practical problems. Among the three, EDAS and the similarity measure methods obtain the final ranking based on the accurate calculation of alternative evaluation values. In addition, the level soft set method makes a decision by roughly comparing the threshold value with alternative evaluation values. In terms of the aggregated neutrosophic set method [37] (p. 11), alternatives are aggregated using the arithmetic average and sorted by TOPSIS.

Note that there are two crucial issues. On one hand, the above methods all make decisions under a single decision maker. In order to successfully apply them to group decision making, this paper employs the weighted average algorithm to the score of alternatives to all decision makers, based on the decision maker determinacy degree of this study. On the other hand, the method in [17] (p. 1) and [37] (p. 11) does not take parameter weights into consideration. Although the EDAS, similarity measure and level soft set methods [25] (p. 2) comprehensively consider objective weights and subjective weights, the subjective weights are directly given determinate values, which cannot reflect the hesitancies of decision makers under uncertainties. Considering this, the comparative analysis applies the subjective weights obtained from this study to the three methods in [25] (p. 2).

The final ranking of the stochastic multi-criteria group decision making problem mentioned in Section 5.1 are presented in Table 2, by utilizing the proposed method and the methods in [17,25,37] (pp. 1–2, 11). By comparison, the results of the proposed method are consistent with those of most comparison methods, which prove the effectiveness of the proposed method.

Table 2. A comparative study with some existing methods.

Method	The Final Ranking	The Optimal Alternative
The proposed method		
Weighted geometric neutrosophic rule	$x_5 \succ x_3 \succ x_4 \succ x_2 \succ x_1$	x_5
Weighted average neutrosophic rule	$x_5 \succ x_3 \succ x_4 \succ x_2 \succ x_1$	x_5
The determinacy degree of decision makers $\psi_t = \{0.3913, 0.2826, 0.3261\}$		
Maji [17]	$x_5 \succ x_4 \succ x_3 \succ x_2 \succ x_1$	x_5
EDAS [25]	$x_5 \succ x_3 \succ x_4 \succ x_2 \succ x_1$	x_5
Similarity [25]	$x_5 \succ x_3 \succ x_4 \succ x_2 \succ x_1$	x_5
Level soft set [25]	$x_5 \succ x_4 \succ x_3 \succ x_2 \succ x_1$	x_5
TOPSIS [37]	$x_5 \succ x_3 \succ x_4 \succ x_2 \succ x_1$	x_5

From Table 2, we can find that the final rankings of the proposed algorithm are different from Maji's method and the level soft set method. The difference can be attributed to two reasons. One is that both methods are approximate comparisons of the alternative evaluation values, and the original evaluation values are not used to the greatest extent. The other is that the threshold value difference of the level soft set method can directly lead to different final rankings. However, decision makers can hardly decide which threshold value to use.

Through comparison, the final rankings of the other three methods are consistent with the proposed method. Among them, EDAS also adopts the aggregation method just as the proposed method. Different from EDAS, the proposed method considers the psychological expectation of decision makers in the borrower selection issue. Thus, in complex group decision making problems, the proposed method can produce more reasonable results than existing methods.

From the above analysis, the main superiorities of the proposed method can be summarized into three aspects. Firstly, this study originally employs neutrosophic soft sets for handling stochastic multi-criteria group decision making problems, which cannot be solved in existing methods. Secondly, the proposed method expresses the weights of subjective weights of parameters by neutrosophic numbers, which can fully reflect the hesitancies of decision makers. Meanwhile, this study presents the weights of decision makers by neutrosophic numbers, which can better incorporate stochastic into the decision making process. Thirdly, the proposed method considers the psychological expectations of decision makers in the borrower selection process. Therefore, it is able to analyze the decision making behavior more objectively.

6. Conclusions

Under uncertain environments, a mass of inconsistent information appears. Neutrosophic soft sets are powerful tools to address these issues involving inconsistent information. Considering this, we develop a generalized stochastic multi-criteria group decision making framework under neutrosophic soft sets, by innovatively integrating the prospect theory and neutrosophic soft sets into our framework. This paper describes the reference point, the psychological expectations of decision makers, in the form of neutrosophic sets. Then, in addition, this study demonstrates how to compute the alternative prospect values as the reference for decision making.

We conduct experiments to test the feasibility and validity of our decision making framework. The main contributions of this paper are fourfold. Firstly, we construct a new algorithm for the stochastic multi-criteria group decision making based on neutrosophic soft sets, which can analyze inconsistent information in decision making effectively. Secondly, the weights of decision makers and parameter subjective weights are both expressed in the form of neutrosophic numbers. Compared with the way directly given determinate values in existing methods [25] (p. 2), the proposed method can embody the stochastic into decision making processes. Thirdly, the research successfully combines the prospect theory with neutrosophic softs sets to construct the stochastic multi-criteria group decision making algorithm. Compared with the existing literatures based on the expected utility theory [16,17,25,26] (pp. 1-2), this research considers the influence of psychological expectations on decision results. Finally, we explore the conflict degree measure of neutrosophic numbers and two aggregation rules of neutrosophic soft sets, and further define the measure of the modified conflict degree to accommodate the multi-criteria group decision making.

The proposed method is not only suitable for credit scoring, but also for decision-making problems in other fields, especially for decisions with inconsistent information. As a suggestion for future researches, we shall integrate more advanced decision theories into neutrosophic soft sets and address stochastic multi-criteria group decision making issues.

Author Contributions: Funding acquisition, Y.D.; Methodology, Y.D. and C.H.; Validation, Y.D. and C.H.; Writing—original draft, Y.D. and C.H.; Writing—review and editing, C.H., Y.P., and K.G.

Acknowledgments: The authors gratefully acknowledge the editor of the Journal and anonymous reviewers for the detailed and valuable comments.

References

1. Zadeh, L.A. Fuzzy sets. *Inf. Control* **1965**, *8*, 338–353. [CrossRef]
2. Pawlak, Z. Rough sets. *Int. J. Comput. Inf. Sci.* **1982**, *11*, 341–356. [CrossRef]
3. Gorzałczany, M.B. A method of inference in approximate reasoning based on interval-valued fuzzy sets. *Fuzzy Sets Syst.* **1987**, *21*, 1–17. [CrossRef]
4. Molodtsov, D. Soft set theory-first results. *Comput. Math. Appl.* **1999**, *37*, 19–31. [CrossRef]
5. Aktaş, H.; Çağman, N. Soft sets and soft groups. *Inf. Sci.* **2007**, *177*, 2726–2735. [CrossRef]
6. Acar, U.; Koyuncu, F.; Tanay, B. Soft sets and soft rings. *Comput. Math. Appl.* **2010**, *59*, 3458–3463. [CrossRef]
7. Min, W.K. A note on soft topological spaces. *Comput. Math. Appl.* **2011**, *62*, 3524–3528. [CrossRef]
8. Çağman, N.; Karataş, S.; Enginoglu, S. Soft topology. *Comput. Math. Appl.* **2011**, *62*, 351–358. [CrossRef]
9. Danjuma, S.; Ismail, M.A.; Herawan, T. An alternative approach to normal parameter reduction algorithm for soft set theory. *IEEE Access* **2017**, *5*, 4732–4746. [CrossRef]
10. Yuksel, S.; Dizman, T.; Yildizdan, G.; Sert, U. Application of soft sets to diagnose the prostate cancer risk. *J. Inequalities Appl.* **2013**, *2013*, 229. [CrossRef]
11. Kamacı, H.; Atagün, A.O.; Sönmezoğlu, A. Row-products of soft matrices with applications in multiple-disjoint decision making. *Appl. Soft Comput.* **2018**, *62*, 892–914. [CrossRef]
12. Fatimah, F.; Rosadi, D.; Hakim, R.B.F.; Alcantud, J.C.R. N-soft sets and their decision making algorithms. *Soft Comput.* **2018**, *22*, 3829–3842. [CrossRef]

13. Feng, F.; Li, C.; Davvaz, B.; Ali, M.I. Soft sets combined with fuzzy sets and rough sets: A tentative approach. *Soft Comput.* **2010**, *14*, 899–911. [CrossRef]
14. Maji, P.K.; Biswas, R.; Roy, A.R. Fuzzy Soft Sets. *J. Fuzzy Math.* **2001**, *9*, 589–602.
15. Jiang, Y.; Tang, Y.; Chen, Q. An adjustable approach to intuitionistic fuzzy soft sets based decision making. *Appl. Math. Model.* **2011**, *35*, 824–836. [CrossRef]
16. Smarandache, F. *A Unifying Field in Logics. Neutrosophy: Neutrosophic Probability, Set and Logic*; American Research Press: Rehoboth, DE, USA, 1999.
17. Maji, P.K. Neutrosophic soft set. *Comput. Math. Appl.* **2013**, *45*, 555–562. [CrossRef]
18. Sahin, R.; Küçük, A. Generalised Neutrosophic Soft Set and its Integration to Decision Making Problem. *Appl. Math. Inf. Sci.* **2014**, *8*, 2751. [CrossRef]
19. Deli, I.; Broumi, S. Neutrosophic soft matrices and NSM-decision making. *J. Intell. Fuzzy Syst.* **2015**, *28*, 2233–2241. [CrossRef]
20. Deli, I. Interval-valued neutrosophic soft sets and its decision making. *Int. J. Mach. Learn. Cybern.* **2017**, *8*, 665–676. [CrossRef]
21. Karaaslan, F. Possibility neutrosophic soft sets and PNS-decision making method. *Appl. Soft Comput.* **2017**, *54*, 403–414. [CrossRef]
22. Karaaslan, F. Correlation coefficients of single-valued neutrosophic refined soft sets and their applications in clustering analysis. *Neural Comput. Appl.* **2017**, *28*, 2781–2793. [CrossRef]
23. Uluçay, V.; Şahin, M.; Hassan, N. Generalized neutrosophic soft expert set for multiple-criteria decision-making. *Symmetry* **2018**, *10*, 437. [CrossRef]
24. Al-Quran, A.; Hassan, N.; Marei, E. A Novel Approach to Neutrosophic Soft Rough Set under Uncertainty. *Symmetry* **2019**, *11*, 384. [CrossRef]
25. Peng, X.; Liu, C. Algorithms for neutrosophic soft decision making based on EDAS, new similarity measure and level soft set. *J. Intell. Fuzzy Syst.* **2017**, *32*, 955–968. [CrossRef]
26. Abu Qamar, M.; Hassan, N. Entropy, measures of distance and similarity of Q-neutrosophic soft sets and some applications. *Entropy* **2018**, *20*, 672. [CrossRef]
27. Kahneman, D.; Tversky, A. Prospect theory: An analysis of decisions under risk. *Econometrica* **1979**, *47*, 263–291. [CrossRef]
28. Peng, X.; Yang, Y. Algorithms for interval-valued fuzzy soft sets in stochastic multi-criteria decision making based on regret theory and prospect theory with combined weight. *Appl. Soft Comput.* **2017**, *54*, 415–430. [CrossRef]
29. Haibin, W.; Smarandache, F.; Zhang, Y.; Sunderraman, R. Single Valued Neutrosophic Sets. *Multispace Multistruct.* **2010**, *4*, 410–413.
30. Liu, C.; Luo, Y.S. Correlated aggregation operators for simplified neutrosophic set and their application in multi-attribute group decision making. *J. Intell. Fuzzy Syst.* **2016**, *30*, 1755–1761. [CrossRef]
31. Ye, J.; Du, S. Some distances, similarity and entropy measures for interval-valued neutrosophic sets and their relationship. *Int. J. Mach. Learn. Cybern.* **2019**, *10*, 347–355. [CrossRef]
32. Sahin, R.; Küçük, A. On similarity and entropy of neutrosophic soft sets. *J. Intell. Fuzzy Syst.* **2014**, *27*, 2417–2430.
33. Alfaro-García, V.G.; Merigó, J.M.; Gil-Lafuente, A.M.; Kacprzyk, J. Logarithmic aggregation operators and distance measures. *Int. J. Intell. Syst.* **2018**, *33*, 1488–1506. [CrossRef]
34. Choi, S.H.; Jung, S.H. Similarity Analysis of Actual Fake Fingerprints and Generated Fake Fingerprints by DCGAN. *Int. J. Fuzzy Log. Intell. Syst.* **2019**, *19*, 40–47. [CrossRef]
35. Biswas, P.; Pramanik, S.; Giri, B.C. Entropy based grey relational analysis method for multi-attribute decision making under single valued neutrosophic assessments. *Neutrosophic Sets Syst.* **2014**, *2*, 102–110.
36. Wu, K.; Jin, J. Attribute recognition method of regional ecological security evaluation based on combined weight on principle of relative entropy. *Sci. Geogr. Sin.* **2008**, *28*, 754–758.
37. Jiang, W.; Zhang, Z.; Deng, X. Multi-Attribute Decision Making Method Based on Aggregated Neutrosophic Set. *Symmetry* **2019**, *11*, 267. [CrossRef]

6

The Asymmetric Alpha-Power Skew-*t* Distribution

Roger Tovar-Falón [1,*], Heleno Bolfarine [2] and Guillermo Martínez-Flórez [1]

1 Departamento de Matemáticas y Estadística, Facultad de Ciencias Básicas, Universidad de Córdoba, Montería 230027, Colombia; guillermomartinez@correo.unicordoba.edu.co

2 Departamento de Estatística, IME, Universidade de São Paulo, São Paulo 1010, Brazil; hbolfar@ime.usp.br

* Correspondence: rjtovar@correo.unicordoba.edu.co

Abstract: In this paper, we propose a new asymmetric and heavy-tail model that generalizes both the skew-*t* and power-*t* models. Properties of the model are studied in detail. The score functions and the elements of the observed information matrix are given. The process to estimate the parameters in model is discussed by using the maximum likelihood approach. Also, the observed information matrix is shown to be non-singular at the whole parametric space. Two applications to real data sets are reported to demonstrate the usefulness of this new model.

Keywords: alpha-power skew-*t* distribution; skew-*t* distribution; power-*t* distribution; asymmetry; Fisher information matrix; maximum likelihood estimation

---

1. Introduction

In recent years, there has been considerable interest in the statistical literature related to flexible families of distributions able of modeling data that present high degree of asymmetry, with kurtosis index greater or smaller than the captured by normal model. In this context, two proposals that have shown a promising behavior in this type of situations are the skew-normal (SN) distribution of Azzalini [1] and the power-normal (PN) distribution of Durrans [2]. The SN distribution has been widely studied by many authors, and its main drawback is that it presents singular Fisher information matrix, implying the inference is useless from the theory of large samples using the maximum likelihood (ML) approach. Although the PN model has a shorter asymmetry range than SN distribution, it presents non-singular information matrix and can easily be extended to censored scenarios, as it has a simple distribution function, see, for example, in Martínez-Flórez et al. [3].

The PN model is part of a wide family of distributions known as alpha-power, which has been widely studied by many authors. In addition to the normal distribution, the Birnbaum–Saunders (BS) distribution [4] has also been considered, see, for example, in Martínez-Flórez et al. [5], who propose an extension of the BS distribution based on the asymmetric alpha-power family of distributions to illustrate the applicability of the new proposal with a data set is related to the lifetimes in cycles $\times 10^{-3}$ $n = 101$ aluminum $6061 - T6$ pieces cut in parallel angle to the rotation direction of rolling at the rate of 18 cycles per second and maximum stress of 21.000 psi. More details of the PN distribution can be found in Gupta and Gupta [6] and Pewsey et al. [7].

An alternative propose for modeling asymmetric data that unifies the two previous approaches was introduced by Martínez-Flórez et al. [8]. The proposed model, which is called alpha-power skew-normal (APSN), has non-singular Fisher information matrix, and it can fit data with much more asymmetry than PN models it can handle. In addition, symmetry can be tested by using the likelihood ratio statistic, as the properties of large samples are satisfied for the ML estimator.

Another set of distributions with non-singular information matrices, useful for modeling asymmetric and heavy-tailed data, are based on generalizations of the Student-*t* distribution, see, for example, in [9–13]. Azzalini and Capitanio [9] for example, introduced a skew-*t* (ST) distribution as an extension of the SN model for modeling asymmetric and heavy-tailed data as follows; The random variable X is said to have the ST distribution with parameter λ and degrees of freedom ν, if X has the probability density function (PDF) given by

$$f_{ST}(x;\lambda,\nu) = 2f_T\left(x;\nu\right)\mathcal{F}_T\left(\lambda\sqrt{\frac{\nu+1}{x^2+\nu}}x;\nu+1\right), \quad x \in \mathbb{R} \tag{1}$$

where $\lambda \in \mathbb{R}$ is a parameter that controls the skewness of the distribution, and $f_T\left(\cdot;\nu\right)$ and $\mathcal{F}_T\left(\cdot;\nu\right)$ denote the PDF and the cumulative distribution function (CDF) of a standard Student-t distribution with ν degree of freedom, respectively. The ST distribution, like an extension of the SN model, inherits the problem of the singularity of the information matrix and before this inconvenience Zhao and Kim [14] proposed the power Student-t (PT) distribution, whose information matrix is non-singular and for a given degree of freedom, the kurtosis range surpasses the kurtosis range of the skew-t model at all times. The PT distribution is defined as follows. The random variable X is said to have the PT distribution with parameter α, and degrees of freedom ν, if X has PDF given by

$$f_{PT}(x;\alpha,\nu) = \alpha f_T\left(x;\nu\right)\left[\mathcal{F}_T\left(x;\nu\right)\right]^{\alpha-1}, \quad x \in \mathbb{R} \tag{2}$$

where $\alpha > 0$ is a parameter that controls the form of the distribution, and, again, $f_T\left(\cdot;\nu\right)$ and $\mathcal{F}_T\left(\cdot;\nu\right)$ denote the PDF and the CDF of a standard Student-t distribution, respectively.

Based on the properties of the ST model, to fit data with high degree of asymmetry and the characteristic of the PN model to capture kurtosis larger than the normal model, in this paper, we introduce a new distribution for modeling asymmetric and heavy-tailed data. The proposed model possess non-singular information matrix, and it is able to fit data with far more asymmetry than ST and PT models can handle and with large sample properties satisfied for the ML estimator. The model introduced in this paper is named as alpha-power skew-t (APST) model and it extends both, ST and PT models. The APSN model by Martínez-Flórez et al. [8] is also a particular case when ν tends to infinite. Note that symmetry can be tested using the likelihood ratio statistics with its large sample chi-square distribution.

The rest of this paper is organized as follows. Section 2 introduces the APST model and some of its properties like moments are studied. In particular, skewness and kurtosis indices are computed showing that their ranges surpass those of the ST and PT models. Section 3 deals with the ML estimation for the location-scale situation and its observed information matrix is derived. The extension to censored data is also presented. Finally, two applications are shown in Section 4, revealing that the model proposed can present much improvement over competitors.

2. The Alpha-Power Skew-t Distribution

Definition 1. *The random variable X is said to have an alpha-power skew-t (APST) distribution, if X has PDF given by*

$$f_{APST}(x;\lambda,\alpha,\nu) = \alpha f_{ST}(x;\lambda,\nu)\left[\mathcal{F}_{ST}(x;\lambda,\nu)\right]^{\alpha-1}, \tag{3}$$

for $x \in \mathbb{R}$, $\lambda \in \mathbb{R}$, and $\alpha,\nu \in \mathbb{R}^+$. Functions $f_{ST}(\cdot)$ and $\mathcal{F}_{ST}(\cdot)$ denote the PDF and the CDF of the standard ST distribution. A random variable having $f_{APST}(x;\lambda,\alpha,\nu)$ distribution is denoted shortly by $X \sim$ APST(λ,α,ν).

Figure 1 displays the form of the APST distribution for some selected values of the parameters λ and α for $\nu = 6$. Note from the figure that the asymmetry and kurtosis of the APST distribution are affected by the parameters α and λ; therefore, the APST model is more flexible to model data that can be highly skewed, as well as heavier tails than ST and PT models.

The following result provides some special cases of the model (3), which occur for different values of λ, α, and ν.

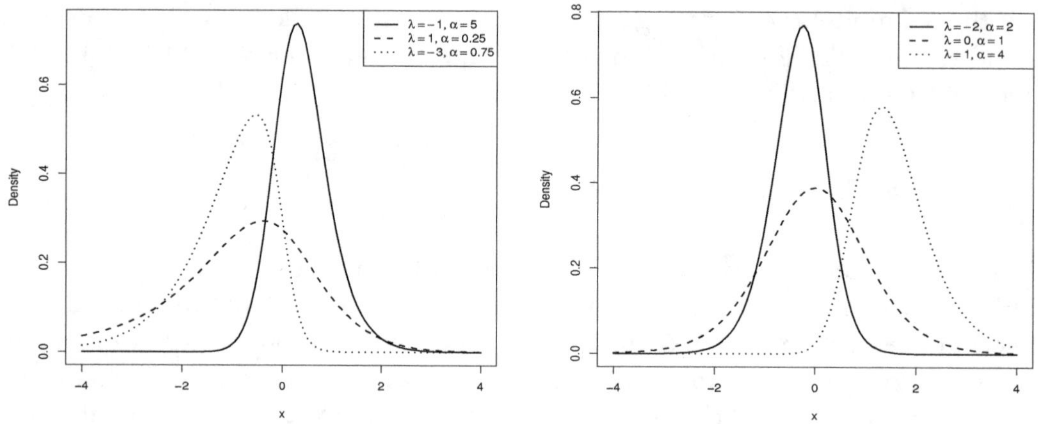

Figure 1. Probability density function of $\mathrm{APST}(\lambda, \alpha, 10)$ for some values of λ and α.

Proposition 1. *Let* $X \sim \mathrm{APST}(\lambda, \alpha, \nu)$,

(i) *if* $\lambda = 0$, *then* $X \sim \mathrm{PT}(\alpha, \nu)$,
(ii) *if* $\alpha = 1$, *then* $X \sim \mathrm{ST}(\lambda, \nu)$,
(iii) *if* $\lambda = 0$ *and* $\alpha = 1$, *then* $X \sim \mathrm{T}(\nu)$, *where* $\mathrm{T}(\nu)$ *denotes the Student-t disribution with* ν *degree of freedom.*
(iv) *if* $\nu \to +\infty$, *then* $X \sim \mathrm{APSN}(\lambda, \alpha)$,
(v) *if* $\lambda = 0$ *and* $\nu \to +\infty$, *then* $X \sim \mathrm{PN}(\alpha)$,
(vi) *if* $\alpha = 1$ *and* $\nu \to +\infty$, *then* $X \sim \mathrm{SN}(\lambda)$,
(vii) *if* $\lambda = 0$, $\alpha = 1$ *and* $\nu \to +\infty$, *then* $X \sim \mathrm{N}(0, 1)$,

Proof. The proof of (i)–(vii) is immediate from the definition of APST distribution. □

2.1. Moments

The following proposition presents an expression to compute the k-th moment of a random variable $\mathrm{APST}(\lambda, \alpha, \nu)$.

Proposition 2. *Let* $X \sim \mathrm{APST}(\lambda, \alpha, \nu)$, *then*

$$\mathbb{E}[X^k] = \mathbb{E}\left[(\mathcal{F}_{ST}^{-1}(Y; \lambda, \nu))^k\right] \tag{4}$$

where Y *follows a* $\mathrm{Beta}(\alpha, 1)$ *distribution and* $\mathcal{F}_{ST}^{-1}(\cdot; \lambda, \nu)$ *is the inverse of the function* $\mathcal{F}_{ST}(\cdot; \lambda, \nu)$.

Proof. We have by definition that

$$\mathbb{E}[X^k] = \int_{\mathbb{R}} x^k \alpha f_{ST}(x) (\mathcal{F}_{ST}(x; \lambda, \nu))^{\alpha - 1} dx$$

thus, letting $y = \mathcal{F}_{ST}(x; \lambda, \nu)$, then $x = \mathcal{F}_{ST}^{-1}(y; \lambda, \nu)$, it follows that

$$\mathbb{E}[X^k] = \int_0^1 \alpha \left(\mathcal{F}_{ST}^{-1}(y; \lambda, \nu)\right)^k y^{\alpha - 1} dy$$

which is the expected value of the function $\left(\mathcal{F}_{ST}^{-1}(Y; \lambda, \nu)\right)^k$, where Y follows a beta distribution with parameters α and 1. □

The indices of skewness ($\sqrt{\beta_1}$) and kurtosis (β_2) of APST distribution can be calculated by using the moments (4) as follows,

$$\sqrt{\beta_1} = \frac{\mu_3 - 3\mu_1\mu_2 + 2\mu_1^3}{(\mu_2 - \mu_1^2)^{3/2}} \quad \text{and} \quad \beta_2 = \frac{\mu_4 - 4\mu_1\mu_3 + 6\mu_2\mu_1^2 - 3\mu_1^4}{(\mu_2 - \mu_1^2)^2}$$

where $\mu_k = \mathbb{E}[X^k]$ for $k = 1, \ldots, 4$. Table 1 presents the ranges of possible values for the indices of asymmetry and kurtosis for $\mathrm{ST}(\lambda, \nu)$, $\mathrm{PT}(\alpha, \nu)$, and $\mathrm{APST}(\lambda, \alpha, \nu)$ distributions, for values of λ between -40 and 40, values of α between 0.5 and 50, and for values of $\nu = 2, 3, 4, 5, 6, 7$. It can seen from Table 1 that the length of the admissible intervals for the skewness and the kurtosis parameters of the APST distribution are larger than the corresponding intervals of the ST and PT distributions. This is an indicator that the APST model is more flexible in terms of asymmetry and kurtosis than the ST and PT models.

Table 1. Skewness and kurtosis for the models $\mathrm{ST}(\lambda, \nu)$, $\mathrm{PT}(\alpha, \nu)$, and $\mathrm{APST}(\lambda, \alpha, \nu)$, for $\lambda \in (-40, 40)$, $\alpha \in (0.5, 50)$ and $\nu = 2, \ldots 7$.

	Skew−t		**Power−t**		**Alpha−Power Skew−t**	
ν	**Skewness**	**Kurtosis**	**Skewness**	**Kurtosis**	**Skewness**	**Kurtosis**
2	(−0.963, 0.963)	(3.170, 3.489)	(−0.119, 3.040)	(1.552, 10.436)	(−2.452, 14.314)	(1.395, 864.385)
3	(−0.950, 0.950)	(3.146, 3.357)	(−0.086, 1.362)	(1.325, 3.223)	(−2.130, 4.902)	(1.628, 114.098)
4	(−1.853, 1.853)	(5.099, 7.824)	(−0.530, 1.178)	(3.461, 5.299)	(−1.898, 3.215)	(3.153, 29.874)
5	(−0.947, 0.947)	(3.051, 3.327)	(−0.475, 0.271)	(1.176, 3.130)	(−1.968, 3.046)	(3.862, 19.925)
6	(−1.681, 1.681)	(4.554, 7.279)	(−0.533, 1.118)	(3.974, 5.173)	(−1.681, 2.145)	(3.892, 11.893)
7	(−0.944, 0.944)	(3.007, 3.367)	(−0.710, 0.243)	(1.264, 3.082)	(−1.535, 2.536)	(3.136, 15.924)

2.2. Distribution Function

Proposition 3. *Let $X \sim \mathrm{APST}(\lambda, \alpha, \nu)$, then the CDF of X, namely, $\mathcal{F}_{APST}(x; \lambda, \alpha, \nu)$ is*

$$\mathcal{F}_{APST}(x; \lambda, \alpha, \nu) = \left[\mathcal{F}_{ST}(x; \lambda, \nu)\right]^{\alpha}, \quad x \in \mathbb{R}. \tag{5}$$

Proof. The proof is immediate and it follows from results of Durrans [2]. □

The inversion method can be used to generate a random variable with APST distribution. Thus, taking $\lambda \in \mathbb{R}$, $\alpha, \nu \in \mathbb{R}^+$ and a random variable with uniform distribution, namely, $U \sim \mathrm{U}(0, 1)$, random variable X with $\mathrm{APST}(\lambda, \alpha, \nu)$ distribution is generated by taking

$$X = \mathcal{F}_{ST}^{-1}\left(U^{1/\alpha}; \lambda, \nu\right).$$

Remark 1. *We consider a truncated $\mathrm{APST}(\lambda, \alpha)$ distribution to obtain a new and useful lifetime distribution. A random variable T has a truncated alpha-power skew-t distribution (at zero), denoted by $TAPST(\lambda, \alpha, \nu)$, if its PDF is given by*

$$f(t) = \frac{\alpha f_{ST}(t, \lambda, \nu)\left[\mathcal{F}_{ST}(t, \lambda, \nu)\right]^{\alpha-1}}{1 - \left[\mathcal{F}_{ST}(0, \lambda, \nu)\right]^{\alpha}}; \quad t > 0 \tag{6}$$

The survival and hazard rate functions of a random variable T following a $\mathrm{TAPST}(\lambda, \alpha, \nu)$ distribution are given by

$$\mathcal{S}_T(t) = P(T > t) = \frac{1 - \left[\mathcal{F}_{ST}(0, \lambda, \nu)\right]^{\alpha} - \left[\mathcal{F}_{ST}(t, \lambda, \nu)\right]^{\alpha}}{1 - \left[\mathcal{F}_{ST}(0, \lambda, \nu)\right]^{\alpha}}; \quad t > 0 \tag{7}$$

and

$$h_T(t) = \frac{\alpha f_{ST}(t, \lambda, \nu)\left[f_{ST}(t, \lambda, \nu)\right]^{\alpha-1}}{1 - \left[\mathcal{F}_{ST}(0, \lambda, \nu)\right]^{\alpha} - \left[\mathcal{F}_{ST}(t, \lambda, \nu)\right]^{\alpha}}; \quad t > 0 \tag{8}$$

respectively.

2.3. Location and Scale Extension

We can also consider a generalization of a APST distribution by adding location and scale parameters. The following definition gives a generalization of the APST model.

Definition 2. *Let* $X \sim \mathrm{APST}(\lambda, \alpha, \nu)$*. The APST density of location and scale is defined as the distribution of* $Y = \mu + \sigma X$*, for* $\mu \in \mathbb{R}$ *and* $\sigma > 0$*. The corresponding PDF is given by*

$$f_{APST}(y; \mu, \sigma, \lambda, \alpha, \nu) = \frac{\alpha}{\sigma} f_{ST}\left(\frac{y-\mu}{\sigma}; \lambda, \nu\right)\left[\mathcal{F}_{ST}\left(\frac{y-\mu}{\sigma}; \lambda, \nu\right)\right]^{\alpha-1}, \quad x \in \mathbb{R}, \tag{9}$$

for $\lambda \in \mathbb{R}$ *and* $\alpha, \nu \in \mathbb{R}^{+}$*. A random variable following a APST distribution of location and scale is denoted by* $Y \sim \mathrm{APST}(\mu, \sigma, \lambda, \alpha, \nu)$*.*

The k-th moment of a random variable $Y \sim \mathrm{APST}(\mu, \sigma, \lambda, \alpha, \nu)$ can be obtained from the formula

$$\mathbb{E}[Y^k] = \sum_{i=0}^{k} \binom{k}{i} \mu^i \sigma^{k-i} \mathbb{E}[X^{k-i}],$$

where $X \sim \mathrm{APST}(\lambda, \alpha, \nu)$.

3. Statistical Inference for APST Distribution

This section concerns likelihood inference about the parameter vector $\boldsymbol{\theta} = (\mu, \sigma, \lambda, \alpha, \nu)^{\top}$ of the location-scale family defined in Equation (9). Let $\mathbf{Y} = (Y_1, \ldots, Y_n)^{\top}$ be a random sample of the distribution $\mathrm{APST}(\mu, \sigma, \lambda, \alpha, \nu)$. The log-likelihood function for $\boldsymbol{\theta} = (\mu, \sigma, \lambda, \alpha, \nu)^{\top}$ can be written as follows,

$$\begin{aligned}
\ell(\boldsymbol{\theta}; \mathbf{Y}) \propto\; & n \log \alpha - n \log \sigma - \frac{n}{2} \log \nu \\
& + n \log \Gamma\left(\frac{\nu+1}{2}\right) - n \log \Gamma\left(\frac{\nu}{2}\right) - \frac{\nu+1}{2} \sum_{i=1}^{n} \log\left(1 + \frac{z_i^2}{\nu}\right) \\
& + \sum_{i=1}^{n} \log \mathcal{F}_T\left(\lambda z_i \sqrt{\frac{\nu+1}{z_i^2+\nu}}; \nu+1\right) + (\alpha-1) \sum_{i=1}^{n} \log \mathcal{F}_{ST}(z_i; \lambda, \nu)
\end{aligned} \tag{10}$$

where $z_i = (y_i - \mu)/\sigma$. Thus, by differentiating the log-likelihood function, we obtain the following score equations,

$$\begin{aligned}
\frac{\partial \ell(\boldsymbol{\theta}; \mathbf{Y})}{\partial \mu} = & \frac{\nu+1}{\sigma \nu} \sum_{i=1}^{n} z_i \left(1 + \frac{z_i^2}{\nu}\right)^{-1} \\
& - \frac{\lambda}{\sigma} \sum_{i=1}^{n} w_i \left(1 + \frac{z_i^2}{\nu}\right)^{-1} \frac{f_T(\lambda z_i w_i; \nu+1)}{\mathcal{F}_T(\lambda z_i w_i; \nu+1)} - \frac{\alpha-1}{\sigma} \sum_{i=1}^{n} \frac{f_{ST}(z_i; \lambda, \nu)}{\mathcal{F}_{ST}(z_i; \lambda, \nu)} = 0
\end{aligned} \tag{11}$$

$$\begin{aligned}
\frac{\partial \ell(\boldsymbol{\theta}; \mathbf{Y})}{\partial \sigma} = & -\frac{n}{\sigma} + \frac{\nu+1}{\sigma \nu} \sum_{i=1}^{n} z_i^2 \left(1 + \frac{z_i^2}{\nu}\right)^{-1} - \frac{\lambda}{\sigma} \sum_{i=1}^{n} z_i w_i \left(1 + \frac{z_i^2}{\nu}\right)^{-1} \frac{f_T(\lambda z_i w_i; \nu+1)}{\mathcal{F}_T(\lambda z_i w_i; \nu+1)} \\
& - \frac{\alpha-1}{\sigma} \sum_{i=1}^{n} z_i \frac{f_{ST}(z_i; \lambda, \nu)}{\mathcal{F}_{ST}(z_i; \lambda, \nu)} = 0
\end{aligned}$$

(12)

$$\frac{\partial\ell(\boldsymbol{\theta};\mathbf{Y})}{\partial\lambda} = \sum_{i=1}^{n} z_i w_i \frac{f_T(\lambda z_i w_i;\nu+1)}{\mathcal{F}_T(\lambda z_i w_i;\nu+1)} - \frac{\alpha-1}{\pi(1+\lambda^2)}\sum_{i=1}^{n}\frac{\left(1+(1+\lambda^2)z_i^2/\nu\right)^{-\frac{\nu}{2}}}{\mathcal{F}_{ST}(z_i;\lambda,\nu)} = 0, \tag{13}$$

$$\frac{\partial\ell(\boldsymbol{\theta};\mathbf{Y})}{\partial\alpha} = \frac{n}{\alpha} + \sum_{i=1}^{n}\log\mathcal{F}_{ST}(z_i;\lambda,\nu) = 0, \tag{14}$$

$$\begin{aligned}\frac{\partial\ell(\boldsymbol{\theta};\mathbf{Y})}{\partial\nu} &= \frac{n\alpha}{2}\left(\psi\left(\frac{\nu+1}{2}\right) - \psi\left(\frac{\nu}{2}\right) - \frac{1}{\nu}\right) - \frac{1}{2}\sum_{i=1}^{n}\log\left(1+\frac{z_i^2}{\nu}\right)\\ &+ \frac{\nu+1}{2\nu^2}\sum_{i=1}^{n} z_i^2\left(1+\frac{z_i^2}{\nu}\right)^{-1}\\ &+ \frac{\lambda}{2\nu(\nu+1)}\sum_{i=1}^{n} z_i^3 w_i\left(1+\frac{z_i^2}{\nu}\right)^{-1}\frac{f_T(\lambda z_i w_i;\nu+1)}{\mathcal{F}_T(\lambda z_i w_i;\nu+1)}\\ &- \frac{\lambda}{2\nu(\nu+1)}\sum_{i=1}^{n} z_i w_i\left(1+\frac{z_i^2}{\nu}\right)^{-1}\frac{f_T(\lambda z_i w_i;\nu+1)}{\mathcal{F}_T(\lambda z_i w_i;\nu+1)}\\ &- \frac{(\alpha-1)}{2\pi(\nu+1)}\frac{\lambda}{(1+\lambda^2)}\sum_{i=1}^{n}\frac{\left(1+(1+\lambda^2)z_i^2/\nu\right)^{-\frac{\nu}{2}}}{\mathcal{F}_{ST}(z_i;\lambda,\nu)} + \frac{\alpha-1}{2}\sum_{i=1}^{n}\frac{g(z_i;\nu)}{\mathcal{F}_{ST}(z_i;\lambda,\nu)} = 0\end{aligned} \tag{15}$$

where $\psi(\cdot)$ is the digamma function, $w_i = \sqrt{\frac{\nu+1}{x_i^2+\nu}}$ for $i = 1,\ldots,n$, and $g(x;\nu)$ is the function defined by

$$\begin{aligned}g(x;\nu) = &\int_{-\infty}^{x}\left\{\frac{(\nu+1)}{\nu^2}s^2\left(1+\frac{s^2}{\nu}\right)^{-1} - \log\left(1+\frac{s^2}{\nu}\right)\right\} f_{ST}(s;\lambda,\nu)ds\\ &- \frac{\lambda}{\pi\nu}\int_{-\infty}^{x} s\left(1+\frac{s^2}{\nu}\right)^{-1}\left\{1+(1+\lambda^2)\frac{s^2}{\nu}\right\}^{-\frac{\nu+2}{2}} ds\end{aligned} \tag{16}$$

Equations (11)–(15) include nonlinear functions; therefore, it is not possible to obtain explicit forms of the maximum likelihood estimators (MLEs), and they must be calculated by using numerical methods. In this work, we used the *maxLik* function of R Development Core Team [15] which uses the Newton–Raphson optimization method. The elements of the observed information matrix are easily obtained after calculating the second derivative of the log-likelihood function and multiplying by -1, that is,

$$j_{\theta_i\theta_k} = -\frac{\partial\ell(\boldsymbol{\theta};\mathbf{Y})}{\partial\theta_i\partial\theta_k}, \quad i,k = 1,2,\ldots,5 \tag{17}$$

where $\boldsymbol{\theta} = (\mu,\sigma,\lambda,\alpha,\nu)^{\top}$. This elements are given in the Appendix A. To find the standard errors (EE) of the MLEs and calculate confidence intervals, the information matrix $\mathbf{I}$ (or Fisher information) must be calculated, which is defined as the expected value of the second derived from the log-likelihood function or less the expected value of the Hessian matrix; from this matrix, we calculate the EE as the diagonal elements of the inverse of this matrix. The elements of the $\mathbf{I}$ matrix are obtained as

$$\mathbf{I}(i,k) = -E\left(\frac{\partial\ell(\boldsymbol{\theta};\mathbf{Y})}{\partial\theta_i\partial\theta_k}\right), \quad i,k = 1,2,\ldots,5 \tag{18}$$

The role of the Fisher information in the asymptotic theory of maximum-likelihood estimation was emphasized by Ronald Fisher following some initial results by Francis Edgeworth, see Lehman and Casella [16] and Frieden [17] for more details. The Fisher-information matrix is used to calculate the covariance matrices associated with maximum-likelihood estimates, and it can also be used in the formulation of test statistics, such as the Wald test.

As the expected value under the APST distribution and the second-order derivatives are not direct, numerical methods must be used to obtain the explicit form of the information matrix $\mathbf{I}$.

Therefore, we use the observed information matrix to calculate the standard errors in the rest of the document.

When ν tends to infinite the ST distribution converges to the SN distribution and we recall that the information matrix of a random variable $X \sim \mathrm{SN}(\mu, \sigma, \lambda)$ which is denoted by $\mathbf{I}_\lambda(\boldsymbol{\varphi})$, where $\boldsymbol{\varphi} = (\mu, \sigma, \lambda)^\top$, is singular for $\lambda = 0$. Therefore, it is convenient to use a centered parameterization of the ST distribution proposed by Arellano-Valle and Azzalini [18].

The centered parameterization of the SN distribution was proposed as an alternative to the problem of singularity of the information matrix of the SN when $\lambda = 0$. Arellano-Valle and Azzalini [19] proposed a second representation of the SN by defining a new random variable X as

$$X = \mu + \sigma\left(\frac{Z - \mathbb{E}[Z]}{\sqrt{\mathrm{Var}[Z]}}\right),$$

where $\mu \in \mathbb{R}$ and $\sigma > 0$ are parameters of the random variable X and $Z \sim \mathrm{SN}(\lambda)$. This representation is called centered parameterization, as $\mathbb{E}[X] = \mu$ and $\mathrm{Var}[X] = \sigma^2$ and it is denoted by $\mathrm{CSN}(\mu, \sigma, \gamma_1)$, where $-0.9953 < \gamma_1 < 0.9953$. Under the centered parameterization model, μ, σ, and $\gamma_1 = \sqrt{\beta_1}$ represent the mean, the standard deviation and the skewness index of X, respectively. If $Z \sim \mathrm{SN}(\lambda)$ then $\mathbb{E}[Z] = b\delta$ and $\mathrm{Var}[Z] = 1 - (b\delta)^2$, where $b = \sqrt{2/\pi}$ and $\delta = \lambda/\sqrt{1+\lambda^2}$; it has that the random variable X can be written as $X = \mu + \sigma Z$ which has $\mathrm{SN}(\lambda_1, \lambda_2, \lambda)$ distribution, where

$$\lambda_1 = \mu - c\sigma\gamma_1^{1/3}, \quad \lambda_2 = \sigma\sqrt{1 + c^2\gamma_1^{2/3}}, \quad \lambda = \frac{c\gamma_1^{1/3}}{\sqrt{b^2 + c^2(b^2-1)\gamma_1^{2/3}}} \tag{19}$$

with $c = \{2/(4-\pi)\}^{1/3}$. Under this denomination, the information matrix can be written as $\mathbf{I}_{\gamma_1} = \mathbf{D}^\top \mathbf{I}_\lambda \mathbf{D}$, where $\mathbf{D}$ is a matrix that represents the derivative of the parameters of the standard representation (λ_1, λ_2 and λ) regarding to the new parameters (μ, σ and γ_1). It also follows that the information matrix converges to a diagonal matrix $\boldsymbol{\Sigma}_c^{-1} = \mathrm{diag}(\sigma^2, \sigma^2/2, 6)$ when $\lambda \to 0$. This guarantees the existence and uniqueness of the MLEs of λ_1 and λ_2 for each fixed value of λ.

Following this same line of thought, we suppose that Y follows the model (1) with location parameter $\mu \in \mathbb{R}$ and scale parameter $\sigma > 0$, that is,

$$f_{ST}(y; \mu, \sigma, \lambda, \nu) = \frac{2}{\sigma} f_T\left(\frac{y-\mu}{\sigma}; \nu\right) \mathcal{F}_T\left(\lambda\sqrt{\frac{\nu+1}{Q_y+\nu}}\left(\frac{y-\mu}{\sigma}\right); \nu+1\right), \quad y \in \mathbb{R} \tag{20}$$

where $\lambda \in \mathbb{R}$ and $Q_y = ((y-\mu)/\sigma)^2$. This representation relates to the direct parameterization of the ST distribution with parameter vector $\boldsymbol{\rho} = (\mu, \sigma, \lambda, \nu)^\top$. It follows that $Z_T = (Y-\mu)/\sigma \sim \mathrm{ST}(\lambda, \nu)$, and by the stochastic representation of the ST distribution is given by $Z_T = Z/\sqrt{V}$, where $Z \sim \mathrm{SN}(\lambda)$ and $V \sim \chi^2_\nu/\nu$. This entails to compute the first four cumulants of Z_T denoted by $\mu_1(\delta, \nu)$, $\mu_2(\delta, \nu)$, $\mu_3(\delta, \nu)$ and $\mu_4(\delta, \nu)$, see [18]. The centered parameterization of the ST distribution of a random variable Y comes by defining

$$\mu_t = \mathbb{E}[Y] = \mu + \sigma\mu_1(\delta, \nu) = \mu + \sigma b_\nu \delta$$

$$\sigma_t^2 = \mathrm{Var}[Y] = \sigma^2\mu_2(\delta, \nu) = \eta^2\left\{\frac{\nu}{\nu-2} - b_\nu^2\delta^2\right\},$$

$$\gamma_{1t} = \frac{\mu_3(\delta,\nu)}{\mu_2(\delta,\nu)^{3/2}} = \frac{b_\nu \delta}{\mu_2(\delta,\nu)^{3/2}} \left\{ \frac{\nu(3-\delta^2)}{\nu-3} - \frac{3\nu}{\nu-2} + 2b_\nu^2\delta^2 \right\}$$

$$\gamma_{2t} = \frac{\mu_4(\delta,\nu)}{\mu_2(\delta,\nu)^2} = \frac{1}{\mu_2(\delta,\nu)^2} \left\{ \frac{3\nu^2}{(\nu-2)(\nu-4)} - \frac{4b_\nu^2\delta^2\nu(3-\delta^2) + \frac{6b_\nu^2\delta^2\nu}{\nu-2} - 4b_\nu^4\delta^4}{\nu-3} \right\} - 3.$$

The new representation is defined as the centered skew-t distribution with parameter vector $\tilde{\boldsymbol{\rho}} = (\mu, \sigma^2, \gamma_1, \gamma_2)^\top$. According to Arellano-Valle and Azzalini [18], the information matrix of this representation can be written as

$$\mathbf{I}(\tilde{\boldsymbol{\rho}}) = \mathbf{B}^\top \mathbf{I}(\boldsymbol{\rho})\mathbf{B},$$

where $\mathbf{B}$ is a matrix representing the derivative of the parameter vector $\boldsymbol{\rho}$ with respect to the new vector $\tilde{\boldsymbol{\rho}}$. It can shown that $b_\nu \to b$ when $\nu \to \infty$, see [18]. Therefore, the parameters of the centered ST model converge to $\mu_t \to \mu$, $\sigma_t^2 \to \sigma^2$, and $\gamma_{1t} \to \gamma_1$ when $\nu \to \infty$, that is, the parameters of the CSN. As $Z_T \to \text{SN}(\lambda)$ when $\nu \to \infty$, it follows that the random variable Y converges to a distribution with information matrix

$$\mathbf{I}(\mu, \sigma^2, \gamma_1, \alpha) = \begin{pmatrix} \mathbf{I}_{\boldsymbol{\theta}_1\boldsymbol{\theta}_1} & \mathbf{I}_{\boldsymbol{\theta}_1,\alpha} \\ \mathbf{I}_{\boldsymbol{\theta}_1,\alpha}^\top & I_{\alpha,\alpha} \end{pmatrix}, \quad (21)$$

where the elements of the diagonal correspond to the information of the parameter vector $\boldsymbol{\theta}_1 = (\mu, \sigma^2, \gamma_1)$ and α, and $\mathbf{I}_{\boldsymbol{\theta}_1,\alpha}$ is the joint information of $\boldsymbol{\theta}_1 = (\mu, \sigma^2, \gamma_1)^\top$ and α. Now, when $\lambda \to 0$ and $\alpha = 1$, it can be shown that $\mathbf{I}_{\boldsymbol{\theta}_1\boldsymbol{\theta}_1} \to \text{diag}(\sigma^2, \sigma^2/2, 6)$, with determinant equal to $0.3333/\sigma^4$, and $\mathbf{I}_{\boldsymbol{\theta}_1,\alpha} = (0.9031/\sigma, -0.5956/\sigma, 0.7206)^\top$; therefore, the determinant $|\mathbf{I}(\mu, \sigma^2, \gamma_1, \alpha)| \neq 0$, and it concludes that the random variable Y converges to a distribution with information matrix non-singular when ν tends to infinite.

3.1. Extension to Censored Data

Based on the goodness of the APST distribution to fit asymmetric and heavy-tailed data, in this section we introduce the censored APST model which we will be denote by CAPST.

Definition 3. *Suppose that the random variable Y follows APST distribution, and consider a random sample $\mathbf{Y} = (Y_1, Y_2, \ldots, Y_n)$ where only the Y_i values greater than a constant k are recorded. In addition, for values $Y_i \leq k$ only the value of k is recorded. Therefore, for $i = 1, 2, \ldots, n$, the observed values Y_i^o can be written as*

$$Y_i^o = \begin{cases} k, & \text{if } Y_i \leq k, \\ Y_i, & \text{if } Y_i > k. \end{cases}$$

The resulting sample is said to be a censored APST, and we say that Y is a censored random variable APST. We will use the notation $Y \sim \text{CAPST}(\boldsymbol{\theta})$, where $\boldsymbol{\theta} = (\mu, \sigma, \lambda, \alpha, \nu)^\top$.

From Definition 3 it follows that $P(Y_i^o = k) = P(Y_i \leq k) = \{\mathcal{F}_{ST}((k-\mu)/\sigma)\}^\alpha$ and for the observations $Y_i^o = Y_i$, the distribution of Y_i^o is the same of Y_i, i.e., $Y_i^o \sim \text{APST}(\boldsymbol{\theta})$. For convenience, we choose to work with the case of left-censored data; however, the followings results can be extended to other types of censorship.

3.2. Properties of the CAPST Model

Let $Y \sim \text{CAPST}(\mu, \sigma, \lambda, \alpha, \nu)$,

1. If $\alpha = 1$, then $Y \sim \text{CST}(\mu, \sigma, \lambda, \nu)$, where CST indicates the censored skew-t model.
2. If $\lambda = 0$, then $Y \sim \text{CPT}(\mu, \sigma, \alpha, \nu)$, where CPT indicates the censored power-t model.
3. If $\alpha = 1$ and $\lambda = 0$, then $Y \sim \text{CT}(\mu, \sigma, \nu)$, that is, the censored Student-t model follows.
4. If $\nu \to +\infty$, then $Y \sim \text{CAPSN}(\mu, \sigma, \lambda, \alpha)$, where CAPSN indicates the censored alpha-power skew-normal model.

5. If $\alpha = 1$ and $\nu \to +\infty$, then $Y \sim \text{CSN}(\mu, \sigma, \lambda)$, that is, the censored skew-normal model follows.
6. If $\lambda = 0$ and $\nu \to +\infty$, then $Y \sim \text{CPN}(\mu, \sigma, \alpha)$, that is, the censored power-normal model follows.
7. If $\alpha = 1$, $\lambda = 0$ and $\nu \to +\infty$, then $Y \sim \text{CN}(\mu, \sigma^2)$, that is, the censored normal model follows.

The estimates of the parameters of the model can be obtained via maximum likelihood method, where the log-likelihood function is given by

$$\begin{aligned}
\ell(\boldsymbol{\theta}; \mathbf{Y}) \propto \alpha \sum_0 \log \mathcal{F}_{ST}\left(\frac{k-\mu}{\sigma}; \lambda, \nu\right) + n_1 \log \alpha - n_1 \log \sigma - \frac{n_1}{2} \log \nu \\
+ n_1 \log \Gamma\left(\frac{\nu+1}{2}\right) - n_1 \log \Gamma\left(\frac{\nu}{2}\right) - \frac{\nu+1}{2} \sum_1 \log\left(1 + \frac{x_i^2}{\nu}\right) \\
+ \sum_1 \log \mathcal{F}_T\left(\lambda x_i \sqrt{\frac{\nu+1}{x_i^2+\nu}}; \nu+1\right) + (\alpha-1) \sum_1 \log \mathcal{F}_{ST}(x_i; \lambda, \nu)
\end{aligned}$$

where $x_i = (y_i - \mu)/\sigma$; $\sum_1$ and $\sum_0$ are the sum over censored individuals and uncensored individuals, respectively; and n_1 is the number of uncensored individuals.

4. Real Data Applications

In this section, we illustrate the applicability of the proposed model in Section 2 by analyzing two data sets. We use the statistical software *R* [15], version 3.5.3 with the package *maxLike* for maximizing the corresponding likelihood functions. For comparing purposes of various models, the AIC Akaike [20], BIC Schwarz [21], and corrected AIC (CAIC) Bozdogan [22] information criteria were used.

4.1. Application 1: Volcano Heights Data

Consider the data set related to heights of 1520 volcanoes in the world which is available in website dx.doi.org/10.5479/si.GVP.VOTW4-2013 [23]. Table 2 presents the summary statistics for the data set. It can be noted that the asymmetry and kurtosis indices seem to indicate that the use of an asymmetric and heavy-tailed model is appropriate to analyze this data set. We analyzed these data by fitting the Student-*t*, ST, PT, and APST distributions.

Table 2. Volcano heights data: Statistical summary.

n	Mean	Variance	$\sqrt{b_1}$	b_2
1520	16.7760	15.6682	0.6461	4.3809

Table 3 shows the parameter estimates, together with their corresponding standard errors (SE). Note that the values of the standard errors of the μ and σ estimates for the APST model are smaller than the corresponding standard errors of the respective parameters for the Student-*t*, ST, and PT models. Table 3 also presents some model selection criteria, together with the values of the log-likelihood. The AIC, BIC, and CAIC criteria indicate that the APST model seems to provide better fit to the volcanoes heights data than the T, ST, and PT models, supporting the asymmetry assertion of the volcano's heights variable. Figure 2 shows the graphs *QQplot* of the fitted models. It can be clearly seen from the figure that the APST model fits the data better than the Student-*t*, ST, and PT models. In addition, we can use the likelihood ratio (LR) test statistic to conform our claim. To do this, we consider the following hypotheses,

$$H_0: (\lambda, \alpha) = (0, 1)\ (\text{T}(\mu, \sigma, \nu)) \quad \text{v.s} \quad H_1: (\lambda, \alpha) \neq (0, 1)\ (\text{APST}(\mu, \sigma, \lambda, \alpha, \nu)),$$

The value of the LR test statistic is $-2\log(\Lambda) = -2(\ell_T(\hat{\theta}) - \ell_{APST}(\hat{\theta}))$ =134.823 and comparing this quantity with χ_2^2 =5.9914, the null hypotheses is rejected. The APST model is also compared with the ST and PT models by considering the hypotheses

$$H_{01} : \alpha = 1 \ (\text{ST}(\mu, \sigma, \lambda, \nu)) \quad \text{v.s} \quad H_{11} : \alpha \neq 1 \ (\text{APST}(\mu, \sigma, \lambda, \alpha, \nu)),$$

and

$$H_{02} : \lambda = 0 \ (\text{PT}(\mu, \sigma, \alpha, \nu)) \quad \text{v.s} \quad H_{12} : \lambda \neq 0 \ (\text{APST}(\mu, \sigma, \lambda, \alpha, \nu)),$$

respectively. The respective values of the LR test statistic are given by $-2\log(\Lambda_1) = -2(\ell_{ST}(\hat{\theta}) - \ell_{APST}(\hat{\theta}))$ =26.620 and $-2\log(\Lambda_2) = -2(\ell_{PT}(\hat{\theta}) - \ell_{APST}(\hat{\theta}))$ =45.660 and comparing these quantities with χ_1^2 =3.8414, both null hypotheses are rejected. Finally, Figure 3left shows the histogram of the volcano heights variable, whereas Figure 3right presents the empirical CDF (solid line) together with the CDF of the fitted APST model (dotted line).

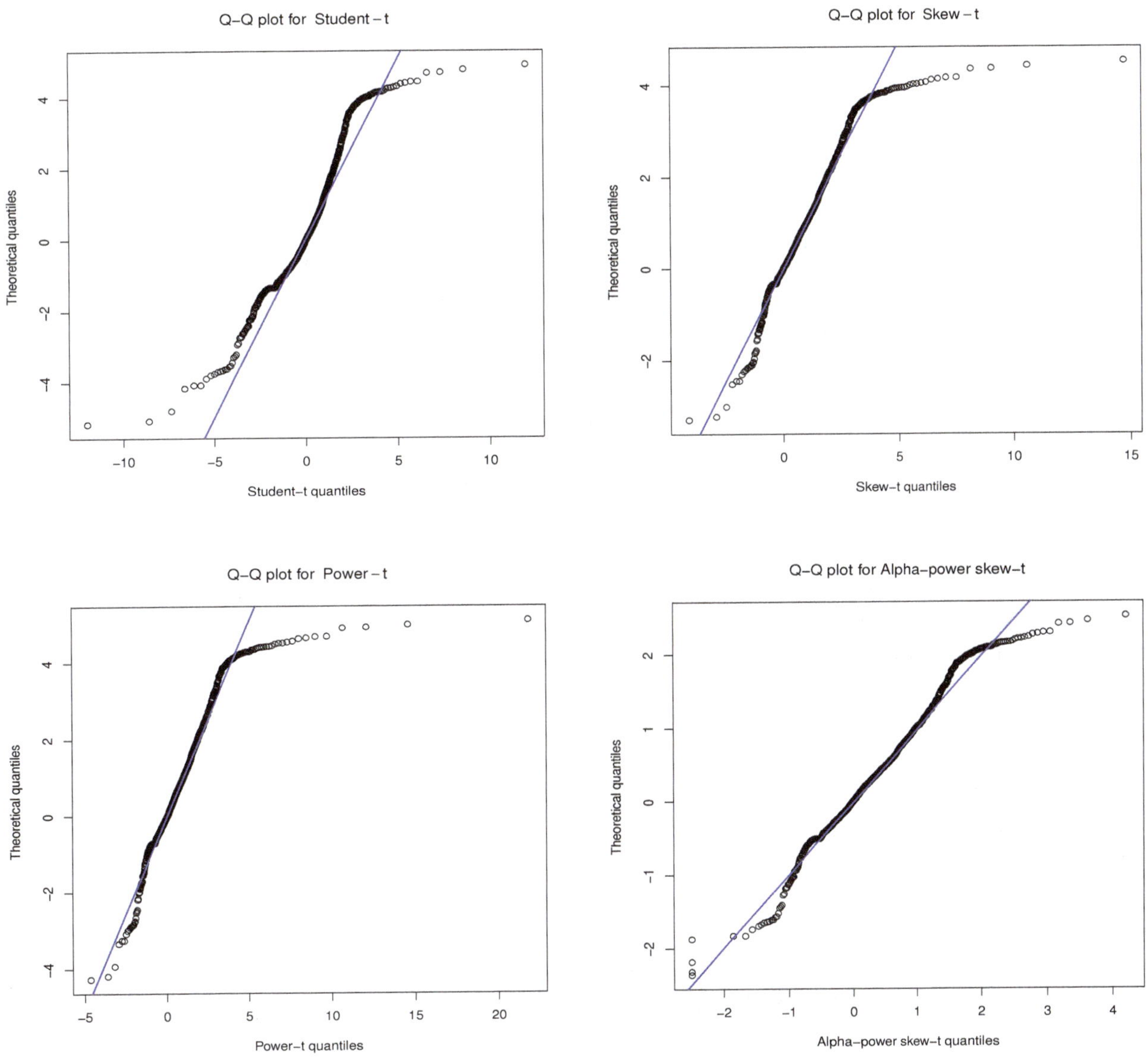

Figure 2. Volcano height data: QQplot for Student-*t*, ST, PT, and APST models.

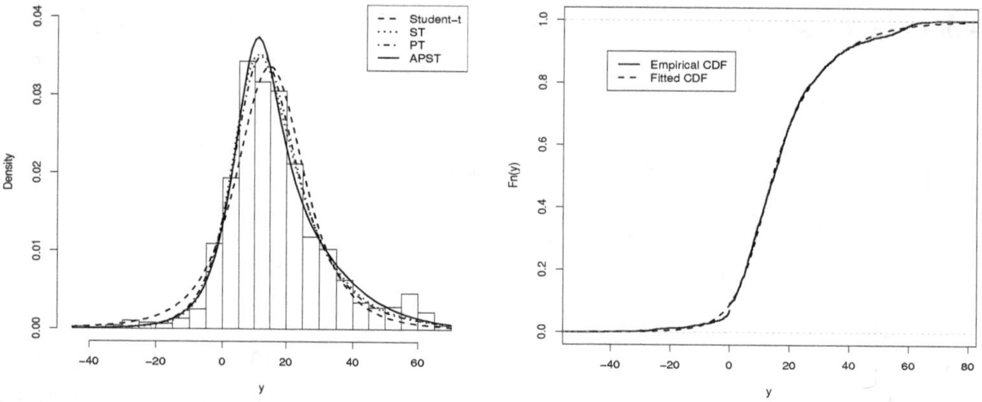

Figure 3. (**Left**) Graph of fitted densities to volcano height data. (**Right**) Empirical CDF and CDF of fitted APST model.

Table 3. Parameter estimates (SE) for the fitted models to the volcano height data.

Estimates	Distribution			
	Student-*t*	**ST**	**PT**	**APST**
$\hat{\mu}$	14.7835(0.3615)	4.7469(0.6892)	8.4027(0.7923)	11.5509(0.1337)
$\hat{\sigma}$	11.0045(0.3975)	14.1532(0.7237)	11.8146(0.4707)	22.6885(0.0792)
$\hat{\lambda}$	–	1.5673(0.1838)	–	5.2347(0.2870)
$\hat{\alpha}$	–	–	1.7912(0.1147)	0.3205(0.0347)
$\hat{\upsilon}$	3.4156(0.3601)	3.4075(0.3454)	2.7473(0.2566)	12.8734(2.9729)
$\hat{\ell}$	−6273.35	−6219.25	−6228.77	−6205.94
AIC	12,552.70	12,446.49	12,465.53	12,421.87
BIC	12,568.68	12,467.79	12,486.53	12,448.50
CAIC	12,571.68	12,471.79	12,490.83	12,453.50

4.2. Application 2: Stellar Abundances Data

The second data set is related to measurements for 68 solar-type stars, which are available in the package *astrodatR* of the software *R* [24] under the name *Stellar abundances*. These data were previously analyzed Mattos et al. [25] by using the Scale Mixture of Skew Normal Censored Regression (SMSNCR) models. We take only the response variable: log *N(Be)*, which represents the log of the abundance of beryllium scaled to Sun's abundance (i.e., the Sun has $\log N(Be) = 0.0$)

In astronomical research, a previously identified sample of objects (stars, galaxies, quasars, X-ray sources, etc.) is observed at some new wavebands. According to Feigelson [24], due to limited sensitivities, some objects may be undetected, leading to upper limits in their derived luminosities. For this dataset we have 12 left-censored data points, i.e., 12 undetected beryllium measurement, that represents 19.35% of observations. Table 4 presents the ML estimates for the parameters of the censored Studen-*t* (CT), censored skew-*t* (CST), censored power-*t* (CPT), and censored alpha-power skew-*t* (CAPST) models, together with their corresponding standard errors. Table 4 also compares the fit of the four models using the model selection criteria (AIC, CAIC and BIC). Note that, again, the CAPST model with heavy tails have better fit than the CT, CST, and CPT models.

To identify atypical observations and/or model mispecification, we analyzed the transformation of the martingale residual, r_{MT_i}, proposed in Barros et al. [26]. These residuals are defined by

$$r_{MTi} = \text{sign}(r_{Mi})\sqrt{-2[r_{Mi} + \delta_i \log(\delta_i - r_{M_i})]}, \qquad i = 1, \ldots, n$$

where $r_{Mi} = \delta_i + \log S(y_i; \hat{\boldsymbol{\theta}})$ is the martingal residual proposed by Ortega et al. [27], where $\delta_i = 0, 1$ indicates whether the i-th observation is censored or not, respectively; $\text{sign}(r_{Mi})$ denotes the sign of

r_{Mi}; and $S(y_i; \hat{\boldsymbol{\theta}}) = P_{\hat{\boldsymbol{\theta}}}(Y_i > y_i)$ represents the survival function evaluated at y_i, where $\hat{\boldsymbol{\theta}}$ are the MLE for $\boldsymbol{\theta}$. The plots of r_{MT_i} with generated confidence envelopes are presented in Figure 4. From this figure, we can see clearly that the CST, CPT, and CAPST models fit better the data than the CT model, since, in that cases, there are not observations which lie outside the envelopes. The Figure 5 shows the graph of the densities of the different models fitted to the stellar abundances data. From the figure, the CAPST model seems to fit better the stellar abundances data than CT, CST and CPT models.

Table 4. Parameter estimates (SE) for the fitted models to the stellar abundances data.

	Distribution			
Estimates	**CT**	**CST**	**CPT**	**CAPST**
$\hat{\mu}$	1.0314(0.0010)	1.2306(0.0018)	1.2098(0.0052)	1.1761(0.0054)
$\hat{\sigma}$	0.1596(0.0012)	0.2712(0.0058)	0.0818(0.0008)	0.0905(0.0020)
$\hat{\lambda}$	–	−3.5655(3.7748)	–	0.6580(0.5031)
$\hat{\alpha}$	–	–	0.1705(0.0208)	0.1518(0.0251)
$\hat{\nu}$	0.9974(0.0884)	1.2501(0.1774)	6.0927(0.7501)	6.0999(0.7326)
$\hat{\ell}$	−29.50743	−18.87016	−17.67113	−14.80241
AIC	65.01487	45.74033	43.34227	39.60482
BIC	71.67339	54.61836	52.22030	50.70236
CAIC	59.38987	38.37525	35.97719	30.57256

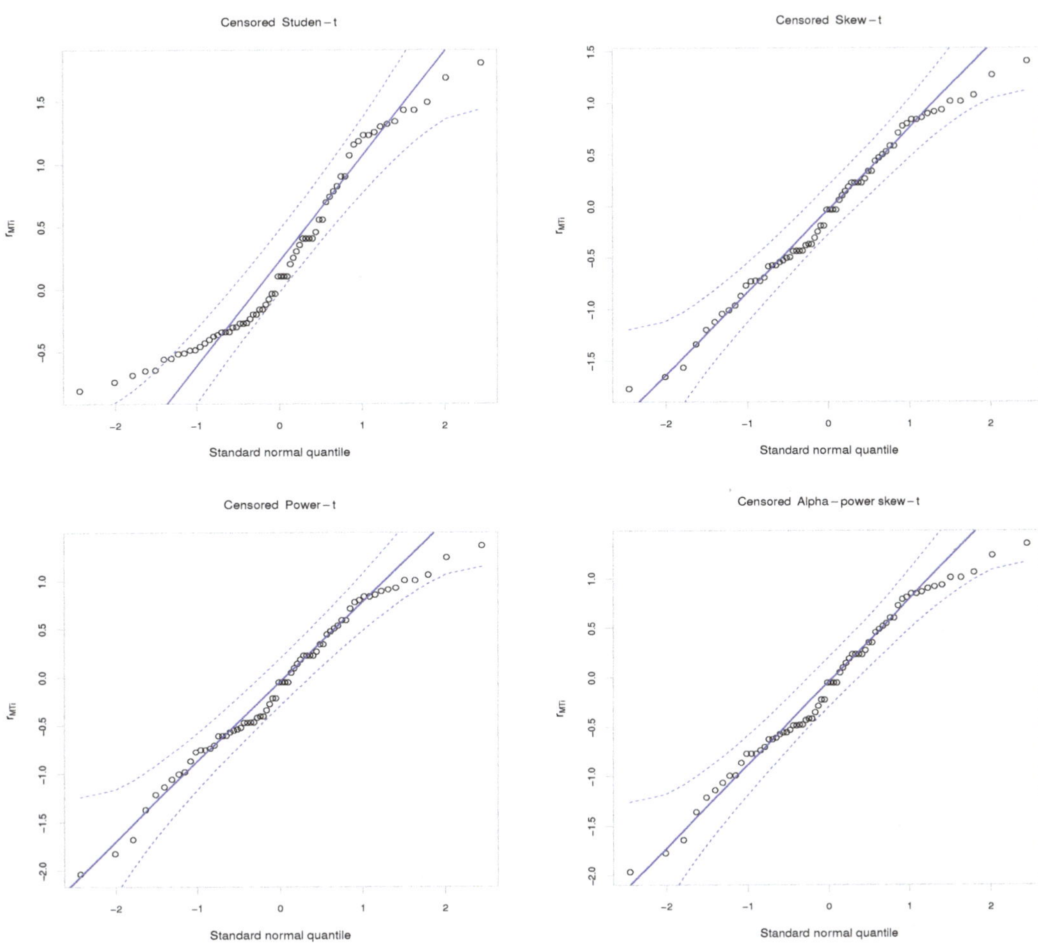

Figure 4. Stellar abundances data. Envelopes of transformed martingale residuals for CT, CST, CPT, and CAPST models.

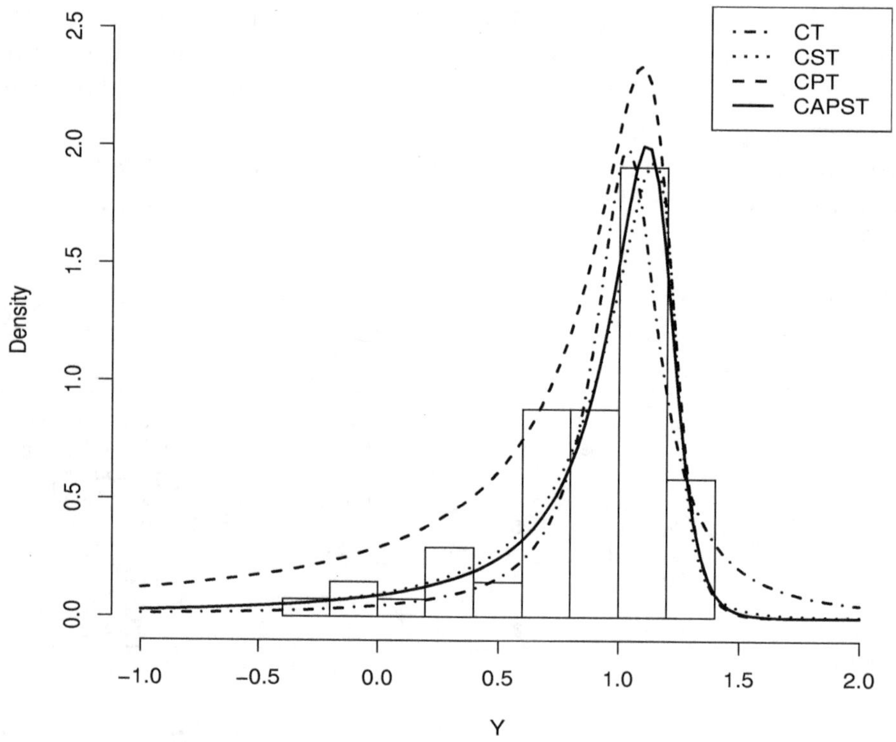

Figure 5. Graph of fitted densities to stellar abundances data.

5. Conclusions

In this work, a new asymmetric model has been introduced. It is based on the combination of skew-*t* [1] and power-*t* [2] models. The new model presents greater ranges of asymmetry and kurtosis, which is very useful for modeling skewed and heavy-tailed data. The problem of estimating the parameters in the model is dealt by using the maximum likelihood approach which is also used for developing large sample properties for the estimators. The elements of the observed information matrix are analytically obtained. The likelihood ratio statistics can be used for testing the APST null hypothesis since the Student-*t*, ST, and PT models are special cases of the model entertained. Two applications to volcano heights data and stellar abundances data indicate that the proposed model can be a useful alternative to the ST and PT models.

Author Contributions: Individual contributions to this article: conceptualization, R.T.-F., H.B., and G.M.-F.; methodology, R.T.-F., H.B., and G.M.-F.; software, R.T.-F., H.B., and G.M.-F.; validation, R.T.-F., H.B., and G.M.-F.; formal analysis, R.T.-F., H.B., and G.M.-F.; investigation, R.T.-F., H.B., and G.M.-F.; resources, R.T.-F., H.B., and G.M.-F.; writing-original draft preparation, R.T.-F., H.B., and G.M.-F.; writing-review and editing, R.T.-F., H.B., and G.M.-F. All authors have read and agreed to the published version of the manuscript.

Acknowledgments: We thank the anonymous referees for helpful suggestions which improved the article.

Appendix A

In this section, expressions for the elements of the observed information matrix of the alpha-power skew-*t* model are provided. Initially we suppose that $Y \sim \text{APST}(\mu, \sigma, \lambda, \alpha, \nu)$, and for $i = 1, \ldots, n$ we define $z_i = (y_i - \mu)/\sigma$, $w_i = \sqrt{(\nu+1)/(z_i^2+\nu)}$, $r_1(z;\nu) = f_T(z;\nu)/\mathcal{F}_T(z;\nu)$, $r_2(z;\lambda,\nu) = f_{ST}(z;\lambda,\nu)/\mathcal{F}_{ST}(z;\lambda,\nu)$, and $r_3(z;\lambda,\nu) = \left(1 + (1+\lambda^2)\frac{z^2}{\nu}\right)^{-\frac{\nu}{2}}/\mathcal{F}_{ST}(z;\lambda,\nu)$. Denoting the elements of the observed information matrix of the APST model by $j_{\mu\mu}, j_{\mu\sigma}, \ldots, j_{\alpha\alpha}$, and after some algebraic manipulations, we obtain

$$
\begin{aligned}
j_{\mu\mu} = & -\frac{1}{\sigma^2}\frac{\nu+1}{\nu^2}\sum_{i=1}^{n} z_i^2\left(1+\frac{z_i^2}{\nu}\right)^{-2} + \frac{1}{\sigma^2}\frac{\nu+1}{\nu}\sum_{i=1}^{n}\left(1+\frac{z_i^2}{\nu}\right)^{-2} \\
& + \frac{\lambda}{\sigma^2}\sum_{i=1}^{n} z_i w_i\left(1+\frac{z_i^2}{\nu}\right)^{-1} r_1(\lambda z_i w_i;\nu+1) \\
& + \frac{2}{\sigma^2}\frac{\lambda}{\nu}\sum_{i=1}^{n} z_i w_i\left(1+\frac{z_i^2}{\nu}\right)^{-2} r_1(\lambda z_i w_i;\nu+1) \\
& + \frac{\lambda^3}{\sigma^2}\frac{\nu+2}{\nu}\sum_{i=1}^{n} z_i w_i\left(1+\frac{z_i^2}{\nu}\right)^{-2}\left(1+(1+\lambda^2)\frac{z_i^2}{\nu}\right)^{-1} r_1(\lambda z_i w_i;\nu+1) \\
& + \frac{\lambda^2}{\sigma^2}\frac{\nu+1}{\nu}\sum_{i=1}^{n}\left(1+\frac{z_i^2}{\nu}\right)^{-3}\left[r_1(\lambda z_i w_i;\nu+1)\right]^2
\end{aligned}
$$

$$
\begin{aligned}
j_{\mu\sigma} = & -\frac{\lambda}{\sigma^2}\frac{\alpha-1}{\pi}\sum_{i=1}^{n}\left(1+\frac{z_i^2}{\nu}\right)^{-1}\left(1+(1+\lambda^2)\frac{z_i^2}{\nu}\right)^{-1} r_3(z_i;\lambda,\nu) \\
& + \frac{\alpha-1}{\sigma^2}\frac{\nu+1}{\nu}\sum_{i=1}^{n} z_i\left(1+\frac{z_i^2}{\nu}\right)^{-1} r_2(z_i;\lambda,\nu) + \frac{\alpha-1}{\sigma^2}\sum_{i=1}^{n}\left[r_2(z_i;\lambda,\nu)\right]^2 \\
& \frac{2}{\sigma^2}\frac{\nu+1}{\nu}\sum_{i=1}^{n} z_i\left(1+\frac{z_i^2}{\nu}\right)^{-2} \\
& + \frac{\lambda^3}{\sigma^2}\frac{\nu+2}{\nu}\sum_{i=1}^{n} z_i^2 w_i\left(1+\frac{z_i^2}{\nu}\right)^{-2}\left(1+(1+\lambda^2)\frac{z_i^2}{\nu}\right)^{-1} r_1(\lambda z_i w_i;\nu+1) \\
& + \frac{2}{\sigma^2}\frac{\lambda}{\nu}\sum_{i=1}^{n} z_i^2 w_i\left(1+\frac{z_i^2}{\nu}\right)^{-2} r_1(\lambda z_i w_i;\nu+1) \\
& - \frac{\lambda}{\sigma^2}\sum_{i=1}^{n} w_i\left(1+\frac{z_i^2}{\nu}\right)^{-2} r_1(\lambda z_i w_i;\nu+1) \\
& + \frac{\lambda^2}{\sigma^2}\frac{\nu+1}{\nu}\sum_{i=1}^{n} z_i\left(1+\frac{z_i^2}{\nu}\right)^{-3}\left[r_1(\lambda z_i w_i;\nu+1)\right]^2 \\
& - \frac{\lambda}{\sigma^2}\frac{\alpha-1}{\pi}\sum_{i=1}^{n} z_i\left(1+\frac{z_i^2}{\nu}\right)^{-1}\left(1+(1+\lambda^2)\frac{z_i^2}{\nu}\right)^{-1} r_3(z_i;\lambda,\nu) \\
& - \frac{\alpha-1}{\sigma^2}\sum_{i=1}^{n} r_2(z_i;\lambda,\nu) + \frac{\alpha-1}{\sigma^2}\frac{\nu+1}{\nu}\sum_{i=1}^{n} z_i^2\left(1+\frac{z_i^2}{\nu}\right)^{-1} r_2(z_i;\lambda,\nu) \\
& + \frac{\alpha-1}{\sigma^2}\sum_{i=1}^{n} z_i\left[r_2(z_i;\lambda,\nu)\right]^2
\end{aligned}
$$

$$\begin{aligned}
j_{\mu\lambda} =& -\frac{\lambda^2}{\sigma}\frac{\nu+2}{\nu}\sum_{i=1}^{n} z_i^2 w_i \left(1+\frac{z_i^2}{\nu}\right)^{-1}\left(1+(1+\lambda^2)\frac{z_i^2}{\nu}\right)^{-1} r_1(\lambda z_i w_i;\nu+1) \\
&+\frac{1}{\sigma}\sum_{i=1}^{n} w_i\left(1+\frac{z_i^2}{\nu}\right)^{-1} r_1(\lambda z_i w_i;\nu+1) \\
&-\frac{\lambda}{\sigma}\frac{\nu+1}{\nu}\sum_{i=1}^{n} z_i\left(1+\frac{z_i^2}{\nu}\right)^{-2}\left[r_1(\lambda z_i w_i;\nu+1)\right]^2 \\
&+\frac{\alpha-1}{\pi\sigma}\sum_{i=1}^{n} z_i^2\left(1+\frac{z_i^2}{\nu}\right)^{-1} r_3(z_i;\lambda,\nu) \\
&+\frac{\alpha-1}{\pi\sigma}\frac{1}{1+\lambda^2}\sum_{i=1}^{n} r_2(z_i;\lambda,\nu) r_3(z_i;\lambda,\nu)
\end{aligned}$$

$$j_{\mu\alpha} = \frac{1}{\sigma}\sum_{i=1}^{n} r_2(z_i;\lambda,\nu)$$

$$\begin{aligned}
j_{\mu\nu} =& -\frac{1}{\sigma\nu}\sum_{i=1}^{n} z_i\left(1+\frac{z_i^2}{\nu}\right)^{-1} + \frac{\nu+1}{\sigma\nu^2}\sum_{i=1}^{n} z_i\left(1+\frac{z_i^2}{\nu}\right)^{-2} \\
&+\frac{\lambda}{\sigma\nu^2}\sum_{i=1}^{n} z_i^2 w_i\left(1+\frac{z_i^2}{\nu}\right)^{-2} r_1(\lambda z_i w_i;\nu+1) \\
&+\frac{\lambda}{2\sigma}\frac{1}{\nu(\nu+1)}\sum_{i=1}^{n}(z_i^2-1) w_i\left(1+\frac{z_i^2}{\nu}\right)^{-2} r_1(\lambda z_i w_i;\nu+1) \\
&+\frac{\lambda^3}{2\sigma}\frac{\nu+2}{\nu^2(\nu+1)}\sum_{i=1}^{n} z_i^2(z_i^2-1) w_i\left(1+\frac{z_i^2}{\nu}\right)^{-2}\left(1+(1+\lambda^2)\frac{z_i^2}{\nu}\right)^{-1} r_1(\lambda z_i w_i;\nu+1) \\
&-\frac{\lambda^2}{2\sigma}\frac{1}{\nu^2}\sum_{i=1}^{n} z_i(z_i^2-1)\left(1+\frac{z_i^2}{\nu}\right)^{-3}\left[r_1(\lambda z_i w_i;\nu+1)\right]^2 \\
&+\frac{\lambda}{2\pi\sigma}\frac{\alpha-1}{\nu+1}\sum_{i=1}^{n} z_i\left(1+(1+\lambda^2)\frac{z_i^2}{\nu}\right)^{-1} r_3(z_i;\lambda,\nu) \\
&+\frac{\alpha-1}{2\pi\sigma(\nu+1)}\frac{\lambda}{1+\lambda^2}\sum_{i=1}^{n} r_2(z_i;\lambda,\nu) r_3(z_i;\lambda,\nu) + \frac{\alpha-1}{2\sigma}\frac{\nu+1}{\nu^2}\sum_{i=1}^{n} z_i^2\left(1+\frac{z_i^2}{\nu}\right)^{-1} r_2(z_i;\lambda,\nu) \\
&-\frac{\alpha-1}{2\sigma}\sum_{i=1}^{n}\log\left(1+\frac{z_i^2}{\nu}\right) r_2(z_i;\lambda,\nu) - \frac{\alpha-1}{2\sigma}\sum_{i=1}^{n}\frac{g(z_i,\nu)}{\mathcal{F}_{ST}(z_i;\lambda,\nu)} r_2(z_i;\lambda,\nu) \\
&+\frac{\lambda}{2\pi\sigma}\frac{\alpha-1}{\nu}\sum_{i=1}^{n} z_i\left(1+\frac{z_i^2}{\nu}\right)^{-1}\left(1+(1+\lambda^2)\frac{z_i^2}{\nu}\right)^{-1} r_3(z_i;\lambda,\nu)
\end{aligned}$$

$$
\begin{aligned}
j_{\sigma\sigma} = &-\frac{n}{\sigma^2} + \frac{1}{\sigma^2}\frac{\nu+1}{\nu}\sum_{i=1}^{n} z_i^2\left(1+\frac{z_i^2}{\nu}\right)^{-1} + \frac{2}{\sigma^2}\frac{\nu+1}{\nu}\sum_{i=1}^{n} z_i^2\left(1+\frac{z_i^2}{\nu}\right)^{-2} \\
&-\frac{2\lambda}{\sigma^2}\sum_{i=1}^{n} z_i w_i\left(1+\frac{z_i^2}{\nu}\right)^{-2} r_1(\lambda z_i w_i;\nu+1) \\
&-\frac{\lambda}{\sigma^2\nu}\sum_{i=1}^{n} z_i^3 w_i\left(1+\frac{z_i^2}{\nu}\right)^{-2} r_1(\lambda z_i w_i;\nu+1) \\
&+\frac{\lambda^3}{\sigma^2}\frac{\nu+2}{\nu}\sum_{i=1}^{n} z_i^3 w_i\left(1+\frac{z_i^2}{\nu}\right)^{-2}\left(1+(1+\lambda^2)\frac{z_i^2}{\nu}\right)^{-1} r_1(\lambda z_i w_i;\nu+1) \\
&+\frac{\lambda^2}{\sigma^2}\frac{\nu+1}{\nu}\sum_{i=1}^{n} z_i^2\left(1+\frac{z_i^2}{\nu}\right)^{-3} r_1(\lambda z_i w_i;\nu+1) \\
&-\frac{\lambda}{\sigma^2}\frac{\alpha-1}{\pi}\sum_{i=1}^{n} z_i^2\left(1+\frac{z_i^2}{\nu}\right)^{-1}\left(1+(1+\lambda^2)\frac{z_i^2}{\nu}\right)^{-1} r_3(z_i;\lambda,\nu) \\
&-\frac{2(\alpha-1)}{\sigma^2}\sum_{i=1}^{n} z_i r_2(z_i;\lambda,\nu) - \frac{\alpha-1}{\sigma^2}\frac{\nu+1}{\nu}\sum_{i=1}^{n} z_i^3\left(1+\frac{z_i^2}{\nu}\right)^{-1} r_2(z_i;\lambda,\nu) \\
&+\frac{\alpha-1}{\sigma^2}\sum_{i=1}^{n} z_i^2 r_2(z_i;\lambda,\nu)
\end{aligned}
$$

$$
\begin{aligned}
j_{\sigma\lambda} = &\frac{1}{\sigma}\sum_{i=1}^{n} z_i w_i\left(1+\frac{z_i^2}{\nu}\right)^{-1} r_1(\lambda z_i w_i;\nu+1) \\
&+\frac{\lambda^2}{\sigma}\frac{\nu+2}{\nu}\sum_{i=1}^{n} z_i^3 w_i\left(1+\frac{z_i^2}{\nu}\right)^{-1}\left(1+(1+\lambda^2)\frac{z_i^2}{\nu}\right)^{-1} r_1(\lambda z_i w_i;\nu+1) \\
&+\frac{\lambda}{\sigma}\frac{\nu+1}{\nu}\sum_{i=1}^{n} z_i\left(1+\frac{z_i^2}{\nu}\right)^{-2}\left[r_1(\lambda z_i w_i;\nu+1)\right]^2 \\
&-\frac{\alpha-1}{\pi\sigma}\sum_{i=1}^{n} z_i^2\left(1+\frac{z_i^2}{\nu}\right)^{-1} r_3(z_i;\lambda,\nu) \\
&+\frac{\alpha-1}{\pi\sigma}\frac{1}{1+\lambda^2}\sum_{i=1}^{n} r_2(z_i;\lambda,\nu) r_3(z_i;\lambda,\nu)
\end{aligned}
$$

$$
j_{\sigma\alpha} = \frac{1}{\sigma}\sum_{i=1}^{n} z_i r_2(z_i;\lambda,\nu)
$$

$$
\begin{aligned}
j_{\sigma\nu} =& -\frac{1}{\sigma\nu}\sum_{i=1}^{n} z_i^2\left(1+\frac{z_i^2}{\nu}\right)^{-1} + \frac{1}{\sigma}\frac{\nu+1}{\nu^2}\sum_{i=1}^{n} z_i^2\left(1+\frac{z_i^2}{\nu}\right)^{-2} \\
&+ \frac{\lambda}{\sigma\nu^2}\sum_{i=1}^{n} z_i^3 w_i\left(1+\frac{z_i^2}{\nu}\right)^{-2} r_1(\lambda z_i w_i;\nu+1) \\
&+ \frac{\lambda}{2\sigma}\frac{1}{\nu(\nu+1)}\sum_{i=1}^{n} z_i(z_i^2-1)w_i\left(1+\frac{z_i^2}{\nu}\right)^{-2} r_1(\lambda z_i w_i;\nu+1) \\
&- \frac{\lambda^3}{2\sigma}\frac{\nu+2}{\nu^2(\nu+1)}\sum_{i=1}^{n} z_i^3(z_i^2-1)w_i\left(1+\frac{z_i^2}{\nu}\right)^{-2}\left(1+(1+\lambda^2)\frac{z_i^2}{\nu}\right)^{-1} r_1(\lambda z_i w_i;\nu+1) \\
&- \frac{\lambda^2}{2\sigma\nu^2}\sum_{i=1}^{n} z_i^2(z_i^2-1)\left(1+\frac{z_i^2}{\nu}\right)^{-3}\left[r_1(\lambda z_i w_i;\nu+1)\right]^2 \\
&+ \frac{\lambda}{2\pi\sigma}\frac{\alpha-1}{\nu(\nu+1)}\sum_{i=1}^{n} z_i^2(z_i^2-1)\left(1+\frac{z_i^2}{\nu}\right)^{-1}\left(1+(1+\lambda^2)\frac{z_i^2}{\nu}\right)^{-1} r_3(z_i;\lambda,\nu) \\
&+ \frac{\alpha-1}{2\pi\sigma(\nu+1)}\frac{\lambda}{1+\lambda^2}\sum_{i=1}^{n} z_i r_2(z_i;\lambda,\nu) r_3(z_i;\lambda,\nu) \\
&+ \frac{\alpha-1}{2\sigma}\frac{\nu+1}{\nu^2}\sum_{i=1}^{n} z_i^3\left(1+\frac{z_i^2}{\nu}\right)^{-1} r_1(\lambda z_i w_i;\nu+1) \\
&- \frac{\alpha-1}{2\sigma}\sum_{i=1}^{n} z_i \log\left(1+\frac{z_i^2}{\nu}\right) r_1(\lambda z_i w_i;\nu+1) \\
&- \frac{\alpha-1}{2\sigma}\sum_{i=1}^{n} z_i \frac{g(z_i,\nu)}{\mathcal{F}_{ST}(z_i,\lambda,\nu)} r_2(z_i;\lambda,\nu)
\end{aligned}
$$

$$
\begin{aligned}
j_{\lambda\lambda} =& \frac{\lambda(\nu+2)}{\nu}\sum_{i=1}^{n} z_i^3 w_i\left(1+(1+\lambda^2)\frac{z_i^2}{\nu}\right)^{-1} r_1(\lambda z_i w_i;\nu+1) \\
&+ \frac{\nu+1}{\nu}\sum_{i=1}^{n} z_i\left(1+\frac{z_i^2}{\nu}\right)^{-1}\left[r_1(\lambda z_i w_i;\nu+1)\right]^2 \\
&- \frac{2(\alpha-1)}{\pi}\frac{\lambda}{(1+\lambda^2)^2}\sum_{i=1}^{n}\left(1+(1+\lambda^2)\frac{z_i^2}{\nu}\right)^{-1} r_3(z_i;\lambda,\nu) \\
&- \frac{\alpha-1}{\pi}\frac{\lambda}{1+\lambda^2}\frac{\nu+2}{\nu}\sum_{i=1}^{n} z_i^2\left(1+(1+\lambda^2)\frac{z_i^2}{\nu}\right)^{-1} r_3(z_i;\lambda,\nu) \\
&+ \frac{\alpha-1}{\pi^2(1+\lambda^2)^2}\sum_{i=1}^{n}\left[r_3(z_i;\lambda,\nu)\right]^2
\end{aligned}
$$

$$j_{\lambda\alpha} = \frac{1}{\pi(1+\lambda^2)} \sum_{i=1}^{n} r_3(z_i;\lambda,\nu)$$

$$\begin{aligned}
j_{\lambda\nu} =& -\frac{1}{2\nu(\nu+1)} \sum_{i=1}^{n} z_i(z_i^2-1)w_i\left(1+\frac{z_i^2}{\nu}\right)^{-1} r_1(\lambda z_i w_i;\nu+1) \\
&+ \frac{\lambda^2}{2\nu^2}\frac{\nu+2}{(\nu+1)^2} \sum_{i=1}^{n} z_i^3(z_i^2-1)w_i\left(1+\frac{z_i^2}{\nu}\right)^{-1}\left(1+(1+\lambda^2)\frac{z_i^2}{\nu}\right)^{-1} r_1(\lambda z_i w_i;\nu+1) \\
&+ \frac{\lambda}{2\nu^2} \sum_{i=1}^{n} z_i^2(z_i^2-1)\left(1+\frac{z_i^2}{\nu}\right)^{-2}\left[r_1(\lambda z_i w_i;\nu+1)\right]^2 \\
&+ \frac{\alpha-1}{2\pi(\nu+1)}\frac{1-\lambda^2}{(1+\lambda^2)^2} \sum_{i=1}^{n} r_3(z_i;\lambda,\nu) + \frac{\alpha-1}{2\pi(\nu+1)}\frac{\lambda}{(1+\lambda^2)^2} \sum_{i=1}^{n}\left[r_3(z_i;\lambda,\nu)\right]^2 \\
&- \frac{\alpha-1}{2\pi(\nu+1)}\frac{\lambda^2}{1+\lambda^2} \sum_{i=1}^{n} z_i^2\left(1+(1+\lambda^2)\frac{z_i^2}{\nu}\right)^{-1} r_3(z_i;\lambda,\nu) \\
&- \frac{\alpha-1}{2\pi(1+\lambda^2)} \sum_{i=1}^{n} \frac{g(z_i,\nu)}{\mathcal{F}_{ST}(z_i,\lambda,\nu)} r_3(z_i;\lambda,\nu) \\
&- \frac{\alpha-1}{2\pi} \sum_{i=1}^{n} \frac{g_1(z_i,\nu)}{\mathcal{F}_{ST}(z_i,\lambda,\nu)}
\end{aligned}$$

$$j_{\alpha\alpha} = \frac{n}{\alpha^2}$$

$$\begin{aligned}
j_{\alpha\nu} =& -\frac{n}{2}\psi\left(\frac{\nu+1}{2}\right) + \frac{n}{2}\psi\left(\frac{\nu}{2}\right) + \frac{n}{2\nu} \\
&+ \frac{1}{2\pi(\nu+1)}\frac{\lambda}{1+\lambda^2} \sum_{i=1}^{n} r_3(z_i;\lambda,\nu) - \frac{1}{2}\sum_{i=1} \frac{g(z_i,\nu)}{\mathcal{F}_{ST}(z_i,\lambda,\nu)}
\end{aligned}$$

$$
\begin{aligned}
j_{\nu\nu} = &- \frac{n\alpha}{2\nu^2} - \frac{n\alpha}{4}\psi_1\Big(\frac{\nu+1}{2}\Big) + \frac{n\alpha}{4}\psi_1\Big(\frac{\nu}{2}\Big) - \frac{\nu-1}{2\nu^3}\sum_{i=1}^{n} z_i^2\Big(1+\frac{z_i^2}{\nu}\Big)^{-1} \\
&+ \frac{\nu+1}{2\nu^3}\sum_{i=1}^{n} z_i^2\Big(1+\frac{z_i^2}{\nu}\Big)^{-2} \\
&+ \frac{\lambda}{4(\nu+1)^2}\frac{1}{\nu^2}\sum_{i=1}^{n} z_i(z_i^2-1)(z_i^2+4\nu+3)w_i\Big(1+\frac{z_i^2}{\nu}\Big)^{-2} r_1(\lambda z_i w_i;\nu+1) \\
&- \frac{\lambda}{4\nu(\nu+1)}\Big(\psi\Big(\frac{\nu+2}{2}\Big) - \psi\Big(\frac{\nu+1}{2}\Big) - \frac{1}{\nu+1}\Big) \\
&\quad \sum_{i=1}^{n} z_i(z_i^2-1)w_i\Big(1+\frac{z_i^2}{\nu}\Big)^{-1} r_1(\lambda z_i w_i;\nu+1) \\
&- \frac{\lambda^3}{4(\nu+1)}\frac{\nu+2}{\nu^3}\sum_{i=1}^{n} z_i^3(z_i^2-1)\Big(1+\frac{z_i^2}{\nu}\Big)^{-2}\Big(1+(1+\lambda^2)\frac{z_i^2}{\nu}\Big)^{-1} r_1(\lambda z_i w_i;\nu+1) \\
&+ \frac{\lambda}{4\nu(\nu+1)}\sum_{i=1}^{n} z_i(z_i^2-1)\log\Big(1+\frac{\lambda^2 z_i^2}{\nu+z_i^2}\Big) r_1(\lambda z_i w_i;\nu+1) \\
&+ \frac{\lambda^2}{4\nu^3(\nu+1)}\sum_{i=1}^{n} z_i^2(z_i^2-1)^2\Big(1+\frac{z_i^2}{\nu}\Big)^{-3}\big[r_1(\lambda z_i w_i;\nu+1)\big]^2 \\
&- \frac{\alpha-1}{2\pi(\nu+1)^2}\frac{\lambda}{1+\lambda^2}\sum_{i=1}^{n} r_3(z_i;\lambda,\nu) \\
&+ \frac{\alpha-1}{4\pi(\nu+1)}\frac{\lambda}{\nu}\sum_{i=1}^{n} z_i^2\Big(1+(1+\lambda^2)\frac{z_i^2}{\nu}\Big)^{-1} r_3(z_i;\lambda,\nu) \\
&- \frac{\alpha-1}{4\pi(\nu+1)}\frac{\lambda}{1+\lambda^2}\sum_{i=1}^{n}\log\Big(1+(1+\lambda^2)\frac{z_i^2}{\nu}\Big) r_3(z_i;\lambda,\nu) \\
&+ \frac{\alpha-1}{4\pi(\nu+1)}\frac{\lambda}{1+\lambda^2}\Big(\psi\Big(\frac{\nu+1}{2}\Big) - \psi\Big(\frac{\nu}{2}\Big) - \frac{1}{\nu}\Big)\sum_{i=1}^{n} r_3(z_i;\lambda,\nu) \\
&- \frac{\alpha-1}{2\pi(\nu+1)}\frac{\lambda}{1+\lambda^2}\sum_{i=1}^{n}\frac{g(z_i,\nu)}{\mathcal{F}_{ST}(z_i,\lambda,\nu)} r_3(z_i;\lambda,\nu) \\
&+ \frac{\alpha-1}{4\pi^2(\nu+1)^2}\frac{\lambda^2}{(1+\lambda^2)^2}\sum_{i=1}^{n}\big[r_3(z_i;\lambda,\nu)\big]^2 \\
&+ \frac{\alpha-1}{4}\sum_{i=1}^{n}\Big(\frac{g(z_i,\nu)}{\mathcal{F}_{ST}(z_i,\lambda,\nu)}\Big)^2 - \frac{\alpha-1}{2}\sum_{i=1}^{n}\frac{g_2(z_i,\nu)}{\mathcal{F}_{ST}(z_i,\lambda,\nu)}
\end{aligned}
$$

where $g(z;\nu)$ is given in Equation (16), and $g_1(z;\nu)$ and $g_2(z;\nu)$ are given in Equations (A1) and (A3), respectively.

$$g_1(x;\nu) = \int_{-\infty}^{x} \left\{ \frac{(\nu+1)s^2}{\nu(s^2+\nu)} - \log\left(1+\frac{s^2}{\nu}\right) \right\} \left\{ 1 + \frac{(1+\lambda^2)s^2}{\nu} \right\}^{-\frac{\nu+2}{2}} s ds$$
$$- \int_{-\infty}^{x} \frac{s}{s^2+\nu} \left\{ 1 + \frac{(1+\lambda^2)s^2}{\nu} \right\}^{-\frac{\nu+2}{2}} ds$$
$$+ \int_{-\infty}^{x} \frac{\lambda^2(\nu+2)s^3}{(s^2+\nu)(\nu+(1+\lambda^2)s^2)} \left\{ 1 + \frac{(1+\lambda^2)s^2}{\nu} \right\}^{-\frac{\nu+2}{2}} ds \tag{A1}$$

$$g_2(x;\nu) = \int_{-\infty}^{x} \left\{ \frac{s^2(s^2\nu - 2\nu - s^2)}{\nu^2(s^2+\nu)^2} + \frac{1}{2}\left[\frac{(\nu+1)s^2}{\nu(s^2+\nu)} - \log\left(1+\frac{s^2}{\nu}\right) \right]^2 \right\} f_{ST}(s;\lambda,\nu) ds$$
$$+ \frac{\lambda}{2\pi(\nu+1)} \int_{-\infty}^{x} \frac{s(s^2-1)}{(s^2+\nu)} \left\{ \frac{(\nu+1)s^2}{\nu(s^2+\nu)} - \log\left(1+\frac{s^2}{\nu}\right) \right\}$$
$$\times \left\{ 1 + \frac{(1+\lambda^2)s^2}{\nu} \right\}^{-\frac{\nu+2}{2}} ds$$
$$+ \frac{\lambda}{\pi} \int_{-\infty}^{x} \frac{s}{(s+\nu)^2} \left\{ 1 + \frac{(1+\lambda^2)s^2}{\nu} \right\}^{-\frac{\nu+2}{2}} ds \tag{A2}$$

$$+ \frac{\lambda}{2\pi} \left(\psi\left(\frac{\nu+1}{2}\right) - \psi\left(\frac{\nu}{2}\right) - \frac{1}{\nu} \right) \int_{-\infty}^{x} \frac{s}{s^2+\nu} \left\{ 1 + \frac{(1+\lambda^2)s^2}{\nu} \right\}^{-\frac{\nu+2}{2}} ds$$
$$- \frac{\lambda}{2\pi} \int_{-\infty}^{x} \frac{s}{s^2+\nu} \left\{ \frac{(\nu+2)(1+\lambda^2)s^2}{\nu(\nu+(1+\lambda^2)s^2)} - \log\left(1+\frac{(1+\lambda^2)s^2}{\nu}\right) \right\}$$
$$\times \left\{ 1 + \frac{(1+\lambda^2)s^2}{\nu} \right\}^{-\frac{\nu+2}{2}} ds \tag{A3}$$

References

1. Azzalini, A. A class of distributions which includes the normal ones. *Scand. J. Stat.* **1985**, *12*, 171–178.
2. Durrans, S.R. Distributions of fractional order statistics in hydrology. *Water Resour. Res.* **1992**, *28*, 1649–1655. [CrossRef]
3. Martínez-Flórez, G.; Bolfarine, H.; Gómez, H.W. The alpha–power tobit model. *Commun. Stat. Theory Methods* **2013**, *42*, 633–643. [CrossRef]
4. Birnbaum, Z.W.; Saunders, S.C. A new family of life distributions. *J. Appl. Probab.* **1969**, *6*, 319–327. [CrossRef]
5. Martínez-Flórez, G.; Bolfarine, H.; Gómez, H.W. An alpha-power extension for the Birnbaum-Saunders distribution. *Statistics* **2014**, *48*, 896–912. [CrossRef]
6. Gupta, R.D.; Gupta, R.C. Analyzing skewed data by power-normal model. *Test* **2008**, *17*, 197–210. [CrossRef]
7. Pewsey, A.; Gómez, H. W.; Bolfarine, H. Likelihood–based inference for power distributions. *Test* **2012**, *21*, 775–789. [CrossRef]
8. Martínez-Flórez, G.; Bolfarine, H.; Gómez, H.W. Skew-normal alpha-power model. *Stat. J. Theor. Appl. Stat.* **2014**, *48*, 1414–1428. [CrossRef]
9. Azzalini, A.; Capitanio, A. Distributions generated by perturbation of symmetry with emphasis on a multivariate skew-*t* distribution. *J. R. Stat. Soc. Ser. B (Stat. Methodol.)* **2003**, *65*, 367–389. [CrossRef]
10. Branco, M.D.; Dey, D.K. General class of multivariate skew-elliptical distributions. *J. Multivar. Anal.* **2001**, *79*, 99–113. [CrossRef]
11. Durrans, S.R. Multivariate skew *t*-distribution. *Stat. J. Theor. Appl. Stat.* **2003**, *37*, 359–363.
12. Sahu, S.K.; Dey, D.K.; Branco, M.D. A new class of multivariate skew distributions with applications to Bayesian regression models. *Can. J. Stat.* **2003**, *31*, 129–150. [CrossRef]
13. Jones, M.C.; Faddy, M.J. A skew extension of the *t*-distribution, with Applications. *J. R. Stat. Soc. Ser. B (Stat. Methodol.)* **2003**, *65*, 159–174. [CrossRef]

14. Zhao, J.; Kim, H.M. Power t distributions. *Commun. Stat. Appl. Methods* **2016**, *23*, 321–334.
15. R Development Core Team. *R: A Language and Environment for Statistical Computing*; R Foundation for Statistical Computing: Vienna, Austria, 2018. Available online: http://www.R-project.org (accessed on 10 October 2019).
16. Lehman, E.L.; Casella, G. *Theory of Point Estimation*, 2nd ed.; Springer: New York, NY, USA, 1998.
17. Frieden, B.R. *Science from Fisher Information: A Unification*; Cambridge Univerisity Press: Cambridge, UK, 2004.
18. Arellano-Valle, R.B.; Azzalini, A. The centered parameterization and related quantities of the skew–t distribution. *J. Multivar. Anal.* **2013**, *113*, 73–90. [CrossRef]
19. Arellano-Valle, R.B.; Azzalini, A. The centered parametrization for the multivariate skew-normal distribution. *J. Multivar. Anal.* **2008**, *99*, 1362–1382. [CrossRef]
20. Akaike, H. A new look at statistical model identification. *IEEE Trans. Autom. Contr.* **1974**, *19*, 716–722. [CrossRef]
21. Schwarz, G. Estimating the dimension of a model. *Ann. Stat.* **1978**, *6*, 461–464. [CrossRef]
22. Bozdogan, H. Model selection and akaike's information criterion (AIC): The general theory and its analytical extensions. *Psychometrika* **2010**, *52*, 345–370. [CrossRef]
23. Siebert, L.; Simkin, T.; Kimberly, P. Global Volcanism Program. In *Volcanoes of the World*; v. 4.6.0.; Venzke, E., Ed.; Smithsonian Institution: Washington, DC, USA, 2013. Available online: https://doi.org/10.5479/si.GVP.VOTW4-2013 (accessed on 10 October 2019).
24. Feigelson, E.D. astrodatR: Astronomical Data. R Package v. 0.1. Available online: http://CRAN.R-project.org/package=astrodatR (accessed on 10 October 2019).
25. Mattos, T.; Garay, A.M.; Lachos, V.H. Likelihood-based inference for censored linear regression models with scale mixtures of skew-normal distributions. *J. Appl. Stat.* **2018**, *45*, 2019–2066. [CrossRef]
26. Barros, M.; Galea, M.; González, M.; Leiva, V. Influence diagnostics in the tobit censored response model. *Stat. Methods Appl.* **2010**, *19*, 379–397. [CrossRef]
27. Ortega, E.M.; Bolfarine, H.; Paula, G.A. Influence diagnostics in generalized log-gamma regression models. *Comput. Stat. Data Anal.* **2003**, *42*, 165–186. [CrossRef]

7

Generation of Julia and Mandelbrot Sets via Fixed Points

Mujahid Abbas [1,2], Hira Iqbal [3] and Manuel De la Sen [4,*]

1 Department of Mathematics, Government College University, Lahore 54000, Pakistan; abbas.mujahid@gmail.com
2 Department of Medical Research, China Medical University No. 91, Hsueh-Shih Road, Taichung 400, Taiwan
3 Department of Sciences and Humanities, National University of Computer and Emerging Sciences, Lahore Campus, Lahore 54000, Pakistan; hira.iqbal@nu.edu.pk
4 Institute of Research and Development of Processes, University of the Basque Country, Campus of Leioa (Bizkaia), P.O. Box 644, Bilbao, Barrio Sarriena, 48940 Leioa, Spain
* Correspondence: manuel.delasen@ehu.eus

Abstract: The aim of this paper is to present an application of a fixed point iterative process in generation of fractals namely Julia and Mandelbrot sets for the complex polynomials of the form $T(x) = x^n + mx + r$ where $m, r \in \mathbb{C}$ and $n \geq 2$. Fractals represent the phenomena of expanding or unfolding symmetries which exhibit similar patterns displayed at every scale. We prove some escape time results for the generation of Julia and Mandelbrot sets using a Picard Ishikawa type iterative process. A visualization of the Julia and Mandelbrot sets for certain complex polynomials is presented and their graphical behaviour is examined. We also discuss the effects of parameters on the color variation and shape of fractals.

Keywords: iteration; fixed points; fractals

MSC: Primary: 47H10; Secondary:47J25

---

1. Introduction

Fixed point theory provides a suitable framework to investigate various nonlinear phenomena arising in the applied sciences including complex graphics, geometry, biology and physics [? ? ? ?]. Complex graphical shapes such as fractals, were discovered as fixed points of certain set maps [?]. Informally, fractals can be treated as self similar mathematical structures which have similarity and symmetry such that considerably small parts of the shape are geometrically akin to the whole shape. Fractals are also known as expanding symmetries or unfolding symmetries. Although, fractals do not have a formal definition, however they are identified through their irregular structure that cannot be found in Euclidean geometry. Julia [?] who is considered as one of the pioneers of fractal geometry, studied iterated complex polynomials and introduced Julia set as a classical example of fractals. Let $\mathbb{C}$ be the complex space, $T : \mathbb{C} \to \mathbb{C}$ be a complex polynomial of degree $n \geq 2$ with complex coefficients and $T^i(x)$ be the ith iterate of x. The behaviour of the iterates $T^i(x)$ for large i determine the Julia set (see [? ? ? ?]).

Definition 1 ([?]). *The set of points in $\mathbb{C}$ whose orbits do not converge to a point at infinity is known as filled Julia set, K_T, that is,*

$$K_T = \left\{ x \in \mathbb{C} : \{|T^i(x)|\}_{i=0}^{\infty} \quad \text{is bounded} \right\}.$$

Julia set of T denoted by J_T is the boundary of filled Julia set, that is, $J_T = \partial K_T$.

Therefore, we may say that $x \in J_T$ if for every neighborhood of x there exist points w and v such that $T^i(w) \to \infty$ and $T^i(v) \nrightarrow \infty$. The complement of a Julia set is a Fatou set.

Let $p \in \mathbb{C}$ be a fixed point of T and $|(T^i)'p| = \rho$, where prime denotes the complex differentiation. A point p is called a periodic point if $p = T^i p$ for some integer $i \geq 0$. Let $\{p, Tp, ..., T^i p, ...\}$ be an orbit of p. The point p is called an attracting point if $0 \leq \rho < 1$ and a repelling point if $\rho > 1$ [? ?]. The following result gives a significant connection between repelling points of a polynomial and the Julia set.

Theorem 1 ([?]). *If T is a complex polynomial, then J_T is the closure of the repelling periodic points of T.*

Let p be an attracting fixed point of T. Then, the set $A(p)$ is called the basin of attraction of p if

$$A(p) = \left\{x \in \mathbb{C} : T^i x \to p \text{ as } i \to \infty\right\}.$$

The basin of attraction of infinity, $A(\infty)$, is defined in the same way. The following lemma is pivotal in determining Julia sets.

Lemma 1. *[?] Let p be an attracting fixed point of T. Then, $J_T = \partial A(p)$.*

Thus, the Julia set is the boundary of the basin of attraction of each attracting fixed point of T, including ∞. The existence of the fixed point p for any complex polynomial is guaranteed by Brouwer fixed point theorem [?]. However, the existence of an attracting fixed point depends on the choice of the parameters. Consider the polynomial $Q_r(x) = x^2 + r$. Then it has two fixed points excluding infinity. In this case, a fixed point p is attracting if $|2p| < 1$ i.e., $|1 - \sqrt{\frac{1}{4} - r}| < 1$. Fix $v_r = \sqrt{\frac{1}{4} - r}$, then the set of parameters r such that Q_r has an attracting fixed point is given by $S = \{r \in \mathbb{C} : |1 - v_r| < 1\}$. Julia sets, J_{Q_r}, on the real axis i.e., $r = 0$ are reflection symmetric while those with complex parameter values, $r \in \mathbb{C}$ demonstrate rotational symmetry.

Mandelbrot [?] extended the idea of Julia sets and presented the notion of fractals. He investigated the graphical behaviour of connected Julia sets and plotted them for complex function, $Q_r(x) = x^2 + r$, where $x \in \mathbb{C}$ is a complex variable and $r \in \mathbb{C}$ is an input parameter. He noted that various geometrical properties involving dimension, symmetry and similarity play consequential role in the study of fractal geometry.

Definition 2 ([?]). *Let T be any complex polynomial of degree $n \geq 2$. A Mandelbrot set M is the set consisting of all parameters r for which the Julia set, J_{Q_r}, is connected, that is,*

$$M = \{r \in \mathbb{C} : J_{Q_r} \quad \textit{is connected}\},$$

or an equivalent definition is

$$M = \{r \in \mathbb{C} : \{|Q_r^n(0)|\} \nrightarrow \infty \quad \textit{as} \quad n \to \infty\}.$$

Mandelbrot [? ?] noted that records of heart beat, irregular coastal structures, variations of traffic flow and many naturally existing textures are examples of fractals.

In order to generate and analyze fractals, various techniques are used such as iterated function systems, random fractals, escape time criterion etc. The escape time algorithm is the stopping criterion that is based on the number of iterations necessary to determine if the orbit sequence tends to infinity or not. This algorithm provides a suitable mechanism used to demonstrate some attributes of dynamic system under iterative process. Generally, the escape criterion for Julia and Mandelbrot sets is given by:

Theorem 2 ([?]). *For $Q_r(x) = x^2 + r$, $x, r \in \mathbb{C}$, if there exists $i \geq 0$ such that*

$$|Q_r^i(x)| > \max\{|r|, 2\},$$

then $Q_r^i(x) \to \infty$ as $i \to \infty$.

The term $\max\{|r|, 2\}$ is also known as escape radius threshold. The escape radius varies in each iteration. The escape radius has a key role in visualizing the fractals.

Historically, Julia and Mandelbrot sets are investigated for the polynomials Q_r but the study has been extended to quadratic, cubic, and n^{th} degree complex polynomials. Lakhtakia et al. [?] explored the Julia sets for general complex function of the form $T(x) = x^n + r$ where $n \in \mathbb{N}$. The superior Julia and superior Mandelbrot sets for such complex polynomials in the context of noises arising in the objects were analyzed by Negi et al. [? ?]. Rochon [?] considered a more generalized form of Mandelbrot sets in bi-complex planes, see also [? ?].

Many authors have utilized various iterative processes to generate fractals. Julia and Mandelbrot sets have usually been studied for quadratic, cubic and higher degree polynomials in Picard orbit [?]. Let $T : \mathbb{C} \to \mathbb{C}$ and $x_0 \in \mathbb{C}$. The Picard orbit [?] is a sequence $\{x_i\}$ which is given by

$$x_{i+1} = T(x_i),$$

where $i \geq 0$.

Since the convergence of Picard process is slow, various faster converging iterative processes have been introduced to generate Julia and Mandelbrot sets. Rani and Kumar [? ?] used one-step Mann iterative process to generate superior Julia and Mandelbrot sets for n^{th} degree complex polynomials of the form $T(x) = x^n + r$. The Mann orbit, for any $x_0 \in \mathbb{C}$, is a sequence $\{x_i\}$ which is given by

$$x_{i+1} = (1 - \alpha)(x_i) + \alpha T(x_i),$$

where $i = 0, 1, \ldots$ and $\alpha \in (0, 1]$.

In 2010, a two-step Ishikawa iteration was used by Rana and Kumar [?] and Chauhan et al. [?] to study relative superior Julia and relative superior Mandelbrot sets, respectively. The dynamics of the nth order complex polynomial for non integer values were investigated in [?]. The authors also obtained new Julia and Mandelbrot sets via Ishikawa orbit. The Ishikawa orbit, for any $x_0 \in \mathbb{C}$, is a sequence $\{x_i\}$ which is given by

$$\begin{cases} x_{i+1} &= (1 - \alpha)x_i + \alpha T y_i, \\ y_i &= (1 - \beta)x_i + \beta T x_i, \end{cases}$$

where $i = 0, 1, \ldots$ and $\alpha, \beta \in (0, 1]$.

Ashish and Rani [?] investigated the three-step Noor iteration process for Julia and Mandelbrot sets. The Noor orbit, for any $x_0 \in \mathbb{C}$, is a sequence $\{x_i\}$ which is given by

$$\begin{cases} x_{i+1} = (1 - \alpha)T x_i + \alpha T y_i, \\ y_i = (1 - \beta)T x_i + \alpha T u_i, \\ u_i = (1 - \gamma)T x_i + \gamma T x_i, \end{cases}$$

where $i = 0, 1, \ldots$ and $\alpha, \beta, \gamma \in (0, 1]$.

The modified Ishikawa process, S-iteration, was employed by Kang et al. [? ?] to study relative superior Mandelbrot sets, tricorn and multicorns. The S-orbit, for any $x_0 \in \mathbb{C}$, is a sequence $\{x_i\}$ given by

$$\begin{cases} x_{i+1} = (1-\alpha)x_i + \alpha T y_i, \\ y_i = (1-\beta)x_i + \alpha T x_i, \end{cases}$$

where $i = 0, 1, \ldots$ and $\alpha, \beta \in (0,1]$.

Kumari et al. [?] used a four-step iterative process which is faster than of Picard, Mann and S-iteration processes and obtained some generalizations of Julia and Mandelbrot sets for quadratic, cubic and higher degree polynomials.

It is noteworthy that for each iterative process the behaviour and dynamics of the Julia and Mandelbrot sets differ. For some thought-provoking and fascinating comparisons, the reader may refer to [? ? ? ? ?] and references therein.

Complex polynomials of the form $T(x) = x^n + mx + r$, where $m, r \in \mathbb{C}$ occur in various engineering problems including digital signal processing. These complex polynomials are used to determine the pole-zero plots for signals and the study of the structure and solutions of linear time invariant (LTI) state-space models, for details see [?]. Thus the study of behaviour of these polynomials and their Julia and Mandelbrot sets has gained immense interest among researchers. Kang et al. [?] introduced Julia and Mandelbrot sets in implicit Jungck Mann and Jungck Ishikawa orbits. Later, several researchers [? ? ? ? ?] employed this implicit iterative process to generate graphs of such complex polynomials. In order to achieve this, they split the polynomial T into two functions $T_1(x) = x^n + r$ and $T_2(x) = mx$. However, the Jungck iterative process and its variants are used to determine the common fixed points of two mappings. Therefore, the question arises whether we can obtain an escape criterion and generate fractals for polynomials of the form T using explicit iterative processes.

The purpose of this paper is to answer this question. In this paper, we discuss the graphical behaviour of the complex polynomial of the form $T(x) = x^n + mx + r$ where $m, r \in \mathbb{C}$ and $n \geq 2$ using Picard Ishikawa type fixed point iteration process for the generation of fractals. Note that the Julia and Mandelbrot sets generated have distinctive shapes for the proposed iterative process as compared to already present iterative processes in the literature. Further, we show the effect of change of parameters on color variation and graph of the sets.

The Picard Ishikawa type iteration process was introduced by Piri et al. [?]. They claimed that this iterative process converges faster than Mann and Ishikawa iteration processes. Let D be a subset of a Banach space and $f : D \to D$ then the three step iteration process is given by

$$\begin{cases} x_1 = x \in D, \\ x_{i+1} = (1-\alpha_i)y_i + \alpha_i f y_i, \\ y_i = f z_i, \\ z_i = f((1-\beta_i)x_i + \beta_i f x_i), & i \geq 0, \end{cases} \tag{1}$$

where $\alpha_i, \beta_i \in (0,1]$.

2. Main Results

In this section, we use a Picard Ishikawa type iterative process and some prove escape criterions to determine the escape radius for this process. Throughout this paper we assume that for any complex polynomial the parameters are chosen in a way that at the least one attracting fixed point exists.

Let $\mathbb{C}$ be a complex space and $T_{\mathbb{C}} : \mathbb{C} \to \mathbb{C}$ be a complex polynomial with complex coefficients. The Picard Ishikawa type orbit around any $x_0 \in \mathbb{C}$, is a sequence $\{x_i\}$ given by

$$\begin{cases} x_{i+1} = (1-\alpha)y_i + \alpha T_{\mathbb{C}} y_i, \\ y_i = T_{\mathbb{C}} z_i, \\ z_i = T_{\mathbb{C}} t_i, \\ t_i = (1-\beta)x_i + \beta T_{\mathbb{C}} x_i, \end{cases} \tag{2}$$

where $i = 0, 1, 2, \ldots$ and $\alpha, \beta \in (0, 1]$.

We need the following escape criterions for the quadratics, cubic and higher degree polynomials.

2.1. *Escape Criterion for Quadratic Complex Polynomials in a Picard Ishikawa Type Orbit*

For the quadratic polynomial $T_{\mathbb{C}}(x) = x^2 + mx + r$ where $m, r \in \mathbb{C}$, we have the following result.

Theorem 3. *Suppose that* $|x| \geq |r| > \max\left\{\frac{2(1+|m|)}{\alpha}, \frac{2(1+|m|)}{\beta}\right\}$, $\alpha, \beta \in (0, 1]$. *Define* $\{x_i\}_{i \in \mathbb{N}}$ *as in* (**??**) *where* $x_0 = x$, $y_0 = y$, $z_0 = z$ *and* $t_0 = t$. *Then,* $|x_i| \to \infty$ *as* $i \to \infty$.

Proof. As, $T_{\mathbb{C}}(x) = x^2 + mx + r$. From (**??**), we have

$$\begin{aligned} |t| &= |(1-\beta)x + \beta T_{\mathbb{C}} x| \\ &= |(1-\beta)x + \beta(x^2 + mx + r)| \\ &\geq |(1-\beta)x + \beta(x^2 + mx)| - \beta|r|. \end{aligned}$$

The assumption $|x| \geq |r|$ yields

$$\begin{aligned} |t| &\geq |(1-\beta)x + \beta(x^2 + mx)| - \beta|x| \\ &\geq \beta|x^2| - (1 - \beta + \beta|m|)|x| - \beta|x| \\ &= \beta|x^2| - (1 + \beta|m|)|x| \\ &= |x|\Big(\beta|x| - (1 + \beta|m|)\Big). \end{aligned}$$

Since $\beta \leq 1$, we obtain $-(1 + \beta|m|) > -(1 + |m|)$ which implies that

$$|t| \geq |x|\Big(\beta|x| - (1 + |m|)\Big).$$

Thus, we have

$$|t| \geq |x|(1 + |m|)\left(\frac{\beta|x|}{1 + |m|} - 1\right).$$

Therefore,

$$\begin{aligned} |t| &\geq \frac{|t|}{(1 + |m|)} \\ &\geq |x|\left(\frac{\beta|x|}{1 + |m|} - 1\right). \end{aligned} \tag{3}$$

From our assumption; $|x| > \max\left\{\frac{2(1+|m|)}{\alpha}, \frac{2(1+|m|)}{\beta}\right\}$, we get

$$\left(\frac{\beta|x|}{1 + |m|} - 1\right) > 1. \tag{4}$$

Now, (**??**) gives that

$$|t| > |x|. \tag{5}$$

As $z = z_0$, (**??**) gives

$$\begin{aligned}|z| =& |T_{\mathbb{C}}(t)| \\ =& |t^2 + mt + r| \geq |t^2 + mt| - |r|.\end{aligned}$$

Since $\beta \leq 1$, it follows from (**??**) and assumption $|x| \geq |r|$ that

$$\begin{aligned}|z| \geq& |t^2 + mt| - |x| \\ \geq& \beta|t^2| - |m||t| - |t| \\ =& |t|\Big(\beta|t| - (1 + |m|)\Big),\end{aligned}$$

which further implies that

$$|z| \geq \frac{|z|}{(1 + |m|)} \geq |t|\Big(\frac{\beta|t|}{(1 + |m|)} - 1\Big). \tag{6}$$

Using (**??**) and (**??**) we have

$$\begin{aligned}|t| >& |x| \\ \Longrightarrow \quad \frac{\beta|t|}{1 + |m|} >& \frac{\beta|x|}{1 + |m|} \\ \Longrightarrow \quad \Big(\frac{\beta|t|}{1 + |m|} - 1\Big) >& \Big(\frac{\beta|x|}{1 + |m|} - 1\Big) > 1.\end{aligned} \tag{7}$$

Consequently, (**??**)–(**??**) yield

$$|z| > |x|. \tag{8}$$

Moreover, let $y = y_0$, $|y| = |T_{\mathbb{C}}(z)| = |z^2 + mz + r|$. Then, by an assumption $|x| \geq |r|$, (**??**) and the fact that $\beta \leq 1$ we obtain

$$\begin{aligned}|y| \geq& |z^2 + mz| - |r| \\ \geq& \beta|z|^2 - |m||z| - |z| \\ \geq& |z|\Big(\beta|z| - (1 + |m|)\Big).\end{aligned}$$

This implies

$$|y| \geq |z|\Big(\frac{\beta|z|}{1 + |m|} - 1\Big).$$

From (**??**) and (**??**) we obtain

$$|y| \geq |x|\Big(\frac{\beta|x|}{1 + |m|} - 1\Big) > |x|. \tag{9}$$

Finally, we have

$$\begin{aligned}|x_1| =& |(1 - \alpha)y + \alpha T_{\mathbb{C}}(y)| \\ =& |(1 - \alpha)y + \alpha(y^2 + my + r)|.\end{aligned}$$

Furthermore, from $|x| \geq |r|$ and (**??**) we get that

$$\begin{aligned}|x_1| =&|(1-\alpha)y + \alpha(y^2 + my + r)|\\ \geq&\alpha|y^2| - (1-\alpha+\alpha|m|)|y| - \alpha|r|\\ \geq&\alpha|y^2| - (1-\alpha+\alpha|m|)|y| - \alpha|y|\\ =&\alpha|y^2| - (1+\alpha|m|)|y|\\ =&|y|\Big(\alpha|y| - (1+\alpha|m|)\Big).\end{aligned}$$

As $\alpha \leq 1$, we obtain

$$\begin{aligned}|x_1| \geq&|y|\Big(\alpha|y| - (1+\alpha|m|)\Big)\\ \geq&|y|\Big(\alpha|y| - (1+|m|)\Big)\\ \geq&|y|(1+|m|)\left(\frac{\alpha|y|}{(1+|m|)} - 1\right).\end{aligned}$$

By (**??**), we have

$$|x_1| \geq |x|\left(\frac{\alpha|x|}{1+|m|} - 1\right).$$

From our given assumption, we have $|x| > \frac{2(1+|m|)}{\alpha}$ and hence $\left(\frac{\alpha|x|}{1+|m|} - 1\right) > 1$. Thus, there exists a real number $\rho > 0$ such that

$$\left(\frac{\alpha|x|}{1+|m|} - 1\right) > 1+\rho.$$

It follows that

$$|x_1| > (1+\rho)|x|.$$

In particular, $|x_1| > |x|$. Continuing in the same manner yields

$$|x_i| > (1+\rho)^i|x|.$$

Therefore, the orbit of x tends to infinity. □

The following corollary is the refinement of the Theorem **??**.

Corollary 1. *Suppose that* $|x_i| > \max\left\{|r|, \frac{2(1+|m|)}{\alpha}, \frac{2(1+|m|)}{\beta}\right\}$ *where* $\alpha, \beta \in (0,1]$ *then* $|x_i| \to \infty$ *as* $i \to \infty$.

2.2. *Escape Criterion for Cubic Complex Polynomials in a Picard Ishikawa Type Orbit*

For the cubic polynomial $T_{\mathbb{C}}(x) = x^3 + mx + r$ where $m, r \in \mathbb{C}$, we have the following result.

Theorem 4. *Suppose* $|x| \geq |r| > \max\left\{\left(\frac{2(1+|m|)}{\alpha}\right)^{\frac{1}{2}}, \left(\frac{2(1+|m|)}{\beta}\right)^{\frac{1}{2}}\right\}$, $\alpha, \beta \in (0,1]$. *Define a sequence* $\{x_i\}_{i\in\mathbb{N}}$ *as in* (**??**) *where* $x_0 = x$, $y_0 = y$, $z_0 = z$ *and* $t_0 = t$. *Then,* $|x_i| \to \infty$ *as* $i \to \infty$.

Proof. As $T_{\mathbb{C}}(x) = x^3 + mx + r$, from (**??**) we have

$$\begin{aligned}|t| =&|(1-\beta)x + \beta T_{\mathbb{C}}(x)| \\ =&|(1-\beta)x + \beta(x^3 + mx + r)| \\ \geq&|(1-\beta)x + \beta(x^3 + mx)| - \beta|r|.\end{aligned}$$

The assumption $|x| \geq |r|$ yields that

$$\begin{aligned}|t| \geq&|(1-\beta)x + \beta(x^3 + mx)| - \beta|x| \\ \geq&\beta|x^3| - (1 - \beta + \beta|m|)|x| - \beta|x| \\ =&\beta|x^3| - (1 + \beta|m|)|x| \\ =&|x|\Big(\beta|x^2| - (1 + \beta|m|)\Big).\end{aligned}$$

As $\beta \leq 1$,

$$|t| \geq |x|\Big(\beta|x^2| - (1 + |m|)\Big).$$

Therefore,

$$\begin{aligned}|t| \geq&\frac{|t|}{(1+|m|)} \\ \geq&|x|\left(\frac{\beta|x^2|}{1+|m|} - 1\right). \end{aligned} \tag{10}$$

The assumption, $|x| > \max\left\{\left(\frac{2(1+|m|)}{\alpha}\right)^{\frac{1}{2}}, \left(\frac{2(1+|m|)}{\beta}\right)^{\frac{1}{2}}\right\}$ implies that

$$\left(\frac{\beta|x^2|}{1+|m|} - 1\right) > 1. \tag{11}$$

It follows from (**??**) that

$$|t| > |x|. \tag{12}$$

As $z = z_0$, by (**??**) we have

$$\begin{aligned}|z| =&|T_{\mathbb{C}}(t)| \\ \geq&|t^3 + mt| - |r|.\end{aligned}$$

As $\beta \leq 1$, from (**??**) and assumption $|x| \geq |r|$ we obtain

$$\begin{aligned}|z| \geq&|t^3 + mt| - |x| \\ =&|t|\Big(\beta|t^2| - (1 + |m|)\Big)\end{aligned}$$

which further implies that

$$|z| \geq |t|\left(\frac{\beta|t^2|}{(1+|m|)} - 1\right). \tag{13}$$

Now by (**??**) and (**??**), we have

$$\left(\frac{\beta|t|^2}{1+|m|} - 1\right) \geq \left(\frac{\beta|x|^2}{1+|m|} - 1\right) > 1. \tag{14}$$

Consequently, (**??**), (**??**) and (**??**) imply that

$$|z| > |x|. \tag{15}$$

Also, $y = y_0$, $|y| = |T_{\mathbb{C}}(z)| = |z^3 + mz + r|$. Then, the given assumption $|x| \geq |r|$, (**??**) and the fact that $\beta \leq 1$ yield

$$\begin{aligned}|y| &\geq |z^3 + mz| - |r| \\ &\geq |z|\Big(\beta|z^2| - (1 + |m|)\Big).\end{aligned}$$

Thus

$$|y| \geq |z|\left(\frac{\beta|z^2|}{1 + |m|} - 1\right).$$

From (**??**) and (**??**), we obtain

$$|y| \geq |x|\left(\frac{\beta|x^2|}{1 + |m|} - 1\right) > |x|. \tag{16}$$

Lastly, we have

$$\begin{aligned}|x_1| &= |(1 - \alpha)y + \alpha T_{\mathbb{C}}y| \\ &= |(1 - \alpha)y + \alpha(y^3 + my + r)|.\end{aligned}$$

From $|x| \geq |r|$, (**??**) and $\alpha \leq 1$, we have

$$\begin{aligned}|x_1| &= |(1 - \alpha)y + \alpha(y^3 + my + r)| \\ &\geq \alpha|y^3| - (1 - \alpha + \alpha|m|)|y| - \alpha|y| \\ &= \alpha|y^2| - (1 + \alpha|m|)|y| \\ &\geq |y|\Big(\alpha|y^2| - (1 + |m|)\Big) \\ &\geq |y|(1 + |m|)\left(\frac{\alpha|y^2|}{(1 + |m|)} - 1\right).\end{aligned}$$

From (**??**), we have

$$|x_1| \geq |x|\left(\frac{\alpha|x^2|}{1 + |m|} - 1\right).$$

By our assumption we have $|x| > \left(\frac{2(1+|m|)}{\alpha}\right)^{\frac{1}{2}}$ and hence $\left(\frac{\alpha|x^2|}{1+|m|} - 1\right) > 1$. Thus, there exists a real number $\rho > 0$ such that

$$\left(\frac{\alpha|x^2|}{1 + |m|} - 1\right) > 1 + \rho.$$

It follows that

$$|x_1| > (1 + \rho)|x|.$$

Continuing in the same manner, we obtain

$$|x_i| > (1 + \rho)^i|x|.$$

Therefore, the orbit of x tends to infinity. □

The following corollary is the refinement of the Theorem **??**.

Corollary 2. *Suppose that* $|x_i| > \max\left\{|r|, \left(\frac{2(1+|m|)}{\alpha}\right)^{\frac{1}{2}}, \left(\frac{2(1+|m|)}{\beta}\right)^{\frac{1}{2}}\right\}$ *where* $\alpha, \beta \in (0,1]$ *then* $|x_i| \to \infty$ *as* $i \to \infty$.

2.3. Escape Criterion for General Complex Polynomials in a Picard Ishikawa Type Orbit

For the general complex polynomial $T_{\mathbb{C}}(x) = x^n + mx + r$ where $m, r \in \mathbb{C}$, we have the following result.

Theorem 5. *Suppose* $|x| \geq |r| > \max\left\{\left(\frac{2(1+|m|)}{\alpha}\right)^{\frac{1}{n-1}}, \left(\frac{2(1+|m|)}{\beta}\right)^{\frac{1}{n-1}}\right\}$, *with* $n \geq 2$ *and* $\alpha, \beta \in (0,1]$. *Define a sequence* $\{x_i\}_{i\in\mathbb{N}}$ *as in* (**??**) *where* $x_0 = x$, $y_0 = y$, $z_0 = z$ *and* $t_0 = t$. *Then,* $|x_i| \to \infty$ *as* $i \to \infty$.

Proof. Let $T_{\mathbb{C}}(x) = x^n + mx + r$. Note that (**??**), assumptions $|x| \geq |r|$ and $\beta \leq 1$ give

$$\begin{aligned}
|t| =& |(1-\beta)x + \beta T_{\mathbb{C}}(x)| \\
\geq& |(1-\beta)x + \beta(x^n + mx)| - \beta|r| \\
\geq& \beta|x^n| - (1-\beta+\beta|m|)|x| - \beta|x| \\
=& |x|\left(\beta|x^{n-1}| - (1+\beta|m|)\right) \\
\geq& |x|\left(\beta|x^{n-1}| - (1+|m|)\right).
\end{aligned}$$

Therefore,

$$|t| \geq |x|\left(\frac{\beta|x^{n-1}|}{1+|m|} - 1\right). \tag{17}$$

By our assumption, we have $|x| > \left(\frac{2(1+|m|)}{\beta}\right)^{\frac{1}{n-1}}$ and hence

$$\left(\frac{\beta|x^{n-1}|}{1+|m|} - 1\right) > 1. \tag{18}$$

It follows from (**??**) that

$$|t| > |x|. \tag{19}$$

Since $z = z_0$, so from (**??**) we obtain

$$|z| \geq |t^n + mt| - |r|.$$

As $\beta \leq 1$, from (**??**) and assumption $|x| \geq |r|$, we have

$$|z| \geq |t|\left(\frac{\beta|t^{n-1}|}{(1+|m|)} - 1\right). \tag{20}$$

Now by (**??**) and (**??**), we have

$$\left(\frac{\beta|t|^{n-1}}{1+|m|} - 1\right) > 1.$$

Hence,

$$|z| > |x|. \tag{21}$$

As $y = y_0$, $|y| = |T_{\mathbb{C}}(z)| = |z^n + mz + r|$, so using the similar arguments as before we obtain

$$|y| \geq |x|\left(\frac{\beta|x^{n-1}|}{1+|m|} - 1\right) > |x|. \tag{22}$$

Also, from $|x| \geq |r|$, (**??**), and $\alpha \leq 1$ we have

$$\begin{aligned}
|x_1| =& |(1-\alpha)y + \alpha(y^n + my + r)| \\
\geq& \alpha|y^n| - (1-\alpha+\alpha|m|)|y| - |r| \\
=& \alpha|y^2| - (1+\alpha|m|)|y| \\
=& |y|\Big(\alpha|y^2| - (1+\alpha|m|)\Big) \\
\geq& |x|\left(\frac{\alpha|x^2|}{1+|m|} - 1\right).
\end{aligned}$$

Furthermore, from our assumption we have $|x| > \left(\frac{2(1+|m|)}{\alpha}\right)^{\frac{1}{n-1}}$ and thus $\left(\frac{\alpha|x^{n-1}|}{1+|m|} - 1\right) > 1$. Thus, there exists a real number $\rho > 0$ such that

$$\left(\frac{\alpha|x^{n-1}|}{1+|m|} - 1\right) > 1 + \rho.$$

Finally, we obtain

$$|x_1| > (1+\rho)|x|.$$

Now, continuing this process

$$|x_i| > (1+\rho)^i|x|.$$

Therefore, the orbit of x tends to infinity. □

The following corollary is the refinement of the Theorem **??**.

Corollary 3. *Suppose that* $|x_i| > \max\left\{|r|, \left(\frac{2(1+|m|)}{\alpha}\right)^{\frac{1}{n-1}}, \left(\frac{2(1+|m|)}{\beta}\right)^{\frac{1}{n-1}}\right\}$ *where* $n \geq 2$ *and* $\alpha, \beta \in (0,1]$ *then* $|x_i| \to \infty$ *as* $i \to \infty$.

Theorem 6. *Suppose that* $\{x_i\}_{i\in\mathbb{N}\cup\{0\}}$ *is a sequence in the Picard Ishikawa type orbit for the complex polynomial* $T_{\mathbb{C}}(x) = x^n + mx + r$ *where* $m, r \in \mathbb{C}$ *with* $n \geq 2$ *such that* $|x_i| \to \infty$ *as* $i \to \infty$, *then* $|x| \geq |r| > \left(\frac{2(1+|m|)}{\alpha}\right)^{\frac{1}{n-1}}$ *and* $|x| \geq |r| > \left(\frac{2(1+|m|)}{\beta}\right)^{\frac{1}{n-1}}$, $\alpha, \beta \in (0,1]$.

Proof. Let $\{x_i\}_{i\in\mathbb{N}}$ be a sequence in Picard Ishikawa type orbit. First, we prove that $|x| \geq |r|$. According to hypothesis, $|x_i| \to \infty$ as $i \to \infty$, the sequence $\{|x_i|\}$ must be unbounded. Hence, $|x_i| \geq |r|$ for all $i \in \mathbb{N}\cup\{0\}$ and therefore $|x| \geq |r|$. Let $T_{\mathbb{C}}(x) = x^n + mx + r$, $m, r \in \mathbb{C}$ where $t_0 = t$, $x_0 = x$, $y_0 = y$ and $z_0 = z$, then $|x| \geq |r|$ implies that

$$\begin{aligned}
|t| =& |(1-\beta)x + \beta T_{\mathbb{C}}x| \\
=& |(1-\beta)x + \beta(x^n + mx + r)| \\
\geq& |\beta x^n| + ((1-\beta) + m\beta)x| - \beta|r| \\
\geq& \beta|x^n| - ((1-\beta) + |m|\beta)|x| - \beta|x| \\
\geq& \beta|x^n| - (1+|m|\beta)|x|.
\end{aligned}$$

Thus,

$$\begin{aligned}|t| \geq& |x|(\beta|x^{n-1}| - (1+|m|)) \\ =& |x|(1+|m|)\left(\frac{\beta|x^{n-1}|}{1+|m|} - 1\right)\end{aligned}$$

implies

$$|t| \geq |x|\left(\frac{\beta|x^{n-1}|}{1+|m|} - 1\right). \tag{23}$$

Here, we have two possibilities; either $\left(\frac{\beta|x^{n-1}|}{1+|m|} - 1\right) \leq 1$ or $\left(\frac{\beta|x^{n-1}|}{1+|m|} - 1\right) > 1$. If $\left(\frac{\beta|x^{n-1}|}{1+|m|} - 1\right) \leq 1$ we have

$$\frac{\beta|x^{n-1}|}{1+|m|} \leq 2$$

which implies that

$$|x^{n-1}| \leq \frac{2(1+|m|)}{\beta}$$

and hence

$$|x| \leq \left(\frac{2(1+|m|)}{\beta}\right)^{\frac{1}{n-1}},$$

a contradiction. Indeed, $\{|x_i|\}$ is not bounded where $i \in \mathbb{N} \cup \{0\}$. Therefore, we must have $\left(\frac{\beta|x^{n-1}|}{1+|m|} - 1\right) > 1$. Thus, $|x| > \left(\frac{2(1+|m|)}{\beta}\right)^{\frac{1}{n-1}}$. Now, inequality (**??**) implies that

$$|t| \geq |x|\left(\frac{\beta|x^{n-1}|}{1+|m|} - 1\right) > |x|.$$

Furthermore, $\beta \leq 1$ and $|x| \geq |r|$ give

$$\begin{aligned}|z| =& |T_{\mathbb{C}}(t)| \\ \geq& |t^n + mt| - |r| \geq \beta|t^n| - |m||t| - |x| \\ =& |t|(\beta|t^{n-1}| - |(1+|m|)).\end{aligned}$$

As $\left(\frac{\beta|x^{n-1}|}{(1+|m|)} - 1\right) > 1$, so we have

$$|t| > |x|\left(\frac{\beta|x^{n-1}|}{1+|m|} - 1\right) > |x|.$$

As a consequence we obtain

$$|z| \geq |x|\left(\frac{\beta|x^{n-1}|}{1+|m|} - 1\right)(1+|m|).$$

Thus,

$$|z| \geq |x|\left(\frac{\beta|x^{n-1}|}{1+|m|} - 1\right) > |x|. \tag{24}$$

Similarly, $|y| = |T_{\mathbb{C}}(z)| = |z^n + mz + r|$, $|x| > |r|$ and $\beta \leq 1$ imply that

$$|y| \geq |x|\left(\frac{\beta|x^{n-1}|}{1+|m|} - 1\right).$$

Consequently,

$$|y| \geq |x|\left(\frac{\beta|x^{n-1}|}{1+|m|} - 1\right) > |x|. \tag{25}$$

Finally, we have

$$\begin{aligned}
|x_1| =&|(1-\alpha)y + \alpha T_{\mathbb{C}}(y)| \\
=&|(1-\alpha)y + \alpha(y^n + my + r)| \\
\geq&\alpha|y^n| - (1-\alpha+\alpha|m|)|y| - \alpha|r| \\
\geq&\alpha|y^n| - (1-\alpha+\alpha|m|)|y| - \alpha|y| \\
\geq&\alpha|y^n| - (1+\alpha|m|)|y| \\
\geq&\alpha|y^n| - (1+|m|)|y| \\
=&|y|(\alpha|y^{n-1}| - (1+|m|)) \\
\geq&|x|(\alpha|x^{n-1}| - (1+|m|)).
\end{aligned}$$

Hence

$$|x_1| \geq |x|\left(\frac{\alpha|x^{n-1}|}{1+|m|} - 1\right).$$

Using arguments similar to those as before, we only have one possibility that $\left(\frac{\alpha|x^{n-1}|}{1+|m|} - 1\right) > 1$. Therefore, $|x| > \left(\frac{2(1+|m|)}{\alpha}\right)^{\frac{1}{n-1}}$. This completes the proof. □

3. Visualization of Fractals

In this section, we present some Julia and Mandelbrot sets for quadratic and higher order polynomials. We found several captivating new fractals having various geometric shapes. However, we have chosen some figures. The color variation occurs due to the change of input parameters. We have also investigated the effect of change of parameters α and β on the shape and the variation of colors. The number of iterations was fixed at 10.

3.1. Generation of Julia sets

Following Algorithm 1 is the pseudocode for the generation of Julia sets. Note that $T'(z)$ represents the iteration process.

Algorithm 1: Generation of Julia Set

Input :complex polynomial–$T : \mathbb{C} \to \mathbb{C}$, parameters–$r, m \in \mathbb{C}$, Area–$A \subset \mathbb{C}$, number of iterations–$N$, colormap with M colors–$colormap[0...M-1]$

Output: $\Re$ is the area for Julia set

```
R = Threshold radius
for c ∈ A
do
    k = 0
    while k ≤ N do
        z = T′(z)
        if |z| > R then
            break
        end
        k = k + 1
    end
    m = ⌊(M − 1) k/N⌋
    color c with colormap[m]
end
```

Now, we present quadratic, cubic and septic Julia sets in Picard Ishikawa type orbit for the complex polynomial, $T_{\mathbb{C}}(x) = x^n + mx + r$.

1. For Figure **??**, we consider the polynomial $T(x) = x^2 + (-0.5 + 0.7i)x + (-0.01 + 0.18i)$ and $A = [-2.5, 2.5] \times [-2.1, 2.1]$. It is easy to see that T has one attracting fixed point, $p = -0.1427 + 0.1019i$. Observe that for $\alpha = 0.2$, $\beta = 0.097$ and $\alpha = 0.11$,$\beta = 0.18$ we obtain different images due to color variation caused by parameters. It is interesting to note that for $\alpha = 1$, $\beta = 1$ and $\alpha = 10^{-10}$, $\beta = 10^{-10}$ we have similar shapes but there is clear variation of colors.

2. For Figure **??**, we consider the polynomial $T(x) = x^3 + (-0.275 + 0.5i)x + (-0.559 + 0.35i)$ and $A = [-1.5, 1.5] \times [-1.8, 1.8]$. The polynomial T has attracting fixed point $p \sim -0.6434 + 0.2687i$ in A. Note that the cubic Julia sets for $\alpha = 0.08$ and $\beta = 0.09$ have more color variation as compared to the Julia sets for $\alpha = 0.1$, and $\beta = 0.2$. Again, for $\alpha = 1$, $\beta = 1$ and $\alpha = 10^{-10}$, $\beta = 10^{-10}$ the shapes are same but there is variability in colors.

3. For Figure **??**, we input $T(x) = x^7 + (0.23 + 1.2i)x + (0.5 + 0.7i)$ and $A = [-1.3, 1.3]^2$. The attracting fixed point of the polynomial is $p \sim -0.2391 + 0.5835i$. We can see that for $\alpha = 0.01$ and $\beta = 0.08$ the shape is spread and stretched while the shape is dense and neatly packed for $\alpha = 0.1$ and $\beta = 0.05$. Note the variation of colors in figures (C) and (D) as well.

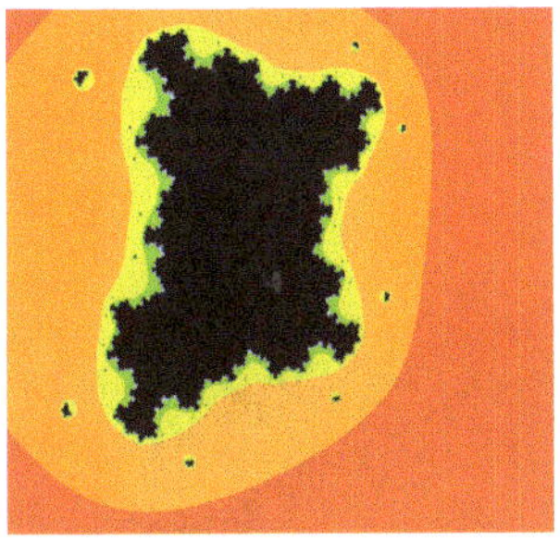

(**a**) $\alpha = 0.2$, $\beta = 0.097$

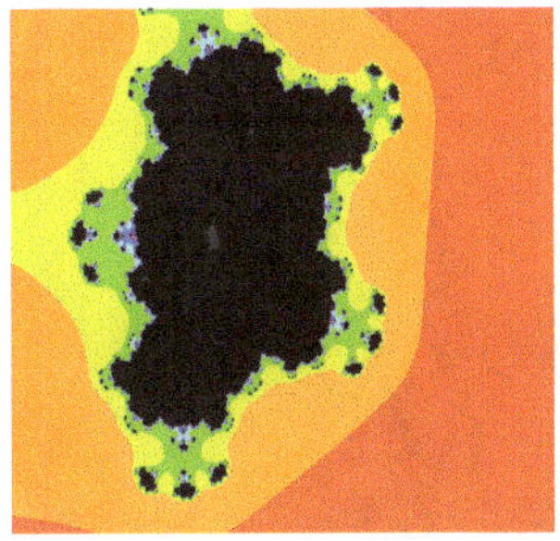

(**b**) $\alpha = 0.11$, $\beta = 0.18$

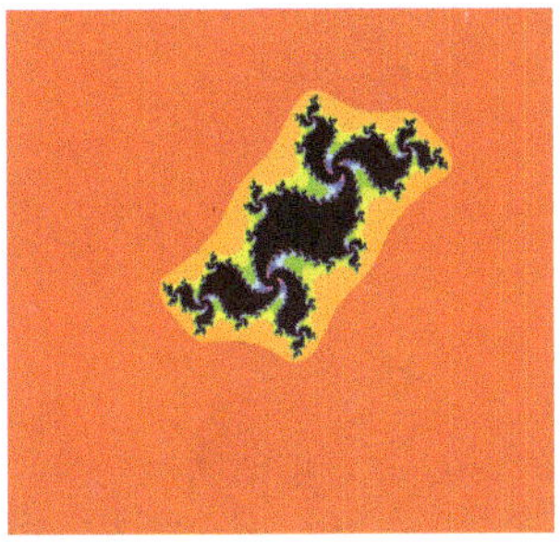

(**c**) $\alpha = 1$, $\beta = 1$

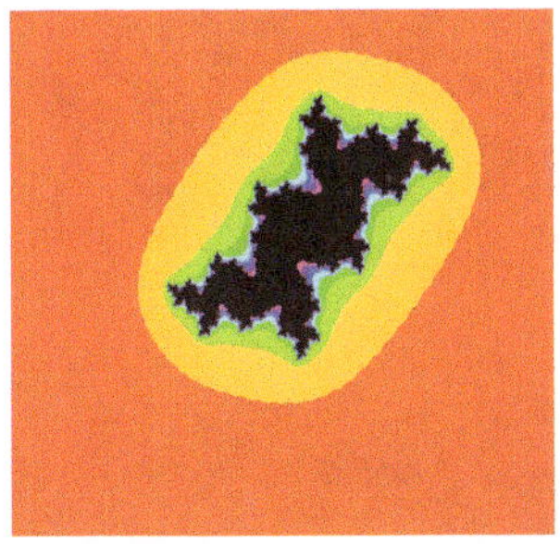

(**d**) $\alpha = 10^{-10}$, $\beta = 10^{-10}$

Figure 1. Quadratic Julia sets.

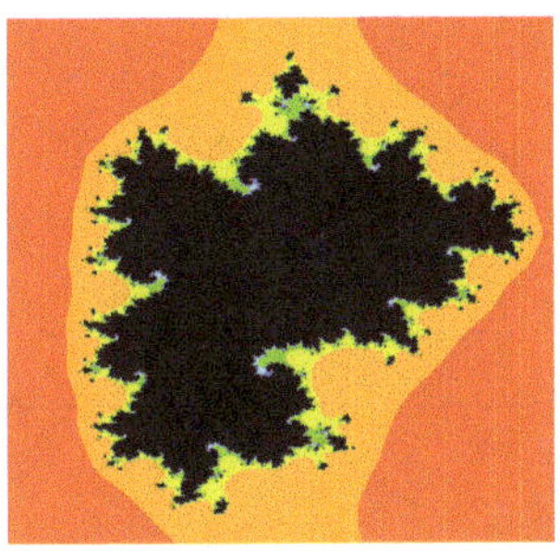

(**a**) $\alpha = 0.08$, $\beta = 0.09$

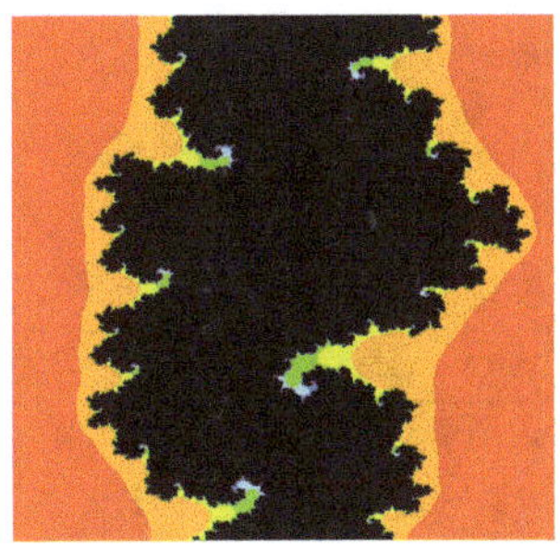

(**b**) $\alpha = 0.1$, $\beta = 0.2$

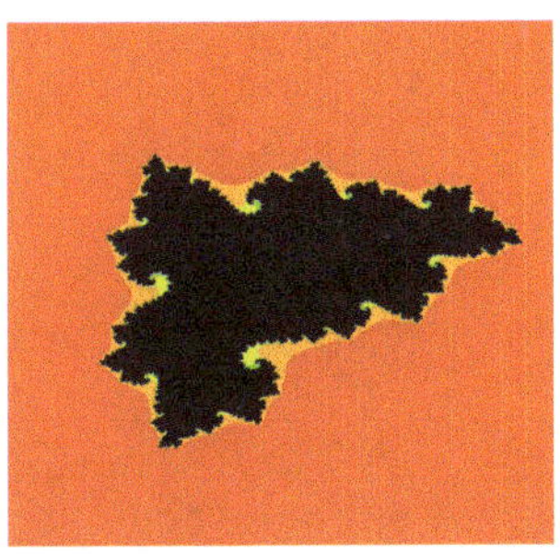

(**c**) $\alpha = 1$, $\beta = 1$

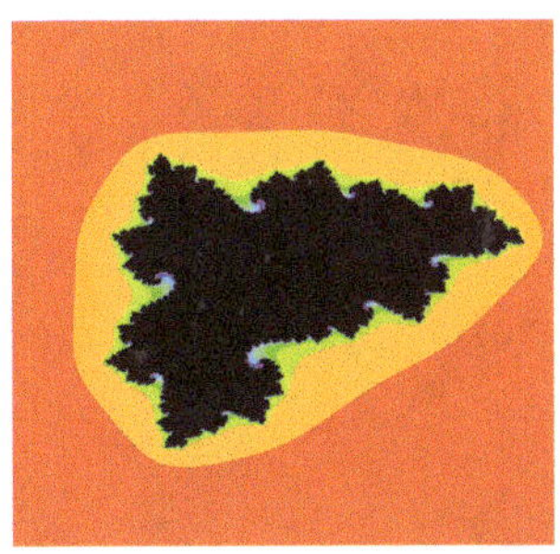

(**d**) $\alpha = 10^{-10}$, $\beta = 10^{-10}$

Figure 2. Cubic Julia sets.

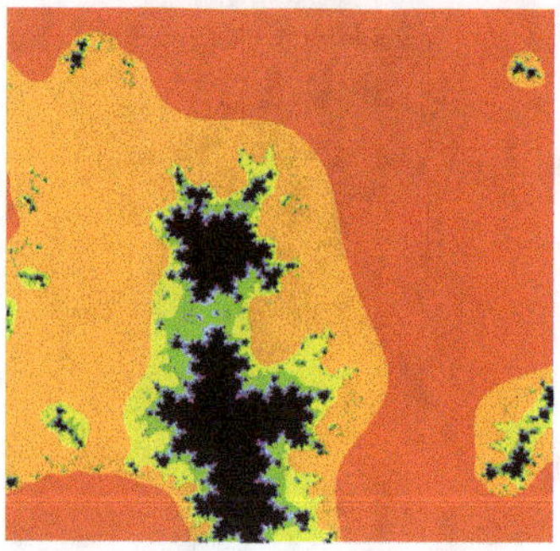

(**a**) $\alpha = 0.01, \beta = 0.08$

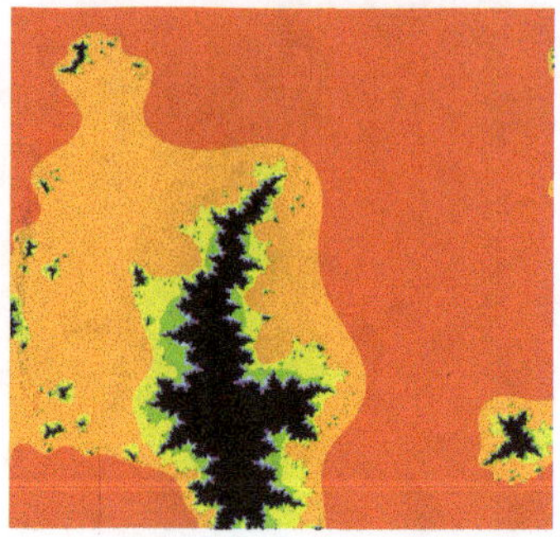

(**b**) $\alpha = 0.1, \beta = 0.05$

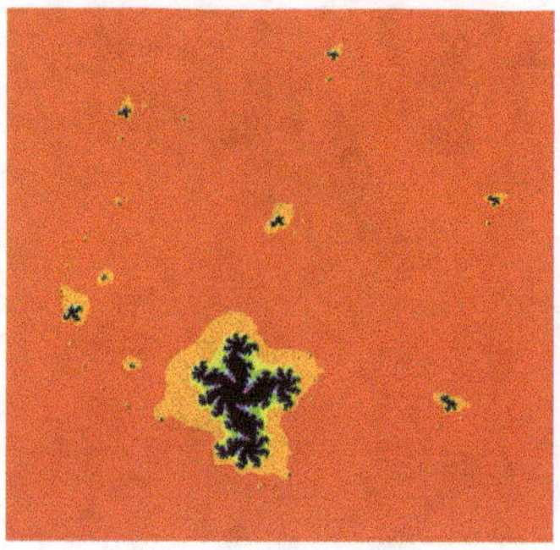

(**c**) $\alpha = 1, \beta = 1$

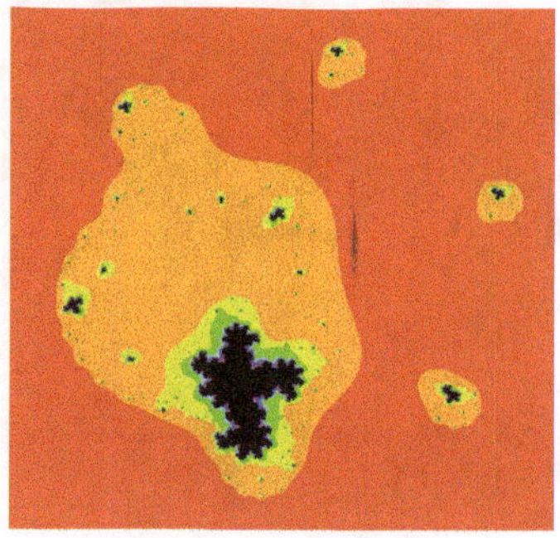

(**d**) $\alpha = 0.009, \beta = 0.009$

Figure 3. Septic Julia sets.

3.2. Generation of Mandelbrot Sets

Following Algorithm 2 is the pseudocode for the generation of Mandelbrot sets. Note that $T'(z)$ represents the iteration process.

Algorithm 2: Generation of Mandelbrot set.

Input : complex polynomial–$T : \mathbb{C} \to \mathbb{C}$, parameters–$r, m \in \mathbb{C}$, Area–$A \subset \mathbb{C}$, number of iterations–$N$, colormap with M colors–$colormap[0...M-1]$

Output: $\Re$ is the area for Mandelbrot set

```
for c ∈ A
do
    R = Threshold radius
    k = 0
    x_0 = critical point of T
    while k ≤ N do
        z = T'(z)
        if |z| > R then
            break
        end
        k = k + 1
    end
    m = ⌊(M − 1) k/N⌋
    color c with colormap[m]
end
```

For Figure **??** we input $A = [-2, 2] \times [-1.2, 2.5]$ and observe that for $\alpha = 0.1$ and $\beta = 0.3$, the shape is stretched and the bulb is wider and for $\alpha = 0.75$ and $\beta = 0.7$ the shape is compact with defined bulb. Notice the variation of colors for Mandelbrot sets for $\alpha = 1$, $\beta = 1$ and $\alpha = 0.009$, $\beta = 0.009$. Also, observe that Mandelbot sets generated are symmetric about origin.

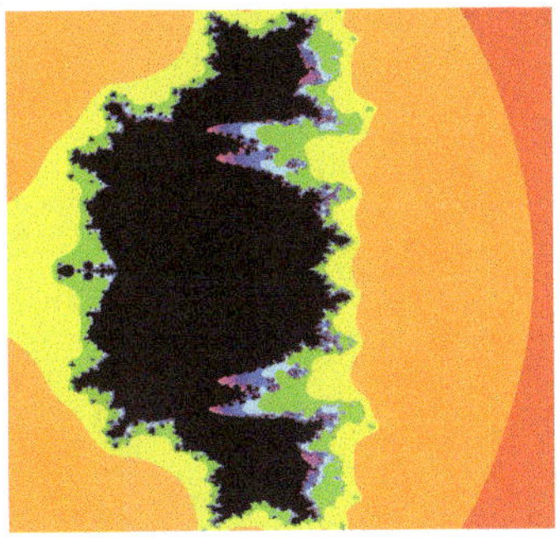

(**a**) $\alpha = 0.1, \beta = 0.3$

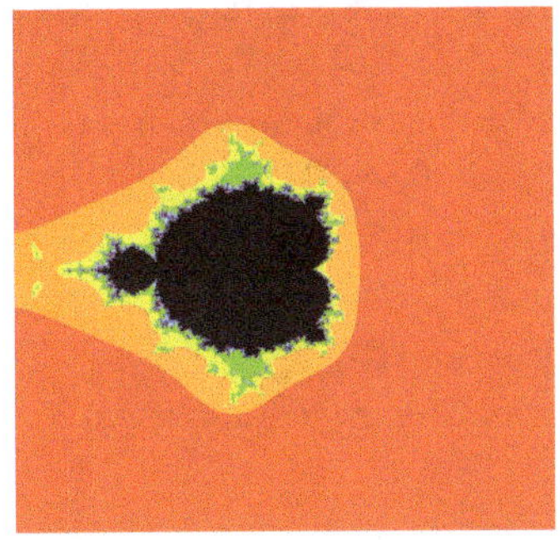

(**b**) $\alpha = 0.75, \beta = 0.7$

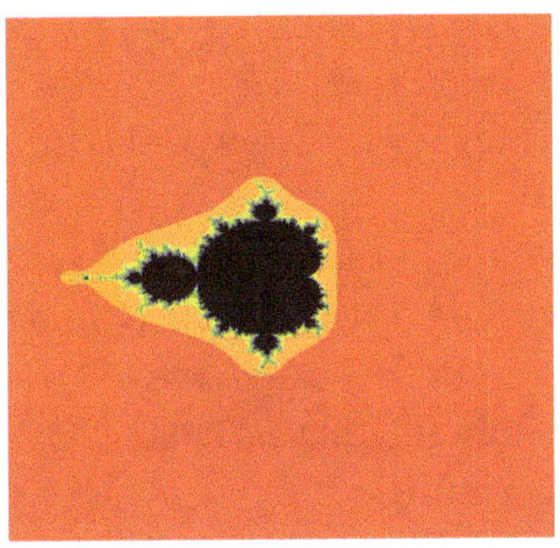

(**c**) $\alpha = 1, \beta = 1$

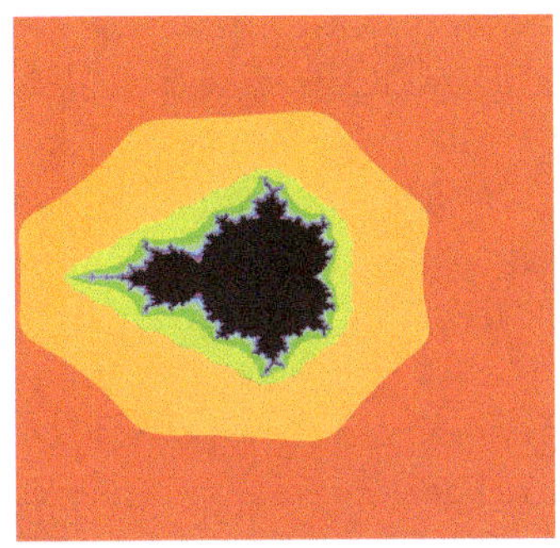

(**d**) $\alpha = 0.009, \beta = 0.009$

Figure 4. Mandelbrot sets.

4. Conclusions

In this paper, a Picard Ishikawa type orbit was used to study the behaviour of complex poylnomials. We obtained escape criterions for complex quadratic, cubic and higher degree polynomials. Some alluring Julia and Mandelbrot sets have been generated. We also observed that the variation of parameters has shown eminent changes in the Julia and Mandelbrot sets. Our results are different from comparable existing results as we obtain escape criterion and fractals for polynomials of the form $T(x) = x^n + mx + r$ where $m, r \in \mathbb{C}$ without using the Jungck iterative process. It is also worth mentioning that the behaviour of the polynomial and shape of the fractal generated under the iterative process (**??**) is different and unique as compared to the iterative process studied before in the literature [**? ? ? ?**].

Author Contributions: Conceptualization, M.A. and H.I.; methodology, M.A. and H.I.; validation, M.A. and M.D.l.S., formal analysis H.I.; investigation, H.I. and M.A.; writing—original draft preparation, H I.; writing—review and editing, H.I.; visualization, M.D.l.S.; supervision, M.A. and M.D.l.S.; project administration, M.D.l.S.; funding acquisition, M.D.l.S. All authors have read and agreed to the published version of the manuscript.

Acknowledgments: All the authors are grateful to the referees for their critical remarks and valuable suggestions which helped to improve the presentation of the paper.

References

1. Barnsley, M. *Fractals Everywhere*, 2nd ed.; Academic Press: San Diego, CA, USA, 1993.
2. Hundertmark-Zauŝkovù, A. On the convergence of fixed point iterations for the moving geometry in a fluid-structure interaction problem. *J. Differ. Equ.* **2019**, *267*, 7002–7046. [CrossRef]
3. Rahmani, M.; Koutsopoulos, H.N.; Jenelius, E. Travel time estimation from sparse floating car data with consistent path inference: A fixed point approach. *Transp. Res. Part C Emerg. Technol.* **2017**, *85*, 628–643. [CrossRef]
4. Strogatz, S.H. *Nonlinear Dynamics and Chaos With Applications to Physics, Biology, Chemistry, and Engineering*, 2nd ed.; CRC Press: Boca Raton, FL, USA, 2018.
5. Julia, G. Memoire sur l'iteration des functions rationnelles. *J. Math. Pures Appl.* **1918**, *8*, 737–747.
6. Devaney, R.L. *A First Course in Chaotic Dynamical Systems: Theory and Experiment*, 2nd ed.; Addison-Wesley: Boston, MA, USA, 1992.
7. Falconer, K. *Fractal Geometry: Mathematical Foundations and Applications*, 2nd ed.; John Wiley & Sons: Chichester, UK, 2004.
8. Frame, M.; Robertson, J. A generalized Mandelbrot set and the role of critical points. *Comput. Graph.* **1992**, *16*, 35–40. [CrossRef]
9. Brouwer, L.E.J. Über Abbildungen von Mannigfaltigkeiten. *Math. Ann.* **1912**, *71*, 97–115. [CrossRef]
10. Mandelbrot, B.B. *The Fractal Geometry of Nature*; W.H. Freeman: New York, NY, USA, 1983; Volume 2.
11. Debnath, L. A brief historical introduction to fractals and fractal geometry. *Int. J. Math. Educ. Sci. Technol.* **2006**, *37*, 29–50. [CrossRef]
12. Lakhtakia, A.; Varadan, W.; Messier, R.; Varadan, V.K. On the symmetries of the Julia sets for the process $z^p + c$. *J. Phys. A Math. Gen.* **1987**, *20*, 3533–3535. [CrossRef]
13. Negi, A.; Rani, M. Midgets of superior Mandelbrot set. *Chaos Solitons Fract.* **2008**, *36*, 237–245. [CrossRef]
14. Negi, A.; Rani, M. A new approach to dynamic noise on superior Mandelbrot set. *Chaos Solitons Fract.* **2008**, *36*, 1089–1096. [CrossRef]
15. Rochon, D. A generalized Mandelbrot set for bicomplex numbers. *Fractals* **2000**, *8*, 355–368. [CrossRef]
16. Wang, X.; Sun, Y. The general quaternionic M-J sets on the mapping $z \leftarrow z^\alpha + c(\alpha \in \mathbb{N})$. *Comput. Math. Appl.* **2007**, *53*, 1718–1732. [CrossRef]
17. Rani, M. Theoretical Framework for Fractal Models under Two-Step Feedback Process. Ph.D. Thesis, Kumaun University, Nainital, India, 2016.
18. Rani, M.; Kumar, V. Superior Julia set. *Res. Math. Educ.* **2004**, *8*, 261–277.
19. Rani, M.; Kumar, V. Superior Mandelbrot set. *Res. Math. Educ.* **2004**, *8*, 279–291.
20. Rani, M.; Chauhan, Y.S.; Negi, A. Non linear dynamics of Ishikawa iteration. *Int. J. Comput. Appl.* **2010**, *7*, 43–49. [CrossRef]
21. Chauhan, Y.S.; Rana, R.; Negi, A. New Julia sets of Ishikawa iterates. *Int. J. Comput. Appl.* **2010**, *7*, 34–42. [CrossRef]
22. Chauhan, Y.S.; Rana, R.; Negi, A. Complex dynamics of Ishikawa iterates for non integer values. *Int. J. Comput. Appl.* **2010**, *9*, 9–16. [CrossRef]
23. Ashish; Rani, M.; Chugh, R. Julia sets and Mandelbrot sets in Noor orbit. *Appl. Math. Comput.* **2014**, *228*, 615–631.
24. Kang, S.M.; Rafiq, A.; Latif, A.; Shahid, A.A.; Ali, F. Fractals through modified iteration scheme. *Filomat* **2016**, *30*, 3033–3046. [CrossRef]
25. Kang, S.M.; Rafiq, A.; Latif, A.; Shahid, A.A.; Kwun, Y.C. Tricorns and multicorns of *S*-iteration scheme. *J. Funct. Spaces* **2015**, *2015*, 417167.
26. Kumari, M.; Ashish, R.C. New Julia and Mandelbrot sets for a new faster iterative process. *Int. J. Pure Appl. Math.* **2016**, *107*, 161–177. [CrossRef]
27. Kang, S.M.; Nazeer, W.; Tanveer, M.; Shahid, A.A. New fixed point results for fractal generation in Jungck Noor orbit with *s*-convexity. *J. Funct. Spaces* **2015**, *2015*, 963016. [CrossRef]
28. Kang, S.M.; Rafiq, A.; Tanveer, M.; Ali, F.; Kwun, Y.C. Julia and Mandelbrot sets in modified Jungck three-step orbit. *Wulfenia J.* **2015**, *22*, 167–185.
29. Kwun, Y.C.; Tanveer, M.; Nazeer, W.; Abbas, M.; Kang, S.M. Fractal generation in modified Jungck-*S* orbit. *IEEE Access* **2019**, *7*, 35060–35071. [CrossRef]

30. Proakis, J.G.; Manolakis, D.G. *Digital Signal Processing: Principles, Algorithms and Applications*, 4th ed.; Pearson: Bengaluru, India, 2007.
31. Mishra, M.K.; Ojha, D.B.; Sharma, D. Fixed point results in tricorn and multicorns of Ishikawa iteration and s-convexity. *IJEST* **2011**, *2*, 157–160.
32. Cho, S.Y.; Shahid, A.A.; Nazeer, W.; Kang, S.M. Fixed point results for fractal generation in noor orbit and s-convexity. *SpringerPlus* **2016**, *5*, 1843. [CrossRef] [PubMed]
33. Nazeer, W.; Kang, S.M.; Tanveer, M.; Shahid, A.A. Fixed point results in the generation of Julia and Mandelbrot sets. *J. Inequal. Appl.* **2015**, *2015*, 298. [CrossRef]
34. Piri, H.; Daraby, B.; Rahrovi, S.; Ghasemi, M. Approximating fixed points of generalized α-nonexpansive mappings Banach spaces by new faster iteration process. *Numer. Algorithms* **2018**, *81*, 1129–1148. [CrossRef]

8

On a Reduced Cost Higher Order Traub-Steffensen-Like Method for Nonlinear Systems

Janak Raj Sharma [1], Deepak Kumar [1,*] and Lorentz Jäntschi [2,*]

1 Department of Mathematics, Sant Longowal Institute of Engineering and Technology, Longowal, Sangrur 148106, India
2 Department of Physics and Chemistry, Technical University of Cluj-Napoca, Cluj-Napoca 400114, Romania
* Correspondence: deepak.babbi@gmail.com (D.K.); lorentz.jantschi@gmail.com (L.L.)

Abstract: We propose a derivative-free iterative method with fifth order of convergence for solving systems of nonlinear equations. The scheme is composed of three steps, of which the first two steps are that of third order Traub-Steffensen-type method and the last is derivative-free modification of Chebyshev's method. Computational efficiency is examined and comparison between the efficiencies of presented technique with existing techniques is performed. It is proved that, in general, the new method is more efficient. Numerical problems, including those resulting from practical problems viz. integral equations and boundary value problems, are considered to compare the performance of the proposed method with existing methods. Calculation of computational order of convergence shows that the order of convergence of the new method is preserved in all the numerical examples, which is not so in the case of some of the existing higher order methods. Moreover, the numerical results, including the CPU-time consumed in the execution of program, confirm the accurate and efficient behavior of the new technique.

Keywords: nonlinear equations; systems; derivative-free methods; fast algorithms; computational efficiency

---

1. Introduction

We are concerned with the problem of solving a system of nonlinear equations

$$F(x) = 0. \tag{1}$$

This problem can precisely be stated as to find a solution vector $\alpha = (\alpha_1, \alpha_2, ..., \alpha_m)^T$ such that $F(\alpha) = 0$, where $\mathrm{F}(x) : D \subset \mathbb{R}^m \longrightarrow \mathbb{R}^m$ is the given nonlinear vector function $F(x) = (f_1(x), f_2(x), ..., f_m(x))^T$ and $x = (x_1, x_2, ..., x_m)^T$. The vector α can be computed as a fixed point of some function $\mathrm{M} : D \subset \mathbb{R}^m \to \mathbb{R}^m$ by means of fixed point iteration

$$\begin{aligned} &x^{(0)} \in D, \\ &x^{(k+1)} = \mathrm{M}(x^{(k)}),\ k \geq 0. \end{aligned} \tag{2}$$

Many applied problems in Science and Engineering are reduced to solve numerically the system $F(x) = 0$ of nonlinear equations (see, for example [1–6]). A plethora of iterative methods are developed in literature for solving such equations. A classical method is cubically convergent Chebyshev's method (see [7])

$$\begin{aligned} &x^{(0)} \in D, \\ &x^{(k+1)} = x^{(k)} - \Big(I + \frac{1}{2}L_F(x^{(k)})\Big)F'(x^{(k)})^{-1}F(x^{(k)}),\ k \geq 0, \end{aligned} \tag{3}$$

where $L_F(x^{(k)}) = F'(x^{(k)})^{-1}F''(x^{(k)})F'(x^{(k)})^{-1}F(x^k)$. This one-point iterative scheme depends explicitly on the first two derivatives of F. In [7], Ezquerro and Hernández present modification in Chebyshev's method that avoids the computation of second derivative F'' while maintaining third-order of convergence. It has the following form:

$$\begin{aligned} &x^{(0)} \in D, \\ &y^{(k)} = x^{(k)} - a\,F'(x^{(k)})^{-1}F(x^{(k)}), \\ &x^{(k+1)} = x^{(k)} - \frac{1}{a^2}F'(x^{(k)})^{-1}\big((a^2+a-1)F(x^{(k)}) + F(y^{(k)})\big),\ k \geq 0. \end{aligned} \tag{4}$$

There is an interest in constructing derivative free iterative processes obtained by considering an approximation of the first derivative of F from a divided difference of first order. One class of such methods is called the class of Secant-type methods which is obtained by replacing F' with the divided difference operator $[x^{(k-1)}, x^{(k)}; F]$. Using this operator a family of derivative free methods is given in [8]. The authors call this family the Chebyshev-Secant-type method and it is defined as

$$\begin{aligned} &x^{(-1)},\ x^{(0)} \in D, \\ &y^{(k)} = x^{(k)} - a\,[x^{(k-1)}, x^{(k)}; F]^{-1}F(x^{(k)}), \\ &x^{(k+1)} = x^{(k)} - [x^{(k-1)}, x^{(k)}; F]^{-1}\big(b\,F(x^{(k)}) + c\,F(y^{(k)})\big),\ k \geq 0, \end{aligned} \tag{5}$$

where a, b and c are non-negative parameters.

Another class of derivative free methods is the class of Steffensen-type processes that replaces F' with operator $[w(x^{(k)}), x^{(k)}; F]$, wherein $w: \mathbb{R}^m \to \mathbb{R}^m$. The work presented in [9] analyzes Steffensen-type iterative method which is given as

$$\begin{aligned} &x^{(0)} \in D, \\ &y^{(k)} = x^{(k)} - a\,[w(x^{(k)}), x^{(k)}; F]^{-1}F(x^{(k)}), \\ &x^{(k+1)} = x^{(k)} - [w(x^{(k)}), x^{(k)}; F]^{-1}\big(b\,F(x^{(k)}) + c\,F(y^{(k)})\big),\ k \geq 0. \end{aligned} \tag{6}$$

For $a = b = c = 1$ and $w(x^{(k)}) = x^{(k)} + \beta F(x^{(k)})$, β is an arbitrary non-zero constant, this method possesses third order convergence. In this case $y^{(k)}$ is Traub-Steffensen iteration [6]. For $\beta = 1$, $y^{(k)}$ belongs to Steffensen iteration [10]. Both of these iterations are quadratically convergent.

The two-step third order Traub-Steffensen-type method, i.e., the case of (6) for $a = b = c = 1$, can be written as

$$\begin{aligned} &x^{(0)} \in D,\ w(x^{(k)}) = x^{(k)} + \beta F(x^{(k)}), \\ &y^{(k)} = \mathrm{M}_{2,1}(x^{(k)}), \\ &x^{(k+1)} = \mathrm{M}_{3,1}(x^{(k)}, y^{(k)}) = y^{(k)} - [w(x^{(k)}), x^{(k)}; F]^{-1}F(y^{(k)}),\ k \geq 0, \end{aligned} \tag{7}$$

where $\mathrm{M}_{2,1}(x^{(k)}) = x^{(k)} - [w(x^{(k)}), x^{(k)}; F]^{-1}F(x^{(k)})$ is the quadratically convergent Traub-Steffensen scheme. Here and in the sequel, the symbol $\mathrm{M}_{p,i}$ is used for denoting an i-th iteration function of convergence order p. It can be observed that the third order scheme (7) is computationally more efficient than quadratically convergent Traub-Steffensen scheme. The reason is that the convergence order is increased from two to three at the cost of only one function evaluation without adding extra inverse operator. We discuss computational efficiency in later sections.

Researchers have always been trying to develop the iterative method with increasing efficiency since different methods converge to the solution with different convergence speed. This can be done either by increasing the convergence order or by decreasing the computational cost or both. In [11], Ren et al. have derived a fourth order derivative-free method that uses three F, three divided differences and two matrix inversions per iteration. Zheng et al. [12] have constructed two families of fourth order derivative-free methods for scalar nonlinear equations, that are extendable to solve systems of nonlinear equations. First family requires to evaluate three F, three divided differences and two matrix inversions, whereas the second family needs three F, three divided differences and three matrix inversions. Grau et al. presented a fourth order derivative-free method in [13] utilizing four F, two divided differences and two matrix inversions. Sharma and Arora [14] presented a fourth order derivative-free method that uses the evaluations of three F, three divided differences and one matrix inversion per each step.

In search of more fast techniques, researchers have also introduced sixth and seventh order derivative-free methods in [13,15–18]. The sixth order method in [13] proposed by Grau et al. requires five F, two divided differences and two matrix inverses. Sharma and Arora [17] also developed a method of at least sixth order which requires evaluation of five functions, two divided difference and one matrix inversion per iteration. The seventh order method proposed by Sharma and Arora [15] utilizes four F, five divided differences and two matrix inversions per iteration. The seventh order methods presented by Wang and Zhang [16] use four F, five divided differences and three matrix inversions. Ahmad et al. [18] proposed eighth order derivative free method without memory which uses six functions evaluations, three divided difference and one matrix inversion.

The main goal in this study is to develop a derivative-free method of high computational efficiency, that means a method with high convergence speed and low computational cost. Consequently, we present a Traub-Steffensen-type method of fifth order of convergence which requires the evaluations four F, two divided differences and only one matrix inversion per step. The scheme of the present contribution is simple and consists of three steps. Of the three steps, the first two are that of cubically convergent Traub-Steffensen-type scheme (7) whereas the third is derivative-free modification of Chebyshev's scheme (3). We show that the proposed method is more efficient than existing methods of similar nature.

The content of the rest of the paper is summarized as follows. Basic definitions relevant to the present work are stated in Section 2. In Section 3, the scheme of fifth order method is introduced and its convergence behavior is studied. In Section 4, the computational efficiency of the new method is examined and also compared with the existing derivative-free methods. In Section 5, the basins of attractors are presented to check the stability and convergence of the new method. Numerical tests are performed in Section 6 to verify the theoretical results as proved in Sections 3 and 4. Section 7 contains the concluding remarks.

2. Preliminary Results

2.1. Computational Order of Convergence

Let α be a solution of the function $F(x) = 0$ and $x^{(k-2)}$, $x^{(k-1)}$, $x^{(k)}$ and $x^{(k+1)}$ be the four consecutive iterations close to α. Then, the computational order of convergence (say, p_c) can be calculated using the formula (see [19])

$$p_c = \frac{\log(\|x^{(k+1)} - x^{(k)}\| / \|x^{(k)} - x^{(k-1)}\|)}{\log(\|x^{(k)} - x^{(k-1)}\| / \|x^{(k-1)} - x^{(k-2)}\|)}. \tag{8}$$

2.2. Divided Difference

Divided difference operator for multivariable function F (see [4,5,20]) is a mapping $[\cdot,\cdot\,;F]$: $D \times D \subset \mathbb{R}^m \times \mathbb{R}^m \to L(\mathbb{R}^m)$ which is defined as

$$[x,y;F](x-y) = F(x) - F(y), \forall\,(x,y)\;\in \mathbb{R}^m. \tag{9}$$

If F is differentiable, we can also define first order divided difference as (see [4,20])

$$[x+h,x\,;F] = \int_0^1 F'(x+th)\,dt,\;\forall\,(x,h)\;\in \mathbb{R}^m. \tag{10}$$

This also implies that

$$[x,x\,;F] = F'(x). \tag{11}$$

It can be seen that the divided difference operator $[x,y\,;F]$ is an $m \times m$ matrix and the definitions (9) and (10) are equivalent (for details see [20]). For computational purpose the following definition (see [5]), is used

$$[x,y\,;F]_{ij} = \frac{f_i(x_1,.....,x_j,y_{j+1},.....,y_m) - f_i(x_1,.....,x_{j-1},y_j,.....,y_m)}{x_j - y_j},\;\; 1 \le i,j \le m. \tag{12}$$

2.3. Computational Efficiency

Computational efficiency of an iterative method for solving $F(x) = 0$ is calculated by the efficiency index $E = p^{1/C}$, (for detail see [21,22]), where p is the order of convergence and C is the total cost of computation. The cost of computation C is measured in terms of the total number of function evaluations per iteration and the number of operations (that means products and quotients) per iteration.

3. The Method and Analysis of Convergence

Let us begin with the following three-step scheme

$$\begin{aligned} y^{(k)} &= \mathrm{M}_{2,1}(x^{(k)}),\\ z^{(k)} &= y^{(k)} - [w^{(k)},x^{(k)}\,;F]^{-1}F(y^{(k)}),\\ x^{(k+1)} &= z^{(k)} - \Big(I + \frac{1}{2}L_F(y^{(k)})\Big)F'(y^{(k)})^{-1}F(z^{(k)}), \end{aligned} \tag{13}$$

where $w^{(k)} = x^{(k)} + \beta F(x^{(k)})$, I is $m \times m$ identity matrix and $L_F(y^{(k)}) = F'(y^{(k)})^{-1}F''(y^{(k)})F'(y^{(k)})^{-1}F(y^k)$.

Note that this is a scheme whose first two steps are that of third order Traub-Steffensen-type method (7) whereas third step is based on Chebyshev's method (3). The scheme requires first and second derivatives of F at $y^{(k)}$. To make this a derivative-free method, we describe an approach as follows:

Consider the Taylor expansion of $F(z^{(k)})$ about $y^{(k)}$,

$$F(z^{(k)}) \approx F(y^{(k)}) + F'(y^{(k)})(z^{(k)} - y^{(k)}) + \frac{1}{2}F''(y^{(k)})(z^{(k)} - y^{(k)})^2. \tag{14}$$

Then, it follows that

$$\frac{1}{2}F''(y^{(k)})(z^{(k)} - y^{(k)})^2 \approx F(z^{(k)}) - F(y^{(k)}) - F'(y^{(k)})(z^{(k)} - y^{(k)}). \tag{15}$$

Using the fact that

$$F(z^{(k)}) - F(y^{(k)}) = [z^{(k)}, y^{(k)}; F](z^{(k)} - y^{(k)}),$$

(see, for example [4,5]), we can write (15) as

$$F''(y^{(k)})(z^{(k)} - y^{(k)}) \approx 2([z^{(k)}, y^{(k)}; F] - F'(y^{(k)})). \tag{16}$$

Then, using the second step of (13) in the above equation, it follows that

$$F''(y^{(k)})[w^{(k)}, x^{(k)}; F]^{-1}F(y^{(k)}) \approx -2([z^{(k)}, y^{(k)}; F] - F'(y^{(k)})). \tag{17}$$

Let us assume $F'(y^{(k)}) \approx [w^{(k)}, x^{(k)}; F]$, then (17) implies

$$F''(y^{(k)})[w^{(k)}, x^{(k)}; F]^{-1}F(y^{(k)}) \approx -2([z^{(k)}, y^{(k)}; F] - [w^{(k)}, x^{(k)}; F]). \tag{18}$$

In addition, we have that

$$\begin{aligned} L_F(y^{(k)}) &= F'(y^{(k)})^{-1}F''(y^{(k)})F'(y^{(k)})^{-1}F(y^k) \\ &\approx [w^{(k)}, x^{(k)}; F]^{-1}F''(y^{(k)})[w^{(k)}, x^{(k)}; F]^{-1}F(y^k). \end{aligned} \tag{19}$$

Using (18) in (19), we obtain that

$$\begin{aligned} L_F(y^{(k)}) &\approx [w^{(k)}, x^{(k)}; F]^{-1}F''(y^{(k)})[w^{(k)}, x^{(k)}; F]^{-1}F(y^k) \\ &\approx -2([w^{(k)}, x^{(k)}; F]^{-1}[z^{(k)}, y^{(k)}; F] - I). \end{aligned} \tag{20}$$

Now, we can write the third-step of (13) in modified form as

$$x^{(k+1)} = z^{(k)} - (2I - [w^{(k)}, x^{(k)}; F]^{-1}[z^{(k)}, y^{(k)}; F])[w^{(k)}, x^{(k)}; F]^{-1}F(z^{(k)}). \tag{21}$$

Thus, we define the following new method:

$$\begin{aligned} y^{(k)} &= \mathrm{M}_{2,1}(x^{(k)}), \\ z^{(k)} &= \mathrm{M}_{3,1}(x^{(k)}, y^{(k)}), \\ x^{(k+1)} &= z^{(k)} - H(x^{(k)})[w^{(k)}, x^{(k)}; F]^{-1}F(z^{(k)}), \end{aligned} \tag{22}$$

wherein $H(x^{(k)}) = 2I - [w^{(k)}, x^{(k)}; F]^{-1}[z^{(k)}, y^{(k)}; F]$.

Since the scheme (22) is composed of Traub-Steffensen like steps, we call it the Traub-Steffensen-like method.

In order to explore the convergence properties of Traub-Steffensen-like method, we recall some important results from the theory of iteration functions. First, we state the following well-known result (see [3,23]):

Lemma 1. *Assume that $M : D \subset \mathbb{R}^m \to \mathbb{R}^m$ has a fixed point $\alpha \in int(D)$ and $M(x)$ is Fréchet differentiable on α. If*

$$\rho(M'(\alpha)) = \sigma < 1, \tag{23}$$

then α is a point of attraction for the iteration $x^{(k+1)} = M(x^{(k)})$, where ρ is a spectral radius of $M'(\alpha)$.

Next, we state a result which has been proven in [24] by Madhu et al. and that shows α is a point of attraction for a general iteration function of the form $M(x) = P(x) - Q(x)R(x)$.

Lemma 2. *Let $F : D \subset \mathbb{R}^m \to \mathbb{R}^m$ be sufficiently Fréchet differentiable at each point of an open convex set D of $\alpha \in D$, which is a solution of the nonlinear system $F(x) = 0$. Suppose that $P, Q, R : D \subset \mathbb{R}^m \to \mathbb{R}^m$ are sufficiently Fréchet differentiable functions (depending on F) at each point in the set D with the properties $P(\alpha) = \alpha$, $Q(\alpha) \neq 0$, $R(\alpha) = 0$. Then, there exists a ball*

$$S = \bar{S}(\alpha, \epsilon) = \{\|\alpha - x\| \leq \epsilon\} \subset D,\ \epsilon > 0,$$

on which the mapping

$$M : S \to \mathbb{R}^m,\ M(x) = P(x) - Q(x)R(x), \forall\, x \in S$$

is well defined. Moreover, $M(x)$ is Fréchet differentiable at α, thus

$$M'(\alpha) = P'(\alpha) - Q(\alpha)R'(\alpha).$$

Let us also recall the definition (10) of divided difference operator. Then, expanding $F'(x + th)$ in (10) by Taylor series at the point x and thereafter integrating, we have that

$$[x + h, x; F] = \int_0^1 F'(x + th)\, dt = F'(x) + \frac{1}{2}F''(x)h + \frac{1}{6}F'''(x)h^2 + \frac{1}{24}F^{(iv)}(x)h^3 + O(h^4), \tag{24}$$

where $h^i = (h, h, \stackrel{i}{\cdots}, h), h \in \mathbb{R}^m$. Let $e^{(k)} = x^{(k)} - \alpha$. Assuming that $\Gamma = F'(\alpha)^{-1}$ exists, then expanding $F(x^{(k)})$ and its first three derivatives in a neighborhood of α by Taylor's series, we have that

$$F(x^{(k)}) = F'(\alpha)(e^{(k)} + A_2(e^{(k)})^2 + A_3(e^{(k)})^3 + A_4(e^{(k)})^4 + A_5(e^{(k)})^5 + O((e^{(k)})^6)), \tag{25}$$

$$F'(x^{(k)}) = F'(\alpha)(I + 2A_2e^{(k)} + 3A_3(e^{(k)})^2 + 4A_4(e^{(k)})^3 + 5A_5(e^{(k)})^4 + O((e^{(k)})^5)), \tag{26}$$

$$F''(x^{(k)}) = F'(\alpha)(2A_2 + 6A_3e^{(k)} + 12A_4(e^{(k)})^2 + 20A_5(e^{(k)})^3 + O((e^{(k)})^4)) \tag{27}$$

and

$$F'''(x^{(k)}) = F'(\alpha)(6A_3 + 24A_4e^{(k)} + 60A_5(e^{(k)})^2 + O((e^{(k)})^3)), \tag{28}$$

where $A_i = \frac{1}{i!}\Gamma F^{(i)}(\alpha) \in L_i(\mathbb{R}^m, \mathbb{R}^m)$ and $(e^{(k)})^i = (e^{(k)}, e^{(k)}, \stackrel{i-times}{\cdots}, e^{(k)})$, $e^{(k)} \in \mathbb{R}^m$.

We are in a situation to analyze the behavior of Traub-Steffensen-like method. Thus, the following theorem is proved:

Theorem 1. *Let $F : D \subset \mathbb{R}^m \to \mathbb{R}^m$ be sufficiently Fréchet differentiable at each point of an open convex set D of $\alpha \in \mathbb{R}^m$, which is a solution of $F(x) = 0$. Assume that $x \in S = \bar{S}(\alpha, \epsilon)$ and $F'(x)$ is continuous and nonsingular at α, and $x^{(0)}$ close to α. Then, α is a point of attraction of the sequence $\{x^{(k)}\}$ generated by the Traub-Steffensen-like method (22). Furthermore, the sequence so developed converges locally to α with order at least 5.*

Proof. First we show that α is a point of attraction of Traub-Steffensen-like iteration. In this case, we have that

$$P(x) = z(x),\ \ Q(x) = H(x)[w, x; F]^{-1},\ \ R(x) = F(z(x)).$$

Now, since $F(\alpha) = 0$, $[\alpha, \alpha; F] = F'(\alpha) \neq O$, we have

$$y(\alpha) = \alpha - [\alpha, \alpha; F]^{-1}F(\alpha) = \alpha - F'(\alpha)^{-1}F(\alpha) = \alpha,$$

$$z(\alpha) = \alpha - [\alpha, \alpha; F]^{-1}F(\alpha) - [\alpha, \alpha; F]^{-1}F(\alpha) = \alpha - F'(\alpha)^{-1}F(\alpha) - F'(\alpha)^{-1}F(\alpha) = \alpha,$$

$$H(\alpha) = 2I - [\alpha, \alpha; F]^{-1}[\alpha, \alpha; F] = I,$$

$$P(\alpha) = z(\alpha),\ P'(\alpha) = z'(\alpha),$$

$$
\begin{aligned}
Q(\alpha) = H(\alpha)[\alpha,\alpha\,;F]^{-1} = I\,[\alpha,\alpha\,;F]^{-1} = [\alpha,\alpha\,;F]^{-1} = F'(\alpha)^{-1} \neq O,\\
R(\alpha) = F(z(\alpha)) = F(\alpha) = 0,\\
R'(\alpha) = F'(z(\alpha))z'(\alpha) = F'(\alpha)z'(\alpha),\\
M'(\alpha) = P'(\alpha) - Q(\alpha)R'(\alpha) = z'(\alpha) - F'(\alpha)^{-1}F'(\alpha)z'(\alpha) = O,
\end{aligned}
$$

so that $\rho(M'(\alpha)) = 0 < 1$ and by Lemma 1, α is a point of attraction of (22).

Let $e_w^{(k)} = w^{(k)} - \alpha = x^{(k)} + \beta F(x^{(k)}) - \alpha = e^{(k)} + \beta F(x^{(k)})$. Then using (25), it follows that

$$
e_w^{(k)} = (I + \beta F'(\alpha))e^{(k)} + \beta F'(\alpha)((A_2(e^{(k)})^2 + A_3(e^{(k)})^3) + O((e^{(k)})^4). \tag{29}
$$

Setting $x + h = w^{(k)}$, $x = x^{(k)}$, $h = e_w^{(k)} - e^{(k)}$ in Equation (24) and then using (26)–(29), we can write

$$
\begin{aligned}
[w^{(k)}, x^{(k)}\,;F] = F'(\alpha)(I + X_1A_2e^{(k)} + (\lambda A_2^2 + X_2A_3)(e^{(k)})^2 + X_1(2\lambda A_2A_3\\
+ X_3A_4)(e^{(k)})^3 + O((e^{(k)})^4)),
\end{aligned} \tag{30}
$$

where $\lambda = \beta F'(\alpha)$, $X_1 = \lambda + 2$, $X_2 = \lambda^2 + 3\lambda + 3$ and $X_3 = \lambda^2 + 2\lambda + 2$.

Expansion of the inverse of preceding divided difference operator is given as

$$
\begin{aligned}
[w^{(k)}, x^{(k)}\,;F]^{-1} = (I - X_1A_2(e^{(k)}) + ((1 + X_2)A_2^2 - X_2A_3)(e^{(k)})^2 - X_1((2 + X_3)A_2^3\\
- 2(1 + X_3)A_2A_3 + X_3A_4)(e^{(k)})^3 + O((e^{(k)})^3))\Gamma.
\end{aligned} \tag{31}
$$

By using (25) and (31) in the first step of method (22), we get

$$
e_y^{(k)} = y^{(k)} - \alpha = (-1 + X_1)A_2(e^{(k)})^2 - (X_3A_2^2 + (1 - X_2)A_3)(e^{(k)})^3 + O((e^{(k)})^4). \tag{32}
$$

Taylor expansion of $F(y^k)$ about α yields,

$$
F(y^{(k)}) = F'(\alpha)(e_y^{(k)} + A_2(e_y^{(k)})^2 + O((e_y^{(k)})^3)). \tag{33}
$$

From the second step of (22), on using (31) and (33), it follows that

$$
\begin{aligned}
e_z^{(k)} &= z^{(k)} - \alpha\\
&= X_1A_2(e^{(k)})e_y^{(k)} - A_2(e_y^{(k)})^2 - ((1 + X_2)A_2^2 - X_2A_3)(e^{(k)})^2e_y^{(k)} + O((e^{(k)})^5).
\end{aligned} \tag{34}
$$

By Taylor expansion of $F(z^k)$ about α,

$$
F(z^{(k)}) = F'(\alpha)(e_z^{(k)} + A_2(e_z^{(k)})^2 + O((e_z^{(k)})^3)). \tag{35}
$$

Equation (24), for $x + h = z^{(k)}$, $x = y^{(k)}$ and $h = e_z^{(k)} - e_y^{(k)}$, yields

$$
\begin{aligned}
[z^{(k)}, y^{(k)}\,;F] &= F'(\alpha)(I + A_2(e_z^{(k)} + e_y^{(k)}) + O((e^{(k)})^3))\\
&= F'(\alpha)(I + (\lambda + 1)A_2^2(e^{(k)})^2 + O((e^{(k)})^3)).
\end{aligned} \tag{36}
$$

From (31) and (36), we have

$$
\begin{aligned}
H(x^{(k)}) &= 2I - [w^{(k)}, x^{(k)}\,;F]^{-1}[z^{(k)}, y^{(k)}\,;F]\\
&= I + X_1A_2e^{(k)} + (X_2A_3 - (X_1 + X_2)A_2^2)(e^{(k)})^2 + O((e^{(k)})^3).
\end{aligned} \tag{37}
$$

Equations (31) and (37) yield

$$H(x^{(k)})[w^{(k)},x^{(k)};F]^{-1} = (I-(\lambda^2+5\lambda+5)A_2^2(e^{(k)})^2+O((e^{(k)})^3))\Gamma. \quad (38)$$

Applying Equations (34), (35) and (38) in the last step of method (22) and then simplifying, we get the error equation

$$e^{(k+1)} = (\lambda+1)(\lambda+2)(\lambda^2+5\lambda+5)A_2^4(e^{(k)})^5+O((e^{(k)})^6). \quad (39)$$

This completes the proof of Theorem 1. □

Thus, the Traub-Steffensen-like method (22) defines a one-parameter (β) family of derivative-free fifth order methods. Now onwards we denote it by $M_{5,1}$. In terms of computational cost $M_{5,1}$ utilizes four functions, two divided difference and one matrix inversion per each step. In the next section we will compare the computational efficiency of the new method with the existing derivative-free methods.

4. Computational Efficiency

In order to find the computational efficiency we will use the definition given in Section 2.3. The various evaluations and arithmetic operations that contribute towards the cost of computation are described as follows. For the computation of F in any iterative function we evaluate m scalar functions f_i, $(1 \le i \le m)$ and when computing a divided difference $[x,y;F]$ (see, Section 2.2) we evaluate $m(m-1)$ scalar functions, wherein $F(x)$ and $F(y)$ are evaluated separately. Furthermore, one has to add m^2 divisions from any divided difference. For the computation of inverse linear operator, a linear system can be solved that requires $m(m-1)(2m-1)/6$ products and $m(m-1)/2$ divisions in the LU decomposition process, and $m(m-1)$ products and m divisions in the resolution of two triangular linear systems. Moreover, we add m products for the multiplication of a vector by a scalar and m^2 products for multiplying a matrix by a vector or of a matrix by a scalar.

The comparison of computational efficiency of the present method $M_{5,1}$ is drawn with second order method $M_{2,1}$; third order method $M_{3,1}$; fourth order methods by Ren et al. [11], Grau et al. [13] and Sharma-Arora [14]; fifth order method by Kumar et al. [25]; sixth order method by Grau et al. [13]; seventh order methods by Sharma-Arora [15] and Wang-Zhang [16]. These methods are expressed as follows:

Fourth order method by Ren et al. ($M_{4,1}$):

$$\begin{aligned} y^{(k)} &= x^{(k)} - [u^{(k)},x^{(k)};F]^{-1}F(x^{(k)}), \\ x^{(k+1)} &= y^{(k)} - ([y^{(k)},x^{(k)};F]+[y^{(k)},u^{(k)};F]-[u^{(k)},x^{(k)};F])^{-1}F(y^{(k)}), \end{aligned}$$

where $u^{(k)} = x^{(k)}+F(x^{(k)})$.

Fourth order method by Grau et al. ($M_{4,2}$):

$$\begin{aligned} y^{(k)} &= x^{(k)} - [u^{(k)},v^{(k)};F]^{-1}F(x^{(k)}) \\ x^{(k+1)} &= y^{(k)} - (2[y^{(k)},x^{(k)};F]-[u^{(k)},v^{(k)};F])^{-1}F(y^{(k)}), \end{aligned}$$

where $u = x+F(x)$ and $v = x-F(x)$.

Sharma-Arora fourth order method ($M_{4,3}$):

$$\begin{aligned} y^{(k)} &= x^{(k)} - [w^{(k)},x^{(k)};F]^{-1}F(x^{(k)}) \\ x^{(k+1)} &= y^{(k)} - (3I-[w^{(k)},x^{(k)};F]^{-1}([y^{(k)},x^{(k)};F]+[y^{(k)},w^{(k)};F])) \\ &\quad \times [w^{(k)},x^{(k)};F]^{-1}F(y^{(k)}), \end{aligned}$$

where $w^{(k)} = x^{(k)} + \beta F(x^{(k)})$, β is a non-zero constant.

Fifth order method by Kumar et al. ($M_{5,2}$):

$$\begin{aligned}
y^{(k)} &= x^{(k)} - [w^{(k)}, x^{(k)}; F]^{-1} F(x^{(k)}) \\
z^{(k)} &= y^{(k)} - [w^{(k)}, x^{(k)}; F]^{-1} F(y^{(k)}) \\
x^{(k+1)} &= z^{(k)} - [x^{(k)}, y^{(k)}; F]^{-1} [w^{(k)}, x^{(k)}; F] [w^{(k)}, y^{(k)}; F]^{-1} F(z^{(k)}),
\end{aligned}$$

where $w^{(k)} = x^{(k)} + F(x^{(k)})$.

Sixth order method by Grau et al. ($M_{6,1}$):

$$\begin{aligned}
y^{(k)} &= x^{(k)} - [u^{(k)}, v^{(k)}; F]^{-1} F(x^{(k)}) \\
z^{(k)} &= y^{(k)} - \left(2[y^{(k)}, x^{(k)}; F] - [u^{(k)}, v^{(k)}; F]\right)^{-1} F(y^{(k)}) \\
x^{(k+1)} &= z^{(k)} - \left(2[y^{(k)}, x^{(k)}; F] - [u^{(k)}, v^{(k)}; F]\right)^{-1} F(z^{(k)}).
\end{aligned}$$

Wang-Zhang seventh order method ($M_{7,1}$):

$$\begin{aligned}
y^{(k)} &= x^{(k)} - [u^{(k)}, x^{(k)}; F]^{-1} F(x^{(k)}), \\
z^{(k)} &= y^{(k)} - \left([y^{(k)}, x^{(k)}; F] + [y^{(k)}, u^{(k)}; F] - [u^{(k)}, x^{(k)}; F]\right)^{-1} F(y^{(k)}) \\
x^{(k+1)} &= z^{(k)} - \left([z^{(k)}, x^{(k)}; F] + [z^{(k)}, y^{(k)}; F] - [y^{(k)}, x^{(k)}; F]\right)^{-1} F(z^{(k)}),
\end{aligned}$$

where $u^{(k)} = x^{(k)} + F(x^{(k)})$.

Sharma-Arora seventh order method ($M_{7,2}$):

$$\begin{aligned}
y^{(k)} &= x^{(k)} - [w^{(k)}, x^{(k)}; F]^{-1} F(x^{(k)}) \\
z^{(k)} &= y^{(k)} - \left(3I - [w^{(k)}, x^{(k)}; F]^{-1}([y^{(k)}, x^{(k)}; F] + [y^{(k)}, w^{(k)}; F])\right) \\
&\qquad \times [w^{(k)}, x^{(k)}; F]^{-1} F(y^{(k)}) \\
x^{(k+1)} &= z^{(k)} - [z^{(k)}, y^{(k)}; F]^{-1}\left([w^{(k)}, x^{(k)}; F] + [y^{(k)}, x^{(k)}; F] - [z^{(k)}, x^{(k)}; F]\right) \\
&\qquad \times [w^{(k)}, x^{(k)}; F]^{-1} F(z^{(k)}).
\end{aligned}$$

Let us denote efficiency indices of the methods $M_{p,i}$ by $E_{p,i}$ and their computational costs by $C_{p,i}$. Then, using the definition of the Section 2.3 taking into account the above considerations of evaluations and operations, we have that

$$C_{2,1} = \frac{1}{3}m^3 + 3m^2 + \frac{2}{3}m \quad \text{and} \quad E_{2,1} = 2^{1/C_{2,1}}. \tag{40}$$

$$C_{3,1} = \frac{1}{3}m^3 + 4m^2 + \frac{5}{3}m \quad \text{and} \quad E_{3,1} = 3^{1/C_{3,1}}. \tag{41}$$

$$C_{4,1} = \frac{2}{3}m^3 + 8m^2 - \frac{2}{3}m \quad \text{and} \quad E_{4,1} = 4^{1/C_{4,1}}. \tag{42}$$

$$C_{4,2} = \frac{2}{3}m^3 + 7m^2 + \frac{4}{3}m \quad \text{and} \quad E_{4,2} = 4^{1/C_{4,2}}. \tag{43}$$

$$C_{4,3} = \frac{1}{3}m^3 + 10m^2 + \frac{2}{3}m \quad \text{and} \quad E_{4,3} = 4^{1/C_{4,3}}. \tag{44}$$

$$C_{5,1} = \frac{1}{3}m^3 + 9m^2 + \frac{8}{3}m \quad \text{and} \quad E_{5,1} = 5^{1/C_{5,1}}. \tag{45}$$

$$C_{5,2} = m^3 + 11m^2 \quad \text{and} \quad E_{5,2} = 5^{1/C_{5,2}}. \tag{46}$$

$$C_{6,1} = \frac{2}{3}m^3 + 8m^2 + \frac{7}{3}m \quad \text{and} \quad E_{6,1} = 6^{1/C_{6,1}}. \tag{47}$$

$$C_{7,1} = m^3 + 13m^2 - 2m \quad \text{and} \quad E_{7,1} = 7^{1/C_{7,1}}. \tag{48}$$

$$C_{7,2} = \frac{2}{3}m^3 + 17m^2 - \frac{2}{3}m \quad \text{and} \quad E_{7,2} = 7^{1/C_{7,2}}. \tag{49}$$

To compare the efficiency of considered iterative methods, say $M_{p,i}$ against $M_{q,j}$, we consider the ratio

$$R_{p,i;q,j} = \frac{\log E_{p,i}}{\log E_{q,j}} = \frac{C_{q,j} \log(p)}{C_{p,i} \log(q)}. \tag{50}$$

It is clear that when $R_{p,i;q,j} > 1$, the iterative method $M_{p,i}$ is more efficient than $M_{q,j}$.

$M_{3,1}$ *versus* $M_{2,1}$ *case*:

For this case the ratio (50) is given by

$$R_{3,1;2,1} = \frac{(\frac{1}{3}m^3 + 3m^2 + \frac{2}{3}m)\log(3)}{(\frac{1}{3}m^3 + 4m^2 + \frac{5}{3}m)\log(2)}.$$

It can be easily shown that $R_{3,1;2,1} > 1$ for $m \geq 2$. This implies that $E_{3,1} > E_{2,1}$ for $m \geq 2$. Thus, $M_{3,1}$ is more efficient than $M_{2,1}$ as we have stated in the introduction section.

$M_{5,1}$ *versus* $M_{2,1}$ *case*:

The ratio (50) is given by

$$R_{5,1;2,1} = \frac{(\frac{1}{3}m^3 + 3m^2 + \frac{2}{3}m)\log(5)}{(\frac{1}{3}m^3 + 9m^2 + \frac{8}{3}m)\log(2)}.$$

It is easy to prove that $R_{5,1;2,1} > 1$ for $m \geq 6$. Thus, we conclude that $E_{5,1} > E_{2,1}$ for $m \geq 6$.

$M_{5,1}$ *versus* $M_{3,1}$ *case*:

The ratio (50) is given by

$$R_{5,1;3,1} = \frac{(\frac{1}{3}m^3 + 4m^2 + \frac{5}{3}m)\log(5)}{(\frac{1}{3}m^3 + 9m^2 + \frac{8}{3}m)\log(3)}.$$

It can be checked that $R_{5,1;3,1} > 1$ for $m \geq 21$. Thus, we have that $E_{5,1} > E_{3,1}$ for $m \geq 21$.

$M_{5,1}$ *versus* $M_{4,1}$ *case*:

In this case the ratio

$$R_{5,1;4,1} = \frac{(\frac{2}{3}m^3 + 8m^2 - \frac{2}{3}m)\log(5)}{(\frac{1}{3}m^3 + 9m^2 + \frac{8}{3}m)\log(4)} > 1,$$

for $m \geq 3$, which implies that $E_{5,1} > E_{4,1}$ for $m \geq 3$.

$M_{5,1}$ *versus* $M_{4,2}$ *case*:

Here the ratio

$$R_{5,1;4,2} = \frac{(\frac{2}{3}m^3 + 7m^2 + \frac{4}{3}m)\log(5)}{(\frac{1}{3}m^3 + 9m^2 + \frac{8}{3}m)\log(4)} > 1,$$

for $m \geq 3$ which implies that $E_{5,1} > E_{4,2}$ for $m \geq 3$.

$M_{5,1}$ *versus* $M_{4,3}$ *case*:

Here the ratio

$$R_{5,1;4,3} = \frac{(\frac{1}{3}m^3 + 10m^2 + \frac{2}{3}m)\log(5)}{(\frac{1}{3}m^3 + 9m^2 + \frac{8}{3}m)\log(4)} > 1,$$

for $m \geq 2$ which implies that $E_{5,1} > E_{4,3}$ for $m \geq 2$.

$M_{5,1}$ *versus* $M_{5,2}$ *case*:

In this case the ratio

$$R_{5,1;5,2} = \frac{m^3 + 11m^2}{\frac{1}{3}m^3 + 9m^2 + \frac{8}{3}m} > 1,$$

for $m \geq 2$ which means $E_{5,1} > E_{5,2}$ for $m \geq 2$.

$M_{5,1}$ *versus* $M_{6,1}$ *case*:

Here the ratio

$$R_{5,1;6,1} = \frac{(\frac{2}{3}m^3 + 8m^2 + \frac{7}{3}m)\log(5)}{(\frac{1}{3}m^3 + 9m^2 + \frac{8}{3}m)\log(6)} > 1,$$

for $m \geq 8$ which means $E_{5,1} > E_{6,1}$ for $m \geq 8$.

$M_{5,1}$ *versus* $M_{7,1}$ *case*:

Here also the ratio

$$R_{5,1;7,1} = \frac{(m^3 + 13m^2 - 2m)\log(5)}{(\frac{1}{3}m^3 + 9m^2 + \frac{8}{3}m)\log(7)} > 1,$$

for $m \geq 2$ which means $E_{5,1} > E_{7,1}$ for $m \geq 2$.

$M_{5,1}$ *versus* $M_{7,2}$ *case*:

Here also the ratio

$$R_{5,1;7,2} = \frac{(\frac{2}{3}m^3 + 17m^2 - \frac{2}{3}m)\log(5)}{(\frac{1}{3}m^3 + 9m^2 + \frac{8}{3}m)\log(7)} > 1,$$

for $m \geq 2$ which means $E_{5,1} > E_{7,2}$ for $m \geq 2$.

The above results are summarized in the following theorem:

Theorem 2. *We have that*

(a) $E_{5,1} > E_{2,1}$ *for* $m \geq 6$.
(b) $E_{5,1} > E_{3,1}$ *for* $m \geq 21$.
(c) $\{E_{5,1} > E_{4,1}\ E_{5,1} > E_{4,2}\}$ *for* $m \geq 3$.
(d) $\{E_{3,1} > E_{2,1},\ E_{5,1} > E_{4,3},\ E_{5,1} > E_{5,2},\ E_{5,1} > E_{7,1},\ E_{5,1} > E_{7,2}\}$ *for* $m \geq 2$.
(e) $E_{5,1} > E_{6,1}$ *for* $m \geq 8$.

5. Complex Dynamics of Methods

Our aim is to analyze the complex dynamics of the new method based on graphical tool 'basins of attraction' of the zeros of polynomial $P(z)$ in complex plane. Visual display of the basins gives important information about the stability and convergence of iterative methods. This idea was introduced initially by Vrscay and Gilbert [26]. In recent times, many authors have used this concept in their work, see, for example [27,28] and references therein. We consider the method (22) to analyze the basins of attraction.

To start with we take the initial point z_0 in a rectangular region $R \in \mathbb{C}$ that contains all the zeros of a polynomial $P(z)$. The iterative method, when starting from point z_0 in a rectangle, either converges to the zero $P(z)$ or eventually diverges. Stopping condition for convergence is considered as 10^{-3} to a maximum of 25 iterations. If the required tolerance is not achieved in 25 iterations, we conclude that the iterative scheme starting at point z_0 does not converge to any root. The strategy adopted is as follows: A color is allocated to each initial point z_0 in the basin of attraction of a zero. If the iteration initiating at z_0 converges, then it represents the attraction basin with that assigned color to it, otherwise in the failing (divergence) situation in 25 iterations the iteration represents the black color.

We analyze the basins of attraction of the new method (for the choices $\beta = 10^{-2}, 10^{-4}, 10^{-8}$) on following three polynomials:

Example 1. *In the first case, consider the polynomial* $P_1(z) = z^2 - 1$ *which has zeros* $\{\pm 1\}$*. A grid of* 400×400 *points in a rectangle* $D \in \mathbb{C}$ *of size* $[-2,2] \times [-2,2]$ *is used for drawing the graphics. We assign the color red to each initial point in the basin of attraction of zero '1' and the color green to the points in the basin of attraction of zero '−1'. The graphics are shown in Figure 1 corresponding to* $\beta = 10^{-2}, 10^{-4}, 10^{-8}$*. Observing the behavior of the basins of the new method, we conclude that the convergence domain becoming wider as parameter* β *assumes smaller values since black zones (divergent points) are getting smaller in size.*

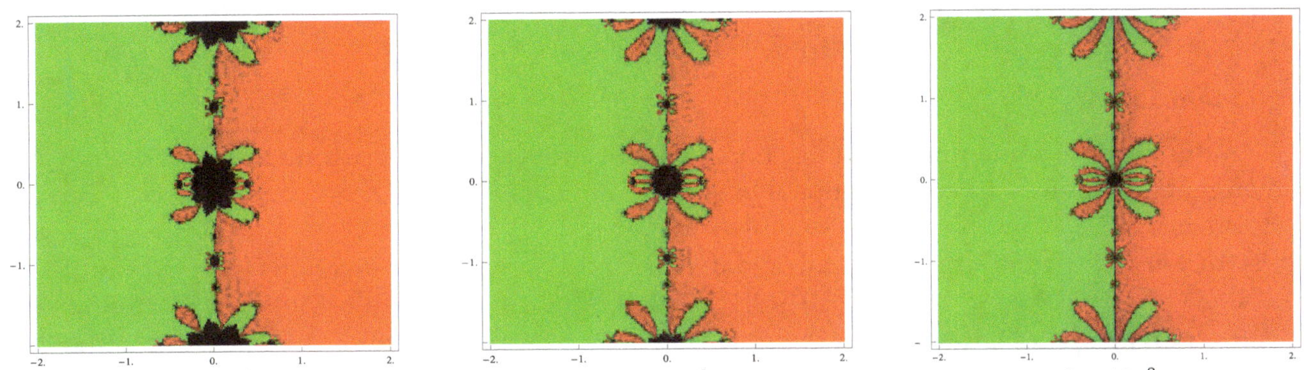

$\beta = 10^{-2}$ $\beta = 10^{-4}$ $\beta = 10^{-8}$

Figure 1. Basins of attraction for polynomial $P_1(z)$.

Example 2. *Let us consider the next polynomial as* $P_2(z) = z^3 - z$ *having zeros* $\{0, \pm 1\}$*. To draw the dynamical view, we select a rectangle* $D = [-2,2] \times [-2,2] \in \mathbb{C}$ *containing* 400×400 *grid points. Then, allocate the colors green, blue and red to each point in the basin of attraction of* 0, 1 *and* -1*, respectively. Basins for this example are exhibited in Figure 2 corresponding to parameter choices* $\beta = 10^{-2}, 10^{-4}, 10^{-8}$ *in the proposed methods. In addition, observe that the basins are becoming larger and larger with the smaller values of* β*.*

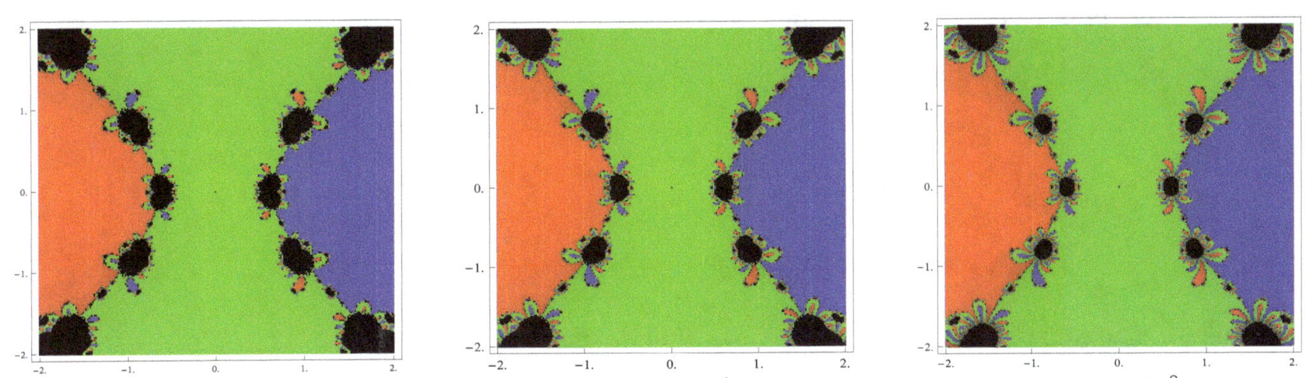

$\beta = 10^{-2}$ $\beta = 10^{-4}$ $\beta = 10^{-8}$

Figure 2. Basins of attraction for polynomial $P_2(z)$.

Example 3. *Lastly, we consider the polynomial as* $P_3(z) = z^5 + 2z - 1$ *having zeros* $\{-0.945068 \pm 0.854518i, 0.701874 \pm 0.879697i, 0.486389\}$*. To draw the dynamical view, we select a rectangle* $D = [-2,2] \times [-2,2] \in \mathbb{C}$ *containing* 400×400 *grid points. Then, allocate the colors green, blue, red, yellow and pink to each point in the basin of attraction of* $0.701874 + 0.879697i$, $-0.945068 - 0.854518i$, $0.701874 - 0.879697i$, 0.486389 *and* $-0.945068 + 0.854518i$*, respectively. Basins for this example are exhibited in Figure 3 corresponding to parameter choices* $\beta = 10^{-2}, 10^{-4}, 10^{-8}$ *in the proposed methods. We observe that the basins are getting larger with the smaller values of* β*.*

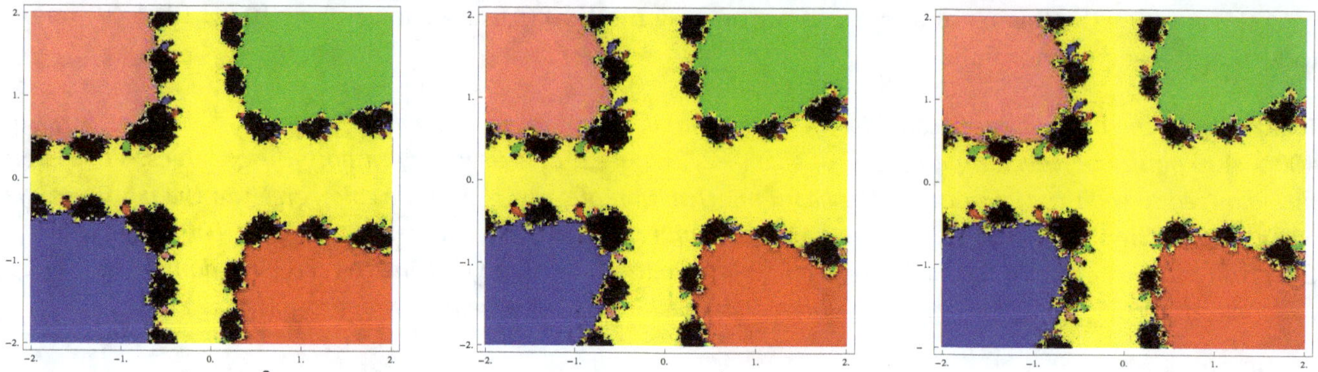

$\beta = 10^{-2}$ $\beta = 10^{-4}$ $\beta = 10^{-8}$

Figure 3. Basins of attraction for polynomial $P_3(z)$.

6. Numerical Tests

In this section, some numerical tests on different problems are performed to demonstrate the convergence behavior and computational efficiency of the method $M_{5,1}$. A comparison between the performance of $M_{5,1}$ with the existing methods $M_{2,1}$, $M_{3,1}$, $M_{4,j}$ $(j = 1, 2, 3)$, $M_{5,2}$, $M_{6,1}$, $M_{7,1}$ and $M_{7,2}$ is also drawn. The programs are performed in the processor with specifications Intel (R) Core (TM) i5-4210U CPU @ 1.70 GHz 2.40 GHz (64-bit Operating System) Microsoft Windows 10 Professional and are complied by *Mathematica* 10.0 using multiple-precision arithmetic. We record the number of iterations (k) required to converge to the solution such that the stopping condition

$$||x^{(k+1)} - x^{(k)}|| + ||F(x^{(k)})|| < 10^{-300}$$

is satisfied. In order to verify the theoretical order of convergence, the computational order of convergence (p_c) is obtained by using the Formula (8). In the comparison of performance of considered methods, we also include the real CPU time elapsed during the execution of program computed by the *Mathematica* command "TimeUsed[ ]".

The methods $M_{2,1}$, $M_{3,1}$, $M_{4,3}$, $M_{5,1}$ and $M_{7,2}$ are tested by using the value 0.01 for the parameter β. In numerical experiments we consider the following five problems:

Example 4. *Let us consider the system of two equations (selected from [29]):*

$$\begin{cases} x^2 + \sin x - e^y = 0, \\ 3x - \cos x - y = 0. \end{cases}$$

The initial guess assumed is $x^{(0)} = \{-1, -2\}^T$ *for obtaining the solution*

$$\alpha = \{-0.90743021707369569\ldots, -3.3380632251862363\ldots\}^T.$$

Example 5. *Now considering the mixed Hammerstein integral equation (see [4]):*

$$x(s) = 1 + \frac{1}{5}\int_0^1 G(s,t)x(t)^3 dt,$$

wherein $x \in C[0,1]$; $s, t \in [0,1]$ *and the kernel G is*

$$G(s,t) = \begin{cases} (1-s)t, & t \leq s, \\ s(1-t), & s \leq t. \end{cases}$$

The above equation is transformed to a finite-dimensional problem by using the Gauss-Legendre quadrature formula

$$\int_0^1 f(t)dt \approx \sum_{j=1}^{m} \varpi_j f(t_j),$$

where the weights ϖ_j and abscissas t_j are obtained for $m = 8$ by Gauss-Legendre quadrature formula. Then, setting $x(t_i) = x_i$, $i = 1, 2, \ldots\ldots, 8$, we obtain the following system of nonlinear equations

$$5\,x_i - 5 - \sum_{j=1}^{8} a_{ij} x_j^3 = 0,$$

where

$$a_{ij} = \begin{cases} \varpi_j t_j (1 - t_i) & \text{if } j \leq i, \\ \varpi_j t_i (1 - t_j) & \text{if } i < j, \end{cases} \qquad i = 1, 2, \ldots\ldots 8.$$

wherein the abscissas t_j and the weights ϖ_j are known and produced in Table 1 for $m = 8$. The initial approximation assumed is

$$x^{(0)} = \{-1, -1, -1, -1, -1, -1, -1, -1\}^T$$

and the solution of this problem is:

$$\alpha = \{1.002096245031..., 1.009900316187..., 1.019726960993..., 1.026435743030..., \\ 1.026435743030..., 1.019726960993..., 1.009900316187..., 1.002096245031...\}^T.$$

Table 1. Weights and abscissas of Gauss-Legendre quadrature formula for $m = 8$.

j	t_j	ϖ_j
1	0.019855071751231884158219565715...	0.050614268145188129576265677154...
2	0.101666761293186630204223031762...	0.111190517226687235272177997213...
3	0.237233795041835507091130475405...	0.156853322938943643668981100993...
4	0.408282678752175097530261928819...	0.181341891689180991482575224638...
5	0.591717321247824902469738071181...	0.181341891689180991482575224638...
6	0.762766204958164492908869524595...	0.156853322938943643668981100993...
7	0.898333238706813369795776968238...	0.111190517226687235272177997213...
8	0.980144928248768115841780434285...	0.050614268145188129576265677154...

Example 6. *Consider the system of 20 equations (see [29]):*

$$\tan^{-1}(x_i) + 1 - 2 \sum_{j=1, j \neq i}^{20} x_j^2 = 0, \quad 1 \leq i \leq 20,$$

This problem has the following two solutions:

$$\alpha_1 = \{0.1757683176158\ldots, 0.1757683176158\ldots, \cdots\cdots, 0.1757683176158\ldots\}^T.$$

and

$$\alpha_2 = \{-0.14968543422\ldots, -0.14968543422, \ldots, \cdots\cdots, -0.14968543422\ldots\}^T.$$

We intend to find the first solution and so choose the initial value: $x^{(0)} = \{0.5, 0.5, 0.5, \cdots\cdots, 0.5\}^T$.

Example 7. *Consider the boundary value problem:*

$$y'' + y^3 = 0, \quad y(0) = 0, \quad y(1) = 1.$$

Assuming the following partitioning of the interval $[0, 1]$*:*
$u_0 = 0 < u_1 < u_2 < \cdots < u_{n-1} < u_n = 1, \quad u_{j+1} = u_j + h, \quad h = 1/n.$

Setting $y_0 = y(u_0) = 0, y_1 = y(u_1), \cdots, y_{n-1} = y(u_{n-1}), y_n = y(u_n) = 1$. *If we discretize the problem by using the finite difference approximation for second derivative*

$$y''_m = \frac{y_{m-1} - 2y_m + y_{m+1}}{h^2}, \quad m = 1, 2, 3, \ldots, n-1,$$

we obtain a system of $n-1$ *equations in* $n-1$ *variables:*

$$y_{m-1} - 2y_m + y_{m+1} + h^2 y_m^3 = 0, \quad m = 1, 2, 3, \ldots, n-1.$$

In particular, let us solve this problem for $n = 51$*, that is for* $m = 50$ *by choosing* $y^{(0)} = \{-1, -1, -1, \cdots, -1\}^T$ *as the initial value. The solution vector* α *of this problem is*

$$\begin{aligned}
\{&0.02071138910\ldots, 0.04142277479\ldots, 0.06213413315\ldots, 0.08284539929\ldots, 0.10355644682\ldots,\\
&0.12426706739\ldots, 0.14497695018\ldots, 0.16568566142\ldots, 0.18639262397\ldots, 0.20709709683\ldots,\\
&0.22779815476\ldots, 0.24849466794\ldots, 0.26918528167\ldots, 0.28986839623\ldots, 0.31054214677\ldots,\\
&0.33120438344\ldots, 0.35185265167\ldots, 0.37248417270\ldots, 0.39309582441\ldots, 0.41368412246\ldots,\\
&0.43424520189\ldots, 0.45477479913\ldots, 0.47526823468\ldots, 0.49572039629\ldots, 0.51612572294\ldots,\\
&0.53647818972\ldots, 0.55677129350\ldots, 0.57699803975\ldots, 0.59715093054\ldots, 0.61722195374\ldots,\\
&0.63720257375\ldots, 0.65708372374\ldots, 0.67685579959\ldots, 0.69650865572\ldots, 0.71603160287\ldots,\\
&0.73541340802\ldots, 0.75464229671\ldots, 0.77370595761\ldots, 0.79259154985\ldots, 0.81128571300\ldots,\\
&0.82977457984\ldots, 0.84804379222\ldots, 0.86607851992\ldots, 0.88386348269\ldots, 0.90138297559\ldots,\\
&0.91862089765\ldots, 0.93556078378\ldots, 0.95218584022\ldots, 0.96847898326\ldots, 0.98442288125\ldots\}^T.
\end{aligned}$$

Example 8. *Consider the following Burger's equation (see [30]):*

$$\frac{\partial^2 f}{\partial u^2} + f\frac{\partial f}{\partial u} - \frac{\partial f}{\partial t} + g(u, t) = 0, \ (u, t) \in [0, 1]^2,$$

where $g(u, t) = -10e^{-2t}[e^t(2 - u + u^2) + 10u(1 - 3u + 2u^2)]$ *and function* $f = f(u, t)$ *satisfies the boundary conditions*

$$f(0, t) = f(1, t) = 0, \ f(u, 0) = 10u(u-1) \text{ and } f(u, 1) = 10u(u-1)/e.$$

Assuming the following partitioning of the domain $[0, 1]^2$*:*

$$\begin{aligned}
0 &= u_0 < u_1 < u_2 < \ldots\ldots\ldots < u_{n-1} < u_n = 1, \ u_{k+1} = u_k + h,\\
0 &= t_0 < t_1 < t_2 < \ldots\ldots\ldots < t_{n-1} < t_n = 1, \ t_{l+1} = t_l + h, \ h = 1/n.
\end{aligned}$$

Let us define $f_{k,l} = f(u_k, t_l)$ *and* $g_{k,l} = g(u_k, t_l)$ *for* $k, l = 0, 1, 2, \ldots\ldots.n$*. Then the boundary conditions would be* $f_{0,l} = f(u_0, t_l) = 0$*,* $f_{n,l} = f(u_n, t_l) = 0$*,* $f_{k,0} = f(u_k, t_0) = 10u_k(u_k - 1)$ *and* $f_{k,n} = f(u_k, t_n) = 10u_k(u_k - 1)/e$*. If we discretize Burger's equation by using the numerical formulas for the partial derivatives*

$$\left(\frac{\partial f}{\partial u}\right)_{i,j} = \frac{f_{i+1,j} - f_{i-1,j}}{2h}, \quad \left(\frac{\partial f}{\partial t}\right)_{i,j} = \frac{f_{i,j+1} - f_{i,j-1}}{2h},$$

$$\left(\frac{\partial^2 f}{\partial u^2}\right)_{i,j} = \frac{f_{i+1,j} - 2f_{i,j} + f_{i-1,j}}{h^2}, \quad i,j = 1,2,\ldots\ldots n-1,$$

then we obtain the following system of $(n-1)^2$ nonlinear equations in $(n-1)^2$ variables:

$$f_{i-1,j}(2 - hf_{i,j}) + h(f_{i,j-1} - f_{i,j+1}) - f_{i,j}(4 - hf_{i+1,j}) + 2f_{i+1,j} + 2h^2 g_{i,j} = 0, \tag{51}$$

where $i,j = 1,2\ldots\ldots n-1$. In particular, we solve this nonlinear system for $n = 11$ so that $m = 100$ by selecting $f_{i,j} = 1$ for $i,j = 1,2\ldots\ldots 10$ as the initial value. The solution of this system of nonlinear equations is given in Table 2.

Table 2. The solution of system (51) with the unknowns $f_{i,j}$ for $i,j = 1,2,\ldots\ldots 10$.

$f_{i,1}$	$f_{i,2}$	$f_{i,3}$	$f_{i,4}$	$f_{i,5}$	$f_{i,6}$	$f_{i,7}$	$f_{i,8}$	$f_{i,9}$	$f_{i,10}$
−0.7546...	−0.6892...	−0.6290...	−0.5750...	−0.5236...	−0.4817...	−0.4306...	−0.4214...	−0.2583...	−0.3068...
−1.3583...	−1.2405...	−1.1322...	−1.0351...	−0.9422...	−0.8675...	−0.7741...	−0.7598...	−0.4358...	−0.5505...
−1.8111...	−1.6541...	−1.5096...	−1.3803...	−1.2559...	−1.1573...	−1.0309...	−1.0110...	−0.6951...	−0.7356...
−2.1130...	−1.9298...	−1.7611...	−1.6106...	−1.4649...	−1.3511...	−1.2014...	−1.1768...	−0.8755...	−0.8606...
−2.2639...	−2.0678...	−1.8869...	−1.7258...	−1.5690...	−1.4485...	−1.2860...	−1.2582...	−0.9783...	−0.9244...
−2.2639...	−2.0678...	−1.8868...	−1.7261...	−1.5686...	−1.4494...	−1.2850...	−1.2558...	−1.0040...	−0.9268...
−2.1129...	−1.9300...	−1.7609...	−1.6112...	−1.4637...	−1.3534...	−1.1987...	−1.1700...	−0.9534...	−0.8672...
−1.8111...	−1.6544...	−1.5093...	−1.3812...	−1.2544...	−1.1604...	−1.0272...	−1.0010...	−0.8270...	−0.7454...
−1.3583...	−1.2408...	−1.1320...	−1.0359...	−0.9407...	−0.8704...	−0.7706...	−0.7492...	−0.6255...	−0.5609...
−0.7546...	−0.6893...	−0.6289...	−0.5755...	−0.5227...	−0.4834...	−0.4285...	−0.4152...	−0.3496...	−0.3128...

In Tables 3–7 we present the numerical results produced for the methods $M_{2,1}$, $M_{3,1}$, $M_{4,j}$ $(j = 1,2,3)$, $M_{5,1}$, $M_{5,2}$, $M_{6,1}$, $M_{7,1}$ and $M_{7,2}$. Displayed in each table are the errors $||x^{(k+1)} - x^{(k)}||$ of first three consecutive approximations to corresponding solution of Examples 4–8, number of iterations (k) needed to converge to the required solution, computational order of convergence p_c, computational cost $C_{p,i}$, computational efficiency $E_{p,i}$ and elapsed CPU-time (e-time) measured in seconds. In each table the meaning of $A(-h)$ is $A \times 10^{-h}$. Numerical values of computational cost and efficiency are obtained according to the corresponding expressions given by (40)–(49). The e-time is calculated by taking the average of 50 performances of the program, where we use $||x^{(k+1)} - x^{(k)}|| + ||F(x^{(k)})|| < 10^{-300}$ as the stopping condition in a single performance of the program.

Table 3. Comparison of performance of methods for Example 4.

Methods	$\lVert x^{(2)} - x^{(1)} \rVert$	$\lVert x^{(3)} - x^{(2)} \rVert$	$\lVert x^{(4)} - x^{(3)} \rVert$	k	p_c	$C_{p,i}$	$E_{p,i}$	**e-Time**
$M_{2,1}(\beta = 0.01)$	9.94(−2)	4.45(−3)	7.14(−6)	9	2.000	16	1.04427	0.2887
$M_{3,1}(\beta = 0.01)$	2.93(−2)	8.14(−6)	1.42(−16)	6	3.000	22	1.05120	0.2630
$M_{4,1}$	3.73(−4)	1.71(−16)	6.94(−66)	5	4.000	36	1.03926	0.3234
$M_{4,2}$	6.17(−2)	7.75(−7)	5.63(−27)	5	4.000	36	1.03926	0.3165
$M_{4,3}(\beta = 0.01)$	5.35(−3)	2.42(−10)	7.56(−40)	5	4.000	44	1.03201	0.3362
$M_{5,1}(\beta = 0.01)$	1.76(−3)	4.72(−15)	4.11(−73)	4	5.000	44	1.03726	0.3297
$M_{5,2}$	1.89(−3)	2.96(−15)	2.04(−74)	4	5.000	52	1.03143	0.3972
$M_{6,1}$	2.97(−2)	8.66(−12)	1.81(−69)	4	6.000	42	1.04358	0.3120
$M_{7,1}$	2.23(−7)	4.55(−52)	0.000	3	7.000	56	1.03536	0.3468
$M_{7,2}(\beta = 0.01)$	7.93(−5)	4.14(−31)	3.40(−215)	3	7.000	72	1.02740	0.4125

Table 4. Comparison of performance of methods for Example 5.

Methods	$\|\|x^{(2)}-x^{(1)}\|\|$	$\|\|x^{(3)}-x^{(2)}\|\|$	$\|\|x^{(4)}-x^{(3)}\|\|$	k	p_c	$C_{p,i}$	$E_{p,i}$	e-Time
$M_{2,1}(\beta=0.01)$	0.202	1.44(−3)	7.18(−8)	9	2.000	368	1.001885	0.3437
$M_{3,1}(\beta=0.01)$	1.73(−3)	1.24(−11)	4.56(−36)	5	3.000	440	1.002500	0.2252
$M_{4,1}$	0.276	6.19(−6)	1.36(−24)	5	4.000	848	1.001636	0.3532
$M_{4,2}$	9.94(−2)	4.12(−8)	1.23(−33)	5	4.000	800	1.001734	0.3562
$M_{4,3}(\beta=0.01)$	1.86(−2)	4.72(−11)	2.06(−45)	5	4.000	816	1.001700	0.2749
$M_{5,1}(\beta=0.01)$	1.20(−5)	3.49(−30)	7.35(−153)	4	5.000	768	1.002098	0.2312
$M_{5,2}$	4.50(−2)	1.98(−11)	3.74(−58)	5	5.000	1216	1.001324	0.4234
$M_{6,1}$	1.39(−2)	3.84(−17)	1.86(−104)	4	6.000	872	1.002057	0.3063
$M_{7,1}$	1.12(−2)	7.70(−21)	6.01(−148)	4	7.000	1328	1.001467	0.3862
$M_{7,2}(\beta=0.01)$	8.66(−5)	4.24(−36)	3.01(−255)	4	7.000	1424	1.001367	0.3625

Table 5. Comparison of performance of methods for Example 6.

Methods	$\|\|x^{(2)}-x^{(1)}\|\|$	$\|\|x^{(3)}-x^{(2)}\|\|$	$\|\|x^{(4)}-x^{(3)}\|\|$	k	p_c	$C_{p,i}$	$E_{p,i}$	e-Time
$M_{2,1}(\beta=0.01)$	0.336	5.87(−2)	2.05(−3)	10	2.000	3880	1.0001787	4.2530
$M_{3,1}(\beta=0.01)$	0.209	4.29(−3)	6.08(−8)	7	3.000	4300	1.0002555	3.3061
* $M_{4,1}$	0.370	2.50(−2)	1.37	18	4.000	8520	1.0001627	15.469
$M_{4,2}$	8.97(−2)	1.02(−5)	2.23(−21)	5	4.000	8160	1.0001699	3.4542
$M_{4,3}(\beta=0.01)$	0.133	2.18(−4)	2.85(−15)	6	4.000	6680	1.0002076	3.8634
$M_{5,1}(\beta=0.01)$	8.15(−2)	3.67(−6)	1.08(−27)	5	5.000	6320	1.0002547	3.3176
$M_{5,2}$	0.434	7.70(−2)	1.33(−2)	7	5.000	12,400	1.0001298	7.2054
$M_{6,1}$	3.16(−2)	1.38(−10)	1.12(−60)	4	6.000	8580	1.0002089	3.3585
* $M_{7,1}$	1.572	3.42(−4)	7.60(−25)	5	7.000	13,160	1.0001478	5.7346
$M_{7,2}(\beta=0.01)$	1.96(−2)	6.66(−13)	3.92(−86)	4	7.000	12,120	1.0001606	4.3594

* The methods $M_{4,1}$ and $M_{7,1}$ converge to the solution α_2.

Table 6. Comparison of performance of methods for Example 7.

Methods	$\|\|x^{(2)}-x^{(1)}\|\|$	$\|\|x^{(3)}-x^{(2)}\|\|$	$\|\|x^{(4)}-x^{(3)}\|\|$	k	p_c	$C_{p,i}$	$E_{p,i}$	e-Time
$M_{2,1}(\beta=0.01)$	3.828	0.681	1.23(−2)	9	2.000	49,200	1.00001409	0.8928
$M_{3,1}(\beta=0.01)$	0.433	9.62(−5)	1.74(−15)	6	3.000	51,750	1.00002123	0.7183
$M_{4,1}$	0.840	9.07(−5)	9.18(−21)	5	4.000	103,300	1.00001342	1.0475
$M_{4,2}$	0.848	9.54(−5)	1.13(−20)	5	4.000	100,900	1.00001374	1.1896
$M_{4,3}(\beta=0.01)$	1.548	3.85(−3)	5.75(−14)	6	4.000	66,700	1.00002078	0.8284
$M_{5,1}(\beta=0.01)$	4.06(−2)	1.22(−12)	2.82(−65)	4	5.000	64,300	1.00002503	0.6102
$M_{5,2}$	8.16(−2)	3.64(−11)	7.86(−58)	5	5.000	152,500	1.00001055	1.6563
$M_{6,1}$	0.159	1.20(−11)	2.24(−72)	6	6.000	103,450	1.00001732	1.0313
$M_{7,1}$	9.92(−2)	1.26(−15)	7.09(−113)	4	7.000	157,400	1.00001236	1.3457
$M_{7,2}(\beta=0.01)$	0.212	6.80(−13)	3.00(−93)	4	7.000	125,800	1.00001547	1.0938

Table 7. Comparison of performance of methods for Example 8.

Methods	$\|\|x^{(2)} - x^{(1)}\|\|$	$\|\|x^{(3)} - x^{(2)}\|\|$	$\|\|x^{(4)} - x^{(3)}\|\|$	k	p_c	$C_{p,i}$	$E_{p,i}$	e-Time
$M_{2,1}(\beta = 0.01)$	1.980	1.78(−2)	2.66(−6)	9	2.000	363,400	1.000001907	7.6572
$M_{3,1}(\beta = 0.01)$	0.951	1.31(−4)	4.43(−16)	6	3.000	373,500	1.000002941	5.9846
$M_{4,1}$	0.158	8.95(−8)	2.27(−25)	6	2.997	746,600	1.000001471	10.503
$M_{4,2}$	0.418	8.99(−6)	2.28(−19)	6	3.000	736,800	1.000001491	10.249
$M_{4,3}(\beta = 0.01)$	0.453	5.14(−6)	9.22(−21)	6	3.001	433,400	1.000002535	6.5627
$M_{5,1}(\beta = 0.01)$	0.137	3.08(−12)	3.28(−65)	4	5.001	423,600	1.000003815	4.8255
$M_{5,2}$	3.12(−2)	5.16(−13)	3.86(−55)	5	3.999	1,110,000	1.000001450	10.876
$M_{6,1}$	8.38(−2)	6.04(−11)	9.18(−46)	5	4.001	746,900	1.000001856	9.2656
$M_{7,1}$	1.75(−4)	1.74(−34)	4.89(−213)	4	5.999	1,129,800	1.000001586	10.235
$M_{7,2}(\beta = 0.01)$	3.83(−3)	5.44(−25)	1.65(−155)	4	5.996	836,600	1.000002142	8.4533

From the numerical results displayed in Tables 3–7, it can be observed that like that of the existing methods the proposed new method shows consistent convergence behavior. Seventh order methods produce approximations with large accuracy due to their higher order of convergence, but they are less efficient. In Example 6, $M_{4,1}$ and $M_{7,1}$ do not converge to the required solution α_1. Instead, they converge to solution α_2 which is far off from initial approximation chosen. Calculation of computational order of convergence shows that the order of convergence of the new method is preserved in all the numerical examples. However, this is not true for some existing methods, e.g., $M_{4,j}$ $(j = 1,2,3)$, $M_{5,2}$, $M_{6,1}$, $M_{7,1}$ and $M_{7,2}$, in Example 8. Values of the efficiency index shown in the penultimate column of each table also verify the theoretical results stated in Theorem 2. The efficiency results are also in complete agreement with the CPU time utilized in the execution of the program since the method with large efficiency uses less computing time than the method with small efficiency. Moreover, the proposed method utilizes less CPU time than existing higher order methods which points to the dominance of the method. In fact, the new method is especially more efficient for large systems of nonlinear equations.

7. Conclusions

In the foregoing study, we have developed a fifth order iterative method for approximating solution of systems of nonlinear equations. The methodology is based on third order Traub-Steffensen method and further developed by using derivative free modification of classical Chebyshev's method. The iterative scheme is totally derivative-free and so particularly suitable to those problems where derivatives are lengthy to compute. To prove the local fifth order of convergence for the new method, a development of first-order divided difference operator and direct computation by Taylor's expansion are used.

We have examined the computational efficiency of the new method. A comparison of efficiencies with that of the existing most efficient methods is also performed. It is proved that, in general, the new algorithm is more efficient. Numerical experiments are performed and the performance is compared with existing derivative-free methods. From numerical results it has been observed that the proposed method has equal or better convergence compared to existing methods. Theoretical results related to convergence order and computational efficiency have also been verified in the considered numerical problems. Similar numerical tests, performed for a variety of other different problems, have confirmed the above drawn conclusions to a good extent.

Author Contributions: Methodology, J.R.S.; writing, review and editing, J.R.S.; investigation, D.K.; data curation, D.K.; conceptualization, L.J; formal analysis, L.J.

References

1. Argyros, I.K. Quadratic equations and applications to Chandrasekhar's and related equations. *Bull. Aust. Math. Soc.* **1985**, *32*, 275–292. [CrossRef]
2. Kelley, C.T. *Iterative Methods for Linear and Nonlinear Equations*; SIAM: Philadelphia, PA, USA, 1995.
3. Ostrowski, A.M. *Solution of Equations and Systems of Equations*; Academic Press: New York, NY, USA, 1960.
4. Ortega, J.M.; Rheinboldt, W.C. *Iterative Solution of Nonlinear Equations in Several Variables*; Academic Press: New York, NY, USA, 1970.
5. Potra, F.-A.; Pták, V. *Nondiscrete Induction and Iterarive Processes*; Pitman Publishing: Boston, MA, USA, 1984.
6. Traub, J.F. *Iterative Methods for the Solution of Equations*; Prentice-Hall: Englewood Cliffs, NJ, USA, 1964.
7. Ezquerro, J.A.; Hernández, M.A. An optimization of Chebyshev's method. *J. Complex.* **2009**, *25*, 343–361. [CrossRef]
8. Argyros, I.K.; Ezquerro, J.A.; Gutiérrez, J.M.; Hernández, M.A.; Hilout, S. On the semilocal convergence of efficient Chebyshev-Secant-type methods. *J. Comput. Appl. Math.* **2011**, *235*, 3195–3206. [CrossRef]
9. Argyros, I.K.; Ren, H. Efficient Steffensen-type algorithms for solving nonlinear equations. *Int. J. Comput. Math.* **2013**, *90*, 691–704. [CrossRef]
10. Steffensen, J.F. Remarks on iteration. *Skand. Aktuar Tidskr.* **1933**, *16*, 64–72. [CrossRef]
11. Ren, H.; Wu, Q.; Bi, W. A class of two-step Steffensen type methods with fourth-order convergence. *Appl. Math. Comput.* **2009**, *209*, 206–210. [CrossRef]
12. Zheng, Q.; Zhao, P.; Huang, F. A family of fourth-order Steffensen-type methods with the applications on solving nonlinear ODEs. *Appl. Math. Comput.* **2011**, *217*, 8196–8203. [CrossRef]
13. Grau-Sánchez, M.; Noguera, M.; Amat, S. On the approximation of derivatives using divided difference operators preserving the local convergence order of iterative methods. *J. Comput. Appl. Math.* **2013**, *237*, 363–372. [CrossRef]
14. Sharma, J.R.; Arora, H. An efficient derivative free iterative method for solving systems of nonlinear equations. *Appl. Anal. Discrete Math.* **2013**, *7*, 390–403. [CrossRef]
15. Sharma, J.R.; Arora, H. A novel derivative free algorithm with seventh order convergence for solving systems of nonlinear equations. *Numer. Algorithms* **2014**, *67*, 917–933. [CrossRef]
16. Wang, X.; Zhang, T. A family of Steffensen type methods with seventh-order convergence. *Numer. Algorithms* **2013**, *62*, 429–444. [CrossRef]
17. Sharma, J.R.; Arora, H. Efficient higher order derivative-free multipoint methods with and without memory for systems of nonlinear equations. *Int. J. Comput. Math.* **2018**, *95*, 920–938. [CrossRef]
18. Ahmad, F.; Soleymani, F.; Haghani, F.K.; Serra-Capizzano, S. Higher order derivative-free iterative methods with and without memory for systems of nonlinear equations. *Appl. Math. Comput.* **2017**, *314*, 199–211. [CrossRef]
19. Cordero, A.; Torregrosa, J.R. Variants of Newton's method for functions of several variables. *Appl. Math. Comput.* **2006**, *183*, 199–208. [CrossRef]
20. Genocchi, A. Relation entre la différence et la dérivée d'un même ordre quelconque. *Arch. Math. Phys. I* **1869**, *49*, 342–345.
21. Cordero, A.; Hueso, J.L.; Martínez, E.; Torregrosa, J.R. A modified Newton-Jarratt's composition. *Numer. Algorithms* **2010**, *55*, 87–99. [CrossRef]
22. Lotfi, T.; Bakhtiari, P.; Cordero, A.; Mahdiani, K.; Torregrosa, J.R. Some new efficient multipoint iterative methods for solving nonlinear systems of equations. *Int. J. Comput. Math.* **2015**, *92*, 1921–1934. [CrossRef]
23. Krasnoselsky, M.A.; Vainikko, G.M.; Zabreiko, P.P.; Rutitskii, J.B.; Stetsenko, V. J. *Approximate Solution of Operator Equations*; Nauka: Moscow, Russia, 1969. (In Russian)
24. Madhu, K.; Babajee, D.K.R.; Jayaraman, J. An improvement to double-step Newton method and its multi-step version for solving system of nonlinear equations and its applications. *Numer. Algorithms* **2017**, *74*, 593–607. [CrossRef]
25. Kumar, M.; Singh, A.K.; Srivastava, A. A new fifth order derivative free Newton-type method for solving nonlinear equations. *Appl. Math. Inf. Sci.* **2015**, *9*, 1507–1513.
26. Vrscay, E.R.; Gilbert, W.J. Extraneous fixed points, basin boundaries and chaotic dynamics for Schröder and König rational iteration functions. *Numer. Math.* **1988**, *52*, 1–16. [CrossRef]

27. Varona, J.L. Graphic and numerical comparison between iterative methods. *Math. Intell.* **2002**, *24*, 37–46. [CrossRef]
28. Scott, M.; Neta, B.; Chun, C. Basin attractors for various methods. *Appl. Math. Comput.* **2011**, *218*, 2584–2599. [CrossRef]
29. Xiao, X.Y.; Yin, H.W. Increasing the order of convergence for iterative methods to solve nonlinear systems. *Calcolo* **2016**, *53*, 285–300. [CrossRef]
30. Sauer, T. *Numerical Analysis*, 2nd ed.; Pearson Education, Inc.: Boston, MA, USA, 2012.

9

Parametric Jensen-Shannon Statistical Complexity and its Applications on Full-Scale Compartment Fire Data

Flavia-Corina Mitroi-Symeonidis [1,2], Ion Anghel [2] and Nicușor Minculete [3,*]

1 Academy of Economic Studies, Department of Applied Mathematics, Calea Dorobantilor 15-17, Sector 1, RO-010552 Bucharest, Romania; fcmitroi@yahoo.com
2 Police Academy "Alexandru Ioan Cuza", Fire Officers Faculty, Str. Morarilor 3, Sector 2, RO-022451 Bucharest, Romania; ion.anghel@academiadepolitie.ro
3 Faculty of Mathematics and Computer Science, Transilvania University of Brașov, Str. Iuliu Maniu50, 500091 Brașov, Romania
* Correspondence: minculete.nicusor@unitbv.ro

Abstract: The order/disorder characteristics of a compartment fire are researched based on experimental data. From our analysis performed by new, pioneering methods, we claim that the parametric Jensen-Shannon complexity can be successfully used to detect unusual data, and that one can use it also as a means to perform relevant analysis of fire experiments. Thoroughly comparing the performance of different algorithms (known as permutation entropy and two-length permutation entropy) to extract the probability distribution is an essential step. We discuss some of the theoretical assumptions behind each step and stress that the role of the parameter is to fine-tune the results of the Jensen-Shannon statistical complexity. Note that the Jensen-Shannon statistical complexity is symmetric, while its parametric version displays a symmetric duality due to the a priori probabilities used.

Keywords: full-scale fire experiment; compartment fire; permutation entropy; two length permutation entropy; time series analysis; parametric Jensen-Shannon statistical complexity; symmetric duality

1. Introduction

We aim to perform a local entropic analysis of the evolution of the temperature during a full-scale fire experiment and seek a straightforward, general, and process-based model of the compartment fire. We propose a new statistical complexity and compare known algorithms dedicated to the extraction of the underlying probabilities, checking their suitability to point out the abnormal values and structure of the experimental time series. For recent research on the fire phenomena performed using entropic tools, see Takagi, Gotoda, Tokuda and Miyano [1] and Murayama, Kaku, Funatsu, and Gotoda [2].

The experimental data was collected during a full-scale fire experiment conducted at the Fire Officers Faculty in Bucharest. We briefly include here the description of the experimental setup (Materials and Methods). Details can be found in [3].

The experiment has been carried out using a container (single-room compartment) having the following dimensions: 12 m × 2.2 m × 2.6 m. A single ventilation opening was available, namely the front door of the container, which remained open during the experiment. Parts of the walls and the ceiling of the container were furnished with oriented strand boards (OSB). The fire source has been a wooden crib made of 36 pieces of wood strips 2.5 cm × 2.5 cm × 30 cm, on which has been poured 500 mL ethanol shortly before ignition. The fire bed was situated in a corner of the compartment, at 1.2 m below the ceiling. The measurement devices consisted of six built-in K-type thermocouples, which were fixed at key locations (see Figure 1) and connected to a data acquisition logger. Flames were observed to impinge on the ceiling and exit through the opening, and we also noted the ignition of crumpled newspaper and stages of fire development that are known as indicators of flashover.

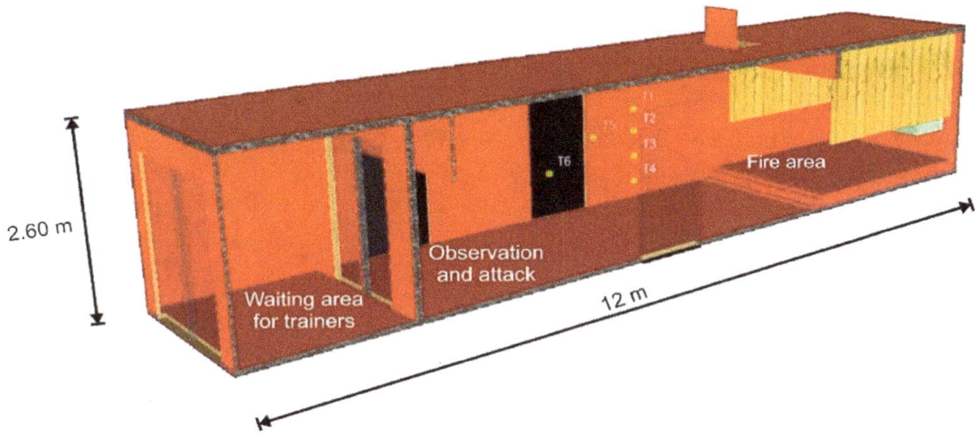

Figure 1. Arrangement of the flashover container.

In Section 2, we present the theoretical background and briefly summarize the approaches that are used to model fire.

Section 3 is dedicated to the results regarding the analysis of the collected raw data.

2. Theoretical Background and Remarks

2.1. Entropy and Statistical Complexity

The natural logarithm is used below, as elsewhere in this paper.

Shannon's entropy [4] is defined as $H(P) = -\sum_{i=1}^{n} p_i \log p_i$, where $P = (p_1, \ldots, p_n)$ is a finite probability distribution. It is nonnegative and its maximum value is $H(U) = \log n$, where $U = \left(\frac{1}{n}, \ldots, \frac{1}{n}\right)$. Throughout the paper, we use the convention $0 \cdot \log 0 = 0$.

The Kullback-Leibler divergence [5] is defined by

$$D(P||R) = \sum_{i=1}^{n} p_i(\log p_i - \log r_i) \quad (1)$$

where $P = (p_1, \ldots, p_n)$ and $R = (r_1, \ldots, r_n)$ are probability distributions. It is nonnegative and it vanishes for $P = R$.

If the value 0 appears in probability distributions $P = (p_1, \ldots, p_n)$ and $R = (r_1, \ldots, r_n)$, it must appear in the same positions for the sake of significance. Otherwise, one usually consider the conventions $0 \log \frac{0}{b} = 0$ for $b \geq 0$ and $a \log \frac{a}{0} = \infty$ for $a > 0$. We remark that these are strong limitations and such conditions rarely occur in practice.

To overcome this issue, the following divergence, well-defined, is used in the literature.

The Jensen-Shannon divergence (see [6,7]) is given by

$$JS(P||R) = \frac{1}{2} D\left(P||\frac{P+R}{2}\right) + \frac{1}{2} D\left(R||\frac{P+R}{2}\right) = H\left(\frac{P+R}{2}\right) - \frac{H(P)+H(R)}{2}. \quad (2)$$

The disequilibrium-based statistical complexity (LMC statistical complexity) introduced in 1995 by López-Ruiz, Mancini, and Calbet in [8] is defined as $C(P) = D(P)\frac{H(P)}{\log n}$, where $D(P)$, which is interpreted as disequilibrium, is the quadratic distance $D(P) = \sum_{i=1}^{n} (p_i - \frac{1}{n})^2$.

Interpreted as entropic non-triviality (in Lamberti et al. [9] and Zunino et al. [10]), the Jensen-Shannon statistical complexity is defined by $C^{(JS)}(P) = Q_{(JS)}(P)\frac{H(P)}{\log n}$, where the disequilibrium $Q_{(JS)}(P)$ is $Q_{(JS)}(P) = k{\cdot}JS(P||U)$. Here, $k = (\max_P JS(P||U))^{-1}$ is the normalizing constant and $U = \left(\frac{1}{n}, \ldots, \frac{1}{n}\right)$. Therefore, we have $0 \le C^{(JS)}(P) \le 1$.

For the convenience of the interested reader, we include the following method to determine the normalizing constant (a result stated for computational purposes, without proof, in [9]).

Proposition 1. *Using the above notation, for the computation of the normalizing constant,* $k = (\max_P JS(P||U))^{-1}$, *the maximum is attained for* P *such that there exists* i, $p_i = 1$.

It holds that $k = \left(\log 2 - \frac{1}{2}\log\frac{n+1}{n} - \frac{\log(n+1)}{2n}\right)^{-1}$.

Proof. We have the following calculations:

$$JS(P||U) = H\left(\frac{P+U}{2}\right) - \frac{H(P)+H(U)}{2} = \frac{1}{2}\sum_{i=1}^{n} p_i \log p_i - \frac{1}{2}\log n - \sum_{i=1}^{n}\left(\frac{p_i}{2} + \frac{1}{2n}\right)\log\left(\frac{p_i}{2} + \frac{1}{2n}\right). \quad (3)$$

$$\frac{\partial JS(P||U)}{\partial p_i} = \frac{1}{2}\log p_i + \frac{1}{2} - \frac{1}{2}\log\left(\frac{p_i}{2} + \frac{1}{2n}\right) - \frac{1}{2} = \frac{1}{2}\log p_i - \frac{1}{2}\log\left(\frac{p_i}{2} + \frac{1}{2n}\right). \quad (4)$$

$$\frac{\partial^2 JS(P||U)}{\partial pi^2} = \frac{1}{2p_i} - \frac{1}{4\left(\frac{p_i}{2} + \frac{1}{2n}\right)} = \frac{1}{2p_i} - \frac{1}{2p_i + \frac{2}{n}} > 0, \quad \frac{\partial^2 JS(P||U)}{\partial p_i \partial p_j} = 0. \quad (5)$$

So, the Hessian of $JS(P||U)$ is everywhere positive definite, whence $JS(P||U)$ is (strictly) convex on the open convex set $\left\{(p_1, \ldots, p_n) : 0 < p_i < 1 \text{ for all } i, \sum_{i=1}^{n} p_i = 1\right\}$. Therefore, $JS(P||U)$ cannot have a maximum inside (otherwise, it would be constant), and the points of maximum must lie on the boundary. See Theorem 3.10.10 in [11] (p. 171). Such points exist, because $JS(P||U)$ is continuous on the compact set $\Delta = \left\{(p_1, \ldots, p_n) : 0 \le p_i \le 1 \text{ for all } i, \sum_{i=1}^{n} p_i = 1\right\}$. The function $JS(P||U)$ is continuous and convex on the compact convex set Δ, so its maximum lies on the set of vertices of Δ (where $p_i = 1$ for one i). See Theorem 3.10.11 in [11] (p. 171). Since $JS(P||U)$ does not depend on the order of the components of P, the maximum value is attained at all vertices, so it can be straightforwardly computed by setting $P = (1, 0, \ldots, 0)$. □

Remark 1. *Note that the maximal value of* $JS(P||U)$ *is* $\log 2 - \frac{1}{2}\log\frac{n+1}{n} - \frac{\log(n+1)}{2n} \nearrow \log 2$, *as* $n \to \infty$. *Since* $JS(P||U)$ *is bounded from above by* $\log 2$, *independently of* n, *the normalization of* $JS(P||U)$ *in the definition of the Jensen-Shannon complexity does not seem to be relevant, and one could simply consider* $JS(P||U)\frac{H(P)}{\log n}$.

Let $\lambda \in [0, 1]$. *The parametric Jensen-Shannon divergence (see for instance, [6]) is given by*

$$\begin{aligned} JS_\lambda(P||R) &= (1-\lambda)D(P||(1-\lambda)P + \lambda R) + \lambda D(R||(1-\lambda)P + \lambda R) \\ &= H((1-\lambda)P + \lambda R) - ((1-\lambda)H(P) + \lambda H(R)). \end{aligned} \quad (6)$$

It is positive and it vanishes for $P = R$ *or* $\lambda = 0$ or 1. *See also Figure 2.*

The values $1-\lambda$ *and* λ *are interpreted as a priori probabilities. Note that* $JS_\lambda(P||R) = JS_{1-\lambda}(R||P)$ *and* JS_λ *is not symmetric, unless* $\lambda = 0.5$.

Mutatis mutandis, from Donald's identity (Lemma 2.12 in [12]), one has

$$JS_\lambda(P||R) + D((1-\lambda)P + \lambda R||Q) = (1-\lambda)D(P||Q) + \lambda D(R||Q) \quad (7)$$

for an arbitrarily fixed $\lambda \in [0,1]$. One needs only straightforward computation to check that it holds. Therefore,

$$JS_\lambda(P\|R) = \min\{(1-\lambda)D(P\|Q) + \lambda D(R\|Q) : Q = \left(q_1,\ldots,q_n\right) \text{ is a finite probability distribution}\Big\}. \tag{8}$$

We introduce the parametric Jensen-Shannon statistical complexity as

$$C_\lambda^{(JS)}(P) \equiv JS_\lambda(P\|U)\frac{H(P)}{\log n}. \tag{9}$$

As in the case of the complexities $C(P), C^{(JS)}(P)$, the new ones, $C_\lambda^{(JS)}(P)$, would be zero (minimum complexity) for $P = U$ or if there exists i such that $p_i = 1$. These two cases describe very different states of the system, both of which are extreme circumstances being considered simple, namely the states with respectively maximum and minimum entropy.

We do not need to normalize $JS_\lambda(P\|U)$ in the definition of the parametric Jensen-Shannon complexity (possibly one can feel more comfortable with its normalized version in other frameworks), but we stress that one can easily prove, following the same recipe as above, that its maximum value is attained for P such that there exists i, $p_i = 1$.

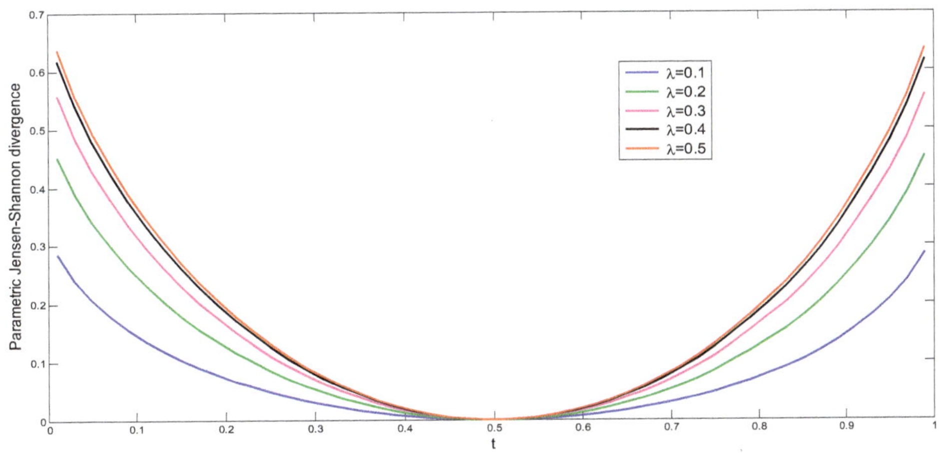

Figure 2. The parametric Jensen-Shannon divergence $JS_\lambda(P\|1-P)$, for $P = (t, 1-t)$, $t \in [0,1]$.

Proposition 2. *Let $\lambda \in [0,1]$. Using the above notation, it holds*

$$\max_P JS_\lambda(P\|U) = -\lambda\log\lambda - (1-\lambda)\log(1-\lambda) - (1-\lambda)\log\left(1+\frac{\lambda}{(1-\lambda)n}\right) - \frac{\lambda}{n}\log\frac{(1-\lambda)n+\lambda}{\lambda}. \tag{10}$$

Moreover, $\max_P JS_\lambda(P\|U) \nearrow -\lambda\log\lambda - (1-\lambda)\log(1-\lambda) \le \log 2$, as $n \to \infty$.

Proof. We omit the computation of $\max_P JS_\lambda(P\|U)$, which is straightforward.

To justify the monotonicity, it is enough to prove that $f(x) = \frac{\lambda}{x}\log\frac{(1g\lambda)x+\lambda}{\lambda}$ is decreasing:

$$f'(x) = -\frac{\lambda}{x^2}\left[\frac{\lambda}{(1-\lambda)x+\lambda} - 1 - \log\frac{\lambda}{(1-\lambda)x+\lambda}\right] < 0, \text{ for } \lambda \in (0,1) \text{ and } x > 0. \tag{11}$$

Furthermore, it is obvious that $(1-\lambda)\log\left(1+\frac{\lambda}{(1-\lambda)n}\right) + \frac{\lambda}{n}\log\frac{(1-\lambda)n+\lambda}{\lambda} \to 0$.

The last inequality follows from Jensen's inequality, which is applied to the concave logarithmic function.

Therefore, $JS_\lambda(P\|U)$ is bounded from above by $\log 2$, independently of n. □

Remark 2. *We split this result into two inequalities (of independent interest), which can be proved by the same technique. Namely, it holds that*

$$\max_P D(P\|(1-\lambda)P+\lambda U) \nearrow -\log(1-\lambda) \tag{12}$$

and

$$\max_P D(U\|(1-\lambda)P+\lambda U) \nearrow -\log\lambda, \tag{13}$$

as $n \to \infty$.

Proposition 3. *Let* $\lambda, \mu \in [0,1]$. *Using the above notation, the following inequality holds:*

$$\min\left\{\frac{1-\lambda}{1-\mu}, \frac{\lambda}{\mu}\right\} JS_{\mu}(P\|R) \leq JS_{\lambda}(P\|R) \leq \max\left\{\frac{1-\lambda}{1-\mu}, \frac{\lambda}{\mu}\right\} JS_{\mu}(P\|R) \tag{14}$$

where $P = (p_1, \ldots, p_n)$ *and* $R = (r_1, \ldots, r_n)$ *are two finite probability distributions.*

Proof. The result is a particular case of Theorem 3.2 from [13]. We include here an alternative proof for the sake of completeness.

It is known that the entropy H is concave; Hence, $H((1-\lambda)P+\lambda R) \geq (1-\lambda)H(P)+\lambda H(R)$. We prove that

$$\min\left\{\frac{1-\lambda}{1-\mu}, \frac{\lambda}{\mu}\right\}[H((1-\mu)P+\mu R)-(1-\mu)H(P)-\mu H(R)] \leq \tag{15}$$

$$H((1-\lambda)P+\lambda R)-(1-\lambda)H(P)-\lambda H(R) \leq \tag{16}$$

$$\max\left\{\frac{1-\lambda}{1-\mu}, \frac{\lambda}{\mu}\right\}[H((1-\mu)P+\mu R)-(1-\mu)H(P)-\mu H(R)]. \tag{17}$$

We consider $0 \leq \frac{1-\lambda}{1-\mu} \leq \frac{\lambda}{\mu}$, so $\lambda \geq \mu$, which implies, by the concavity of H, that

$$(1-\lambda)H(P)+\lambda H(R)+\min\left\{\frac{1-\lambda}{1-\mu}, \frac{\lambda}{\mu}\right\}[H((1-\mu)P+\mu R)-(1-\mu)H(P)-\mu H(R)] = \tag{18}$$

$$(1-\lambda)H(P)+\lambda H(R)+\frac{1-\lambda}{1-\mu}[H((1-\mu)P+\mu R)-(1-\mu)H(P)-\mu H(R)] = \tag{19}$$

$$\frac{\lambda-\mu}{1-\mu}H(R)+\frac{1-\lambda}{1-\mu}H((1-\mu)P+\mu R) \leq H\left(\frac{\lambda-\mu}{1-\mu}R+\frac{1-\lambda}{1-\mu}((1-\mu)P+\mu R)\right) =$$

$$H((1-\lambda)P+\lambda R), \tag{20}$$

because it holds $\frac{\lambda-\mu}{1-\mu}+\frac{1-\lambda}{1-\mu} = 1$ and $\lambda-\mu \geq 0$.

For the second inequality, we have

$$(1-\lambda)H(P)+\lambda H(R)+\max\left\{\frac{1-\lambda}{1-\mu}, \frac{\lambda}{\mu}\right\}[H((1-\mu)P+\mu R)-(1-\mu)H(P)-\mu H(R)] = \tag{21}$$

$$(1-\lambda)H(P)+\lambda H(R)+\frac{\lambda}{\mu}[H((1-\mu)P+\mu R)-(1-\mu)H(P)-\mu H(R)] =$$

$$-\frac{\lambda-\mu}{\mu}H(P)+\frac{\lambda}{\mu}H((1-\mu)P+\mu R) \geq H((1-\lambda)P+\lambda R), \tag{22}$$

because it holds that $\frac{\lambda}{\mu}H((1-\mu)P+\mu R) \geq H((1-\lambda)P+\lambda R)+\frac{\lambda-\mu}{\mu}H(P)$, which is equivalent to

$$H((1-\mu)P+\mu R) \geq \frac{\mu}{\lambda}H((1-\lambda)P+\lambda R)+\frac{\lambda-\mu}{\lambda}H(P). \tag{23}$$

For $0 \leq \frac{\lambda}{\mu} \leq \frac{1-\lambda}{1-\mu}$, the proof is similar. □

Remark 3. *For* $\lambda \in [0,1]$, $\mu \in (0,1)$, *and* $R = U$ *in Equation (1), then the following inequality holds:*

$$\min\left\{\frac{1-\lambda}{1-\mu}, \frac{\lambda}{\mu}\right\} JS_{\mu}(P||U) \leq JS_{\lambda}(P||U) \leq \max\left\{\frac{1-\lambda}{1-\mu}, \frac{\lambda}{\mu}\right\} JS_{\mu}(P||U). \tag{24}$$

For $\mu = \frac{1}{2}$ *in Equation (3), we obtain:*

$$2\min\{1-\lambda, \lambda\} JS(P||U) \leq JS_{\lambda}(P||U) \leq 2\max\{1-\lambda, \lambda\} JS(P||U). \tag{25}$$

Multiplying by $\frac{H(P)}{\log n}$ *in Equation (4), we deduce the following inequality related to the parametric Jensen-Shannon statistical complexity:*

$$2k\min\{1-\lambda, \lambda\} C^{(JS)}(P) \leq C_{\lambda}^{(JS)}(P) \leq 2k\max\{1-\lambda, \lambda\} C^{(JS)}(P), \tag{26}$$

where $k = \log 2 - \frac{1}{2}\log\frac{n+1}{n} - \frac{\log(n+1)}{2n}$.

2.2. Extraction of the Underlying Probability Distribution

The permutation entropy (PE) [14] quantifies randomness and the complexity of a time series based on the appearance of ordinal patterns, that is on comparisons of neighboring values of a time series. For other details on the PE algorithm applied to the present experimental data, see [3].

Let $T = (t_1, \ldots, t_n)$ be a time series with distinct values.

Step 1. The increasing rearranging of the components of each j-tuple $(t_i, \ldots, t_{i+j-1})$ as $(t_{i+r_1-1}, \ldots, t_{i+r_j-1})$ yields a unique permutation of order j denoted by $\pi = (r_1, \ldots, r_j)$, which is an encoding pattern that describes the up-and-downs in the considered j-tuple.

Simple numerical examples may help clarify the concepts throughout this section.

Example 1. *For the five-tuple* (2.3, 1, 3.1, 6.1, 5.2), *the corresponding permutation (encoding) is* (2, 1, 3, 5, 4).

Step 2. The absolute frequency of this permutation (the number of j-tuples which are associated to this permutation) is

$$k_{\pi} \equiv \#\{i : i \leq n-(j-1), (t_i, \ldots, t_{i+j-1}) \text{ is of type } \pi\}. \tag{27}$$

These values have the sum equal to the number of all the consecutive j-tuples; that is, $n-(j-1)$.

Step 3. The permutation entropy of order j is defined as $PE(j) \equiv -\sum_{\pi} p_{\pi}\log p_{\pi}$, where $p_{\pi} = \frac{k_{\pi}}{n-(j-1)}$ is the relative frequency.

In [14], the measured values of the time series are considered distinct. The authors neglect equalities and propose to break them by adding small random perturbations (random noise) to the original series.

Another known approach is to rank the equalities according to their order of emergence (to rank the equalities with their sequential/chronological order, see for instance [15,16]). We use this method throughout the paper to compute PE(j) for j = 3, 4, 5.

Applying the PE algorithm for experimental fire data, $C_{\lambda}^{(JS)}(P)$ cannot be zero. The number of the encoding patterns that occur is >1, and these patterns are not equiprobable: some patterns may be rare or locally forbidden (that is, one encounters such patterns at some thermocouples, but not in all six time series), as discussed in [3].

We briefly describe now the encoding steps in the TLPE algorithm (Two-Length Permutation Entropy algorithm) given by Watt and Politi in [17]; other details are provided in [3].

Step 1 given the j-tuple $T = (t_1, \ldots, t_j)$, we start encoding the last $k \leq j$ elements $(t_{j-k+1}, \ldots, t_j)$ according to the ordinal position of each element; that is, every t_s is replaced by a symbol which indicates the position occupied by t_s within the increasing rearranging of the considered k-tuple.

Next, we proceed by encoding each previous element t_m up to $m = 1$ according to the symbol provided by **Step 1** applied to the k-tuple $(t_m, \ldots, t_{m+k-1})$.

Example 2. *Encoding obtained by the chronological ordering of equal values* (4.1, 4.1, 4.1, 5, 2.1) → (1, 1, 2, 3, 1) *for* $k = 3$ *and* $j = 5$.

Step 2 and **Step 3**, they coincide with **Step 2** and **Step 3** in the PE algorithm above.

This algorithm leads, after computing the relative frequencies of the encoding sequences, to the two-length permutation entropy (TLPE (k, j)).

Given the pair (k, j) of values, the number of symbolic (encoding) sequences of length j is $k!k^{j-k}$, which is a number that can be much smaller than j!, so this algorithm is faster, it involves a simplified computation, and sometimes it makes the results more relevant for big values of j.

We deal with the equal values by using the same method as for PE; that is, we consider them ordered chronologically.

In the next section, we apply the above techniques and observe their capability to discern the changes of the parametric Jensen-Shannon statistical complexity of the experimental data.

3. Raw Data Analysis

The raw data set under consideration consists of measured temperatures during a compartment fire: six thermocouples T1, ... , T6 measure the temperatures every second during the experiment. Hence, we get six time series consisting of 3046 entries (data points), and we aim to a better understanding of these results by modeling the time series using information theory, and to assess the performance of the discussed statistical complexities.

We plot the parametric Jensen-Shannon statistical complexity against the parameter (for $\lambda \in \{0, 0.2, \ldots, 1\}$). We notice the unusual plotting for the time series at T5, which is definitely not caused by the position of this thermocouple. The graph corresponding to the time series at T5 is far from the rest of the graphs for the other thermocouples; hence, smaller values were obtained for the statistical complexities, with no apparent experiment related or mathematical reason. See Figures 3–7.

We conclude that the PE and TLPE algorithms can be successfully used to detect unusual data collected in fire experiments: different embedding dimensions and different algorithms used to determine the underlying probabilities provide the same conclusion, the hierarchy among the statistical complexities established for the thermocouples T1–T5 is the same, and T5 is always at a bigger distance from the rest of them. The position of the thermocouple T5 does not justify this big difference (see Figure 1). This also agrees with the smaller values provided at T5 by the LMC statistical complexity in Figure 8.

It is not clear in [9] why only the disequilibrium provided by JS has been considered and why JS_λ has been avoided. Using experimental data, we have verified that the parametric Jensen-Shannon complexity can be used for the analysis of the time series related to the fire dynamics: except for the trivial cases $\lambda = 0$ or 1, the results are not altered by the non-symmetry of JS_λ for $\lambda \neq 0.5$ (however, the embedding dimension j has to be adequate to the amount of data), so one can draw similar conclusions as for $\lambda = 0.5$. See Figure 8 (the plots obtained by PE(3), PE(4), TLPE(3,5), and TLPE(2,5) look similar, so we do not include them here). We have limitations for the choice of the embedding dimension j, since the factorial increases fast, and one then requires a bigger amount of data n. So, as a guideline for choosing the embedding dimension, the value of the statistical complexities remains relevant for j such that $n \gg j!$. See also [18].

Moreover, the proposed parametric Jensen-Shannon statistical complexities complement and validate the information provided bythe usual LMC and Jensen-Shannon statistical complexities. See Figures 8 and 9 for a quick comparison to the descriptions provided by the Jensen-Shannon and

LMC statistical complexities. According to our findings, the parametric Jensen-Shannon statistical complexity is a valid tool for the analysis of the evolution of the temperature in compartment fire data. The slight differences that appear between the upper line (corresponding to $\lambda = 0.5$) in Figure 9 and the one in Figure 10 are because the Jensen-Shannon complexity [9] is defined using the normalized Jensen-Shannon divergence, while we introduced the parametric Jensen-Shannon divergence in the LMC style, that is without normalizing the disequilibrium.

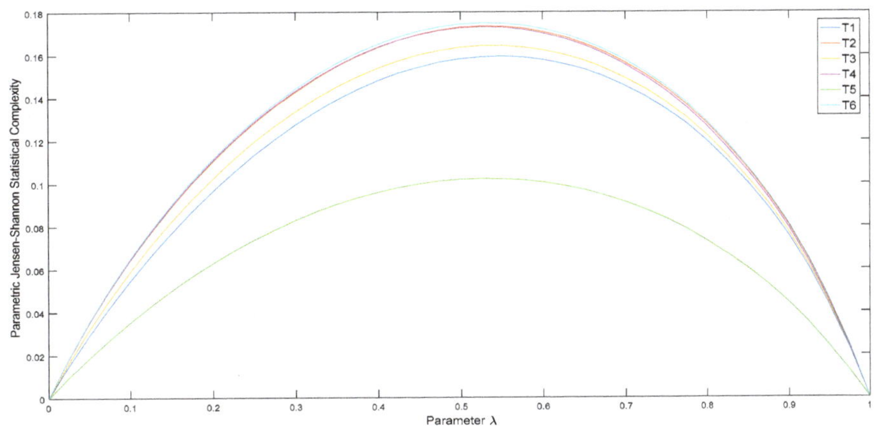

Figure 3. Plot obtained using the PE(5) algorithm. PE: permutation entropy.

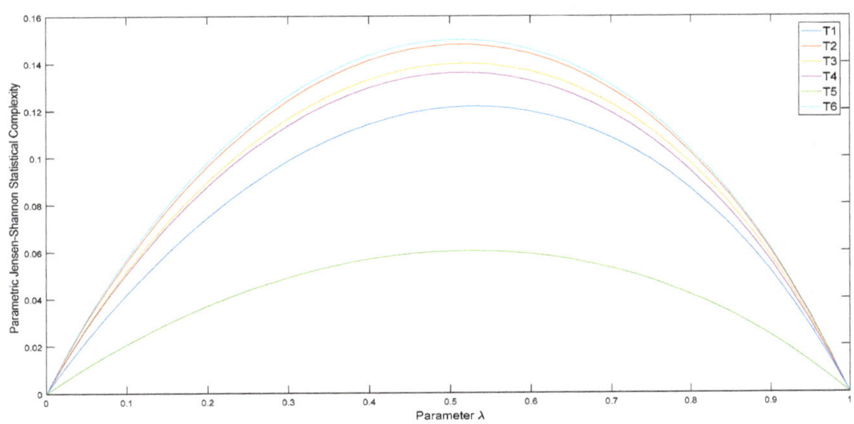

Figure 4. Plot obtained using the PE(4) algorithm.

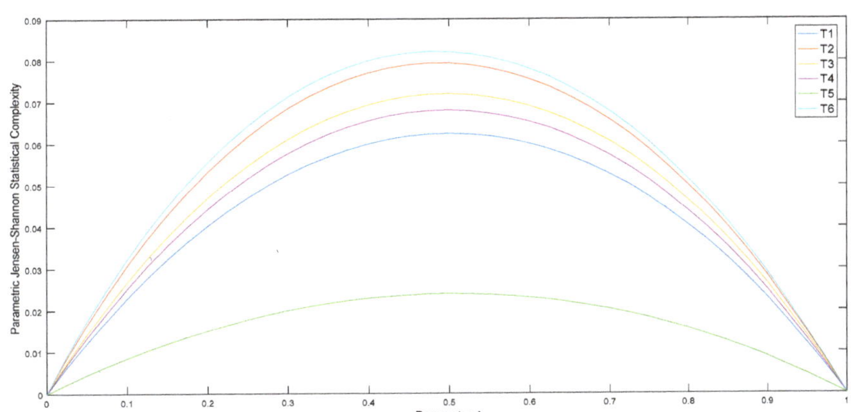

Figure 5. Plot obtained using the PE(3) algorithm.

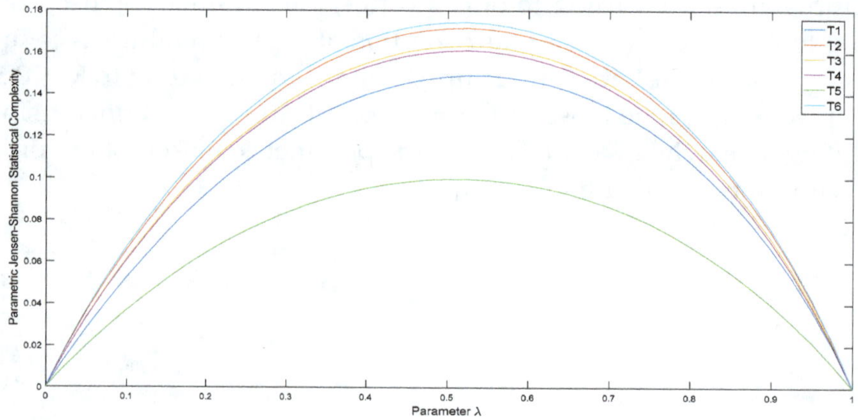

Figure 6. Plot obtained using the TLPE(3,5) algorithm.TLPE: Two-Length Permutation Entropy algorithm.

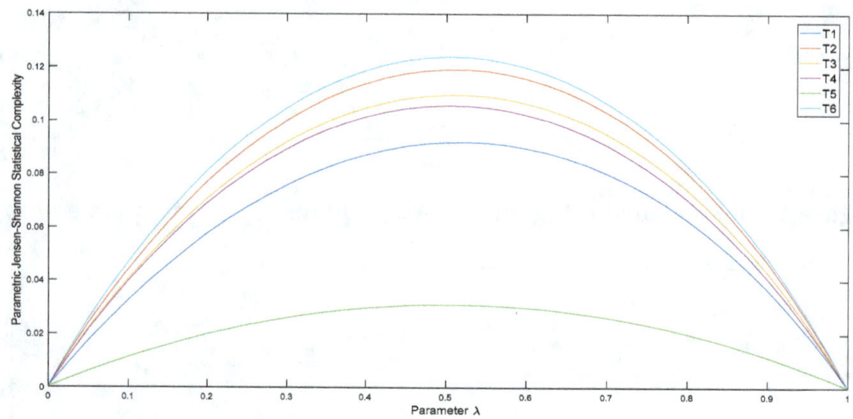

Figure 7. Plot obtained using the TLPE(2,5) algorithm.

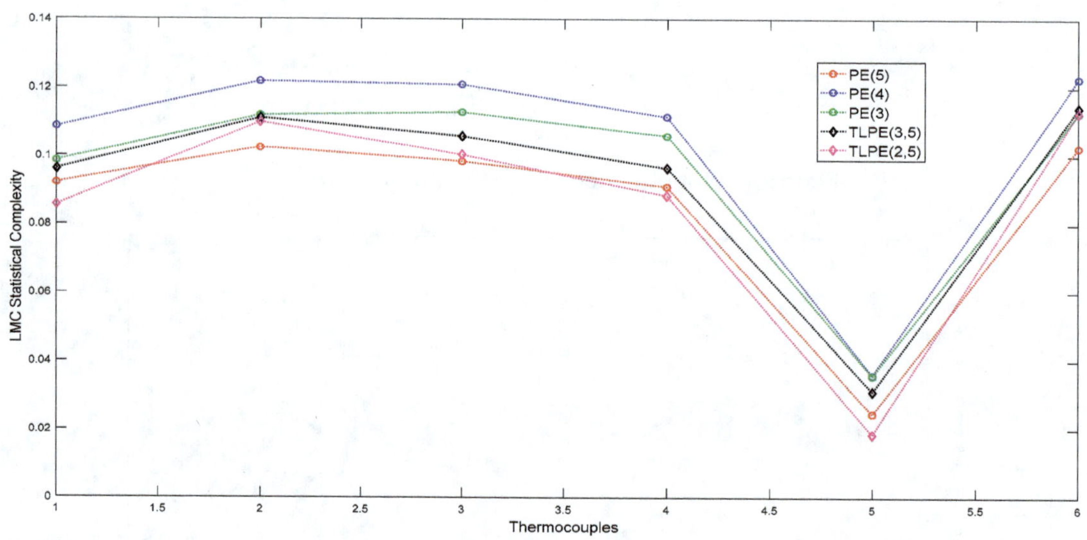

Figure 8. Statistical complexity.

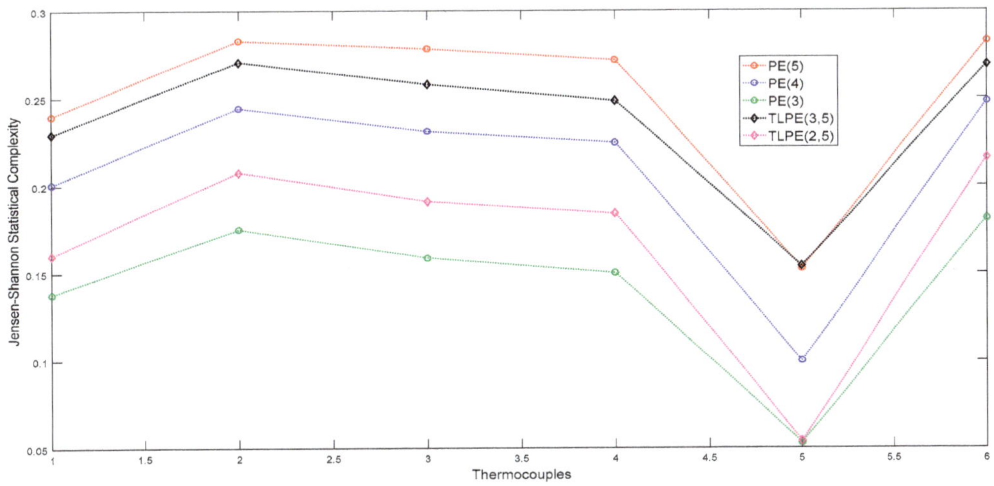

Figure 9. Jensen-Shannon statistical complexity ($\lambda = 1/2$).

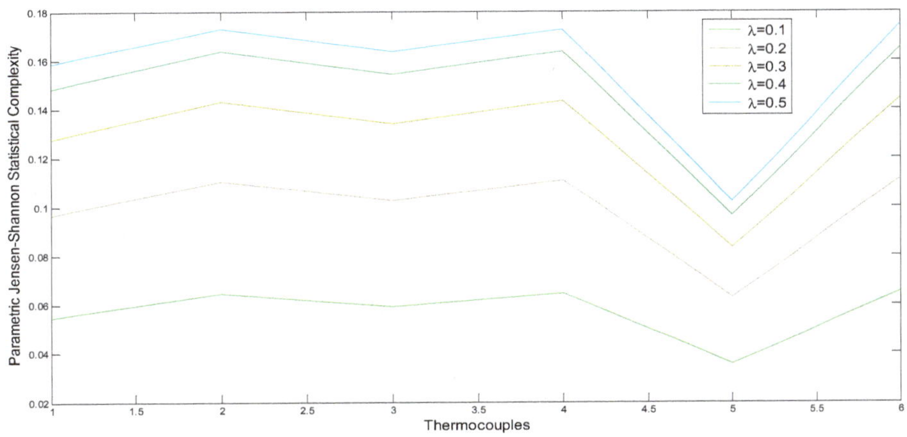

Figure 10. Obtained for the parametric Jensen-Shannon complexity using the PE(5)algorithm.

The most relevant aspect is that by applying the formula for the parametric Jensen-Shannon complexity, one gets similar plots, regardless of the embedding dimension and the encoding type algorithms used to determine the probability distribution, so the analysis is coherent and not misleading. Similarities with other in use complexity formulae would certainly improve the whole picture and bring us one step closer to the understanding of their ability to capture the behavior of various phenomena, in this particular case the fire dynamics. See Figures 8–10. We remark that these types of similarities might yield further mathematical results stating relationships among these mathematical notions: the (parametric) Jensen-Shannon and the LMC complexities.

4. Concluding Remarks on the Limitations of Our Study

The newly proposed complexities are used to analyze a full-scale experimental data set collected from a compartment fire.

For various algorithms and various embedding dimensions, more comparisons can be performed from this point onwards. We could not answer the questions about the merits and demerits of the known statistical complexities: such aspects are not yet clear in the literature, even in other frameworks where the permutation entropy has already been used by many researchers. Therefore, we discussed the relevance of the use of statistical complexities in the framework of fire data: small changes in the algorithms or choosing different embedding dimensions does not affect the interpretation of the results and the conclusions. This means that this new mathematical tool (the parametric Jensen-Shannon complexity) is informally staying "stable" in the framework of fire data. The accuracy of the interpretations can definitely be improved by the choice of the parameters, but the degree of its change cannot be estimated out of the data gathered in just one experiment: further research is required.

Other recent results on the analysis of this data set can be found in [19]. To understand this material the reader is referred to [20]. For the use of the permutation entropy in another framework see [21,22].

Our results might also indicate a turbulenceor a malfunction of the thermocouple T5 (an improperly calibrated scale); however, it is beyond the scope of the present paper to discuss it in detail.

Author Contributions: The work presented here was carried out in collaboration between all authors. All authors contributed equally and significantly in writing this article. All authors have contributed to the manuscript. All authors have read and agreed to the published version of the manuscript.

Acknowledgments: The authors thank Eleutherius Symeonidis for the argumentation of the proof of Proposition 1.

References

1. Takagi, K.; Gotoda, H.; Tokuda, I.T.; Miyano, T. Dynamic behavior of temperature field in a buoyancy-driven turbulent fire. *Phys. Lett. A* **2018**, *382*, 3181–3186. [CrossRef]
2. Murayama, S.; Kaku, K.; Funatsu, M.; Gotoda, H. Characterization of dynamic behavior of combustion noise and detection of blowout in a laboratory-scale gas-turbine model combustor. *Proc. Combust. Inst.* **2019**, *37*, 5271–5278. [CrossRef]
3. Mitroi-Symeonidis, F.-C.; Anghel, I.; Lalu, O.; Popa, C. The permutation entropy and its applications on fire tests data. *arXiv* **2019**, arXiv:1908.04274.
4. Shannon, C.E. A mathematical theory of communication. *Bell Syst. Tech. J.* **1948**, *27*, 379–423. [CrossRef]
5. Kullback, S.; Leibler, L.A. On information and sufficiency. *Ann. Math. Stat.* **1951**, *22*, 79–86. [CrossRef]
6. Lin, J. Divergence measures based on the Shannon entropy. *IEEE-Trans. Inf. Theory* **1991**, *37*, 145–151. [CrossRef]
7. Rao, C.R.; Nayak, T.K. Cross entropy, dissimilarity measures, and characterizations of quadratic entropy. *IEEE Trans. Inform. Theory* **1985**, *31*, 589–593. [CrossRef]
8. López-Ruiz, R.; Mancini, H.L.; Calbet, X. A statistical measure of complexity. *Phys. Lett. A* **1995**, *209*, 321–326. [CrossRef]
9. Lamberti, P.W.; Martin, M.T.; Plastino, A.; Rosso, O.A. Intensive entropic non-triviality measure. *Phys. A Stat. Mech. Appl.* **2004**, *334*, 119–131. [CrossRef]
10. Zunino, L.; Soriano, M.C.; Rosso, O.A. Distinguishing chaotic and stochastic dynamics from time series by using a multiscale symbolic approach. *Phys. Rev. E* **2012**, *86*, 046210. [CrossRef]
11. Niculescu, C.P.; Persson, L.-E. *Convex Functions and their Applications. A Contemporary Approach*, 2nd ed.; CMS Books in Mathematics; Springer-Verlag: New York, NY, USA, 2018; Volume 23.
12. Donald, M.J. Further results on the relative entropy. *Math. Proc. Camb. Philos. Soc.* **1987**, *101*, 363–373. [CrossRef]
13. Mitroi-Symeonidis, F.-C. About the precision in Jensen-Steffensen inequality. *Ann. Univ. Craiova Ser. Mat. Inform.* **2010**, *37*, 73–84.
14. Bandt, C.; Pompe, B. Permutation entropy: A natural complexity measure for time series. *Phys. Rev. Lett.* **2002**, *88*, 174102. [CrossRef] [PubMed]
15. Cao, T.; Tung, W.W.; Gao, J.B.; Protopopescu, V.A.; Hively, L.M. Detecting dynamical changes in time series using the permutation entropy. *Phys. Rev. E* **2004**, *70*, 046217. [CrossRef]
16. Duan, S.; Wang, F.; Zhang, Y. Research on the biophoton emission of wheat kernels based on permutation entropy. *Optik* **2019**, *178*, 723–730. [CrossRef]
17. Watt, S.J.; Politi, A. Permutation entropy revisited. *Chaos Solitons Fractals* **2019**, *120*, 95–99. [CrossRef]
18. Riedl, M.; Müller, A.; Wessel, N. Practical considerations of permutation entropy. *Eur. Phys. J. Spec. Top.* **2013**, *222*, 249–262. [CrossRef]
19. Mitroi-Symeonidis, F.-C.; Anghel, I.; Furuichi, S. Encodings for the calculation of the permutation hypoentropy and their applications on full-scale compartment fire data. *Acta Tech. Napoc. Ser. Appl. Math. Mech. Eng.* **2019**, *62*, 607–616.

20. Furuichi, S.; Mitroi-Symeonidis, F.-C.; Symeonidis, E. On some properties of Tsallis hypoentropies and hypodivergences. *Entropy* **2014**, *16*, 5377–5399. [CrossRef]
21. Araujo, F.H.A.; Bejan, L.; Rosso, O.A.; Stosic, T. Permutation entropy and statistical complexity analysis of Brazilian agricultural commodities. *Entropy* **2019**, *21*, 1220. [CrossRef]
22. Song, Y.; Ju, Y.; Du, K.; Liu, W.; Song, J. Online road detection under a shadowy traffic image using a learning-based illumination-independent image. *Symmetry* **2018**, *10*, 707. [CrossRef]

10

Volume Preserving Maps Between p-Balls

Adrian Holhoş * and Daniela Roşca

Department of Mathematics, Technical University of Cluj-Napoca, str. Memorandumului 28, RO-400114 Cluj-Napoca, Romania; daniela.rosca@math.utcluj.ro

* Correspondence: Adrian.Holhos@math.utcluj.ro

Abstract: We construct a volume preserving map $\mathcal{U}_p$ from the p-ball $\mathcal{B}_p(r) = \{\mathbf{x} \in \mathbb{R}^3,\ \|\mathbf{x}\|_p \le r\}$ to the regular octahedron $\mathcal{B}_1(r')$, for arbitrary $p > 0$. Then we calculate the inverse $\mathcal{U}_p^{-1}$ and we also deduce explicit expressions for $\mathcal{U}_\infty$ and $\mathcal{U}_\infty^{-1}$. This allows us to construct volume preserving maps between arbitrary balls $\mathcal{B}_p(r)$ and $\mathcal{B}_{p'}(\tilde{r})$, and also to map uniform and refinable grids between them. Finally we list some possible applications of our maps.

Keywords: equal volume projection; hierarchical grid

---

1. Introduction

The p-norms in $\mathbb{R}^3$ have applications in many branches of mathematics, physics and computer science. For $p \ge 1$, the p-norm of the vector $\mathbf{x} = (x, y, z) \in \mathbb{R}^3$ (also called L_p-norm) is defined as

$$\|\mathbf{x}\|_p = (|x|^p + |y|^p + |z|^p)^{1/p}. \tag{1}$$

For $p = 2$, we arrive at the Euclidean norm, and when $p \to \infty$ the norm is called the infinity norm or the maximum norm and is given by

$$\|\mathbf{x}\|_\infty = \max(|x|, |y|, |z|).$$

When $p \in (0, 1)$, Formula (1) does not define a norm, because the triangle inequality is not satisfied.

2. Preliminaries

For $p > 0$, let $\mathcal{B}_p(r)$ be the 3D p-ball of radius $r > 0$ centered at the origin, defined by

$$\mathcal{B}_p(r) = \left\{\mathbf{x} \in \mathbb{R}^3, \|\mathbf{x}\|_p \le r\right\}.$$

For finite p the parametric equations of $\mathcal{B}_p(r)$ are

$$\begin{aligned}
x &= \rho\,|\cos\theta|^{2/p}\,|\sin\varphi|^{2/p}\,\mathrm{sgn}(\cos\theta)\,\mathrm{sgn}(\sin\varphi),\\
y &= \rho\,|\sin\theta|^{2/p}\,|\sin\varphi|^{2/p}\,\mathrm{sgn}(\sin\theta)\,\mathrm{sgn}(\sin\varphi),\\
z &= \rho\,|\cos\varphi|^{2/p}\,\mathrm{sgn}(\cos\varphi),
\end{aligned}$$

with $\rho \in [0, r]$, $\theta \in [0, 2\pi)$, $\varphi \in [0, \pi]$.

For $p = 1$ the ball $\mathcal{B}_1(r)$ is the regular octahedron with the vertices on the axes, at distance r from the origin. For $p = \infty$, the set $\mathcal{B}_\infty(r)$ is the cube with edge of length $2r$ and for $p = 2$ the region $\mathcal{B}_2(r)$ represents the Euclidean ball. For $p > 2$ the balls are called *superellipsoids* and they are used in computer graphics (see [1,2], where the author uses the name *superquadrics* to refer to both superellipsoids and supertoroids). Some examples of balls $\mathcal{B}_p(r)$, for different values of p are given in Figure 1.

Figure 1. Some balls $\mathcal{B}_p(r)$ for $p = 0.5$, $p = 0.75$, $p = 1$ (**first line**) and $p = 1.2$, $p = 2$ and $p = 2.5$ (**second line**), respectively.

The volume of the 3D p-ball is

$$\begin{aligned}\mathrm{Vol}(\mathcal{B}_p(r)) &= 8\int_0^r \int_0^{(r^p-x^p)^{1/p}} \int_0^{(r^p-x^p-y^p)^{1/p}} \mathrm{d}z\,\mathrm{d}y\,\mathrm{d}x \\ &= 8r^3\frac{\Gamma^3(1/p+1)}{\Gamma(3/p+1)}.\end{aligned}$$

We notice that the radius r' of the regular octahedron $\mathcal{B}_1(r')$ with the same volume as the p-ball $\mathcal{B}_p(r)$ must be

$$r' = rc_p, \quad \text{with } c_p = \sqrt[3]{6}\frac{\Gamma(1/p+1)}{\sqrt[3]{\Gamma(3/p+1)}}.$$

We will construct a map $\mathcal{U}_p : \mathcal{B}_p(r) \to \mathcal{B}_1(r')$ which preserves the volume, i.e., $\mathcal{U}_p$ satisfies

$$\mathrm{Vol}(D) = \mathrm{Vol}(\mathcal{U}_p(D)), \qquad \text{for all domains } D \subseteq \mathcal{B}_p(r). \tag{2}$$

Consider the bijections $F_{1,p}, F_{2,p} : [0,1] \to [0,1]$, which are particular cases of the regularized incomplete Beta function (also known in statistics as cumulative beta distribution functions)

$$\begin{aligned}F_{1,p}(t) &= \frac{1}{\int_0^1 [u(1-u)]^{\frac{1}{p}-1}\,\mathrm{d}u}\int_0^t u^{\frac{1}{p}-1}(1-u)^{\frac{1}{p}-1}\,\mathrm{d}u, \quad \text{for } t \in [0,1], \\ F_{2,p}(t) &= \frac{1}{\int_0^1 u^{\frac{2}{p}-1}(1-u)^{\frac{1}{p}-1}\,\mathrm{d}u}\int_0^t u^{\frac{2}{p}-1}(1-u)^{\frac{1}{p}-1}\,\mathrm{d}u, \quad \text{for } t \in [0,1].\end{aligned}$$

In the standard notation we have $F_{1,p}(t) = I_t(1/p, 1/p)$ and $F_{2,p}(t) = I_t(2/p, 1/p)$, where I_t is the so-called regularized incomplete beta function defined as $I_t(\alpha, \beta) = B(t; \alpha, \beta)/B(1; \alpha, \beta)$, with

$$B(t; \alpha, \beta) = \int_0^t u^{\alpha-1}(1-u)^{\beta-1} du, \quad \text{for } \alpha, \beta > 0.$$

One has $F_{1,p}(0) = F_{2,p}(0) = 0$ and $F_{1,p}(1) = F_{2,p}(1) = 1$, further $F_{1,p}, F_{2,p}$ are increasing functions. Let $G_{1,p}, G_{2,p} : [0,1] \to [0,1]$ be the inverses (in Mathematica one can use the command `InverseBetaRegularized` for the inverses $G_{1,p}$ and $G_{2,p}$) of the functions $F_{1,p}$ and $F_{2,p}$, respectively.

For $a \in (0, \pi/2)$, let

$$\mathcal{B}_{p,a}(r) = \{ (x, y, z) \in \mathcal{B}_p(r),\ x, y, z \geq 0,\ x \tan a \geq y \}.$$

Lemma 1. *For $a \in (0, \pi/2)$ we have*

$$\mathrm{Vol}(\mathcal{B}_{p,a}(r)) = \frac{1}{8} F_{1,p}\left(\frac{\tan^p a}{1 + \tan^p a}\right) \mathrm{Vol}(\mathcal{B}_p(r)).$$

Proof. The volume of $\mathcal{B}_{p,a}(r)$ can be computed using the double integral

$$\mathrm{Vol}(\mathcal{B}_{p,a}(r)) = \iint_D (r^p - x^p - y^p)^{1/p}\, dx\, dy,$$

where $D = \{ (x, y) \in \mathbb{R}^2,\ x^p + y^p \leq r^p,\ 0 \leq y \leq x \tan a \}$. With the change of variables $x = (\rho \cos t)^{2/p}$ and $y = (\rho \sin t)^{2/p}$ the Jacobian is

$$J = \left(\frac{2}{p}\right)^2 \rho^{\frac{4}{p}-1} (\cos t)^{\frac{2}{p}-1} (\sin t)^{\frac{2}{p}-1}$$

and the new domain of integration is

$$\Delta = \left\{ (\rho, t) \in \mathbb{R}^2,\ 0 \leq \rho \leq r^{p/2},\ 0 \leq t \leq \arctan(\tan^{p/2} a) \right\}.$$

The volume of $\mathcal{B}_{p,a}(r)$ is

$$\mathrm{Vol}(\mathcal{B}_{p,a}(r)) = \frac{4}{p^2} \int_0^{r^{p/2}} (r^p - \rho^2)^{\frac{1}{p}} \rho^{\frac{4}{p}-1}\, d\rho \int_0^{\arctan(\tan^{\frac{p}{2}} a)} (\cos t)^{\frac{2}{p}-1} (\sin t)^{\frac{2}{p}-1}\, dt.$$

With the change of variables $u = \rho^2/r^p$ and $v = \sin^2 t$ in the two independent integrals we get

$$\begin{aligned}
\mathrm{Vol}(\mathcal{B}_{p,a}(r)) &= \frac{r^3}{p^2} \int_0^1 (1-u)^{\frac{1}{p}} u^{\frac{2}{p}-1}\, du \int_0^{\frac{\tan^p a}{1+\tan^p a}} v^{\frac{1}{p}-1} (1-v)^{\frac{1}{p}-1}\, dv \\
&= \frac{r^3}{p^2} B(1/p+1, 2/p) B(1/p, 1/p) F_{1,p}\left(\frac{\tan^p a}{1+\tan^p a}\right) \\
&= r^3 \frac{\Gamma^3(1/p+1)}{\Gamma(3/p+1)} F_{1,p}\left(\frac{\tan^p a}{1+\tan^p a}\right).
\end{aligned}$$

□

3. Construction of the Volume Preserving Map $\mathcal{U}_p : \mathcal{B}_p(r) \to \mathcal{B}_1(r')$ and Its Inverse

Of course, there is no unique map $\mathcal{U}_p$ with the volume preserving property. In this section, we will construct a map $\mathcal{U}_p : \mathcal{B}_p(r) \to \mathcal{B}_1(r')$ satisfying the following conditions:

(a) $\mathcal{U}_p$ has the volume preserving property (2);

(b) $\mathcal{U}_p$ is continuous on $\mathcal{B}_p(r)$ and has continuous partial derivatives at every point of $\mathcal{B}_p(r)$, except the points of the coordinate planes;

(c) $\mathcal{U}_p$ has the symmetry property

$$\mathcal{U}_p(x,y,z) = (\text{sgn}(x)\overline{X}, \text{sgn}(y)\overline{Y}, \text{sgn}(z)\overline{Z}), \quad \text{where } (\overline{X},\overline{Y},\overline{Z}) = \mathcal{U}_p(|x|,|y|,|z|);$$

(d) $\mathcal{U}_p$ maps every $\mathcal{B}_{p,a}(\widetilde{r})$ onto some $\mathcal{B}_{1,b}(c_p\widetilde{r})$.

Theorem 2. *The map $\mathcal{U}_p = (X,Y,Z)$ with the properties (a)–(d) is defined by*

$$X = \text{sgn}(x)c_p\left(|x|^p + |y|^p + |z|^p\right)^{\frac{1}{p}}\left[1 - F_{1,p}\left(\frac{|y|^p}{|x|^p+|y|^p}\right)\right]\sqrt{F_{2,p}\left(\frac{|x|^p+|y|^p}{|x|^p+|y|^p+|z|^p}\right)},$$

$$Y = \text{sgn}(y)c_p\left(|x|^p + |y|^p + |z|^p\right)^{\frac{1}{p}}\ F_{1,p}\left(\frac{|y|^p}{|x|^p+|y|^p}\right)\sqrt{F_{2,p}\left(\frac{|x|^p+|y|^p}{|x|^p+|y|^p+|z|^p}\right)},$$

$$Z = \text{sgn}(z)c_p\left(|x|^p + |y|^p + |z|^p\right)^{\frac{1}{p}}\left[1 - \sqrt{F_{2,p}\left(\frac{|x|^p+|y|^p}{|x|^p+|y|^p+|z|^p}\right)}\right],$$

when $|x|^p + |y|^p > 0$, and $(X,Y,Z) = (0,0,c_p z)$ when $|x|^p + |y|^p = 0$.

Proof. Let $(x,y,z) \in \mathcal{B}_p(r)$. Then $(X,Y,Z) = \mathcal{U}_p(x,y,z) \in \mathcal{B}_1(r')$. Consider first the case $x,y,z > 0$. From condition (d) for the limit case $a = \frac{\pi}{2}$ and using (a) and (c) we deduce that $\text{Vol}(\mathcal{B}_p(r)) = \text{Vol}(\mathcal{B}_1(c_p r))$. This relation gives us

$$X + Y + Z = c_p(x^p + y^p + z^p)^{1/p}. \tag{3}$$

From conditions (a) and (d) there is some $b > 0$ such that

$$\text{Vol}(\mathcal{B}_{p,a}(\widetilde{r})) = \text{Vol}(\mathcal{B}_{1,b}(c_p\widetilde{r})).$$

From Lemma 1 we have

$$F_{1,p}\left(\frac{\tan^p a}{1+\tan^p a}\right)\text{Vol}(\mathcal{B}_p(\widetilde{r})) = F_{1,1}\left(\frac{\tan b}{1+\tan b}\right)\text{Vol}(\mathcal{B}_1(c_p\widetilde{r})).$$

Since $\mathcal{B}_p(\widetilde{r})$ and $\mathcal{B}_1(c_p\widetilde{r})$ have the same volume and $F_{1,1}(t) = t$ we obtain

$$F_{1,p}\left(\frac{\tan^p a}{1+\tan^p a}\right) = \frac{\tan b}{1+\tan b}.$$

Further, since $\tan a = y/x$ and $\tan b = Y/X$, this equality can be written as

$$F_{1,p}\left(\frac{y^p}{x^p+y^p}\right) = \frac{Y}{X+Y}. \tag{4}$$

From conditions (a) and (b) the Jacobian of $\mathcal{U}_p$ must be 1, i.e.

$$\begin{vmatrix} \frac{\partial X}{\partial x} & \frac{\partial X}{\partial y} & \frac{\partial X}{\partial z} \\ \frac{\partial Y}{\partial x} & \frac{\partial Y}{\partial y} & \frac{\partial Y}{\partial z} \\ \frac{\partial Z}{\partial x} & \frac{\partial Z}{\partial y} & \frac{\partial Z}{\partial z} \end{vmatrix} = 1. \tag{5}$$

Further, taking into account Formulas (3) and (4) we have

$$Z = c_p(x^p + y^p + z^p)^{1/p} - X - Y,$$
$$Y = XF_{1,p}\left(\frac{y^p}{x^p + y^p}\right)\left(1 - F_{1,p}\left(\frac{y^p}{x^p + y^p}\right)\right)^{-1},$$

then we calculate the partial derivatives of Y and Z with respect to x, y and z and introduce them in (5). After some calculations, we find that X must be solution of the following first order partial differential equation

$$\frac{\partial X}{\partial x}xz^{p-1} + \frac{\partial X}{\partial y}yz^{p-1} - \frac{\partial X}{\partial z}(x^p + y^p) = \frac{(x^p + y^p)^{\frac{2}{p}}\left[1 - F_{1,p}\left(\frac{y^p}{x^p+y^p}\right)\right]^2 B\left(\frac{1}{p}, \frac{1}{p}\right)}{c_p p X(x^p + y^p + z^p)^{\frac{1}{p}-1}}.$$

With $U = X^2$ the equation is rewritten

$$\frac{\partial U}{\partial x}xz^{p-1} + \frac{\partial U}{\partial y}yz^{p-1} - \frac{\partial U}{\partial z}(x^p + y^p) = 2\frac{(x^p + y^p)^{\frac{2}{p}}\left[1 - F_{1,p}\left(\frac{y^p}{x^p+y^p}\right)\right]^2 B\left(\frac{1}{p}, \frac{1}{p}\right)}{c_p p(x^p + y^p + z^p)^{\frac{1}{p}-1}}.$$

We have to solve the symmetric system

$$\frac{dx}{xz^{p-1}} = \frac{dy}{yz^{p-1}} = \frac{dz}{-(x^p + y^p)} = \frac{c_p p(x^p + y^p + z^p)^{\frac{1}{p}-1}du}{2(x^p + y^p)^{\frac{2}{p}}\left[1 - F_{1,p}\left(\frac{y^p}{x^p+y^p}\right)\right]^2 B\left(\frac{1}{p}, \frac{1}{p}\right)}.$$

The first equality gives us $y = xC_1$, for some constant C_1. Replacing this in the equality

$$\frac{dx}{xz^{p-1}} = \frac{dz}{-(x^p + y^p)}$$

we get $x^p + y^p + z^p = C_2$, for some constant C_2. Replacing these two relations in the equality

$$\frac{dx}{xz^{p-1}} = \frac{c_p p(x^p + y^p + z^p)^{\frac{1}{p}-1}du}{2(x^p + y^p)^{\frac{2}{p}}\left[1 - F_{1,p}\left(\frac{y^p}{x^p+y^p}\right)\right]^2 B\left(\frac{1}{p}, \frac{1}{p}\right)},$$

integrating and using that the plane $x = 0$ is mapped onto $U = 0$ (this follows from the conditions (b) and (c) of the map), we obtain

$$U = \frac{2C_2^{\frac{2}{p}} B(1/p, 1/p)B(2/p, 1/p)}{p^2 c_p}\left[1 - F_{1,p}\left(\frac{C_1^p}{1 + C_1^p}\right)\right]^2 F_{2,p}\left(\frac{x^p(1 + C_1^p)}{C_2}\right),$$

which is equivalent to

$$X = c_p(x^p + y^p + z^p)^{1/p}\left[1 - F_{1,p}\left(\frac{y^p}{x^p + y^p}\right)\right]\sqrt{F_{2,p}\left(\frac{x^p + y^p}{x^p + y^p + z^p}\right)}. \tag{6}$$

Then,

$$Y = c_p(x^p + y^p + z^p)^{1/p} F_{1,p}\left(\frac{y^p}{x^p + y^p}\right)\sqrt{F_{2,p}\left(\frac{x^p + y^p}{x^p + y^p + z^p}\right)}, \tag{7}$$

$$Z = c_p(x^p + y^p + z^p)^{1/p}\left[1 - \sqrt{F_{2,p}\left(\frac{x^p + y^p}{x^p + y^p + z^p}\right)}\right]. \tag{8}$$

In the case when $z = 0$ and also in the case when $x = 0$ or $y = 0$ but $x + y > 0$ we use Formulas (6)–(8) to define the map $\mathcal{U}_p$. In the case when $x = y = 0$, we define $\mathcal{U}_p(0,0,z) = (0,0,c_p z)$, for all $z \geq 0$, using the continuity property of the map $\mathcal{U}_p$.

Finally, for the points (x, y, z) in the other seven octants, the map $\mathcal{U}_p$ will be defined as

$$\mathcal{U}_p(x,y,z) = (\text{sgn}(x)\overline{X}, \text{sgn}(y)\overline{Y}, \text{sgn}(z)\overline{Z}), \quad \text{where } (\overline{X},\overline{Y},\overline{Z}) = \mathcal{U}_p(|x|,|y|,|z|).$$

□

Remark. *Not all the partial derivatives of the map $\mathcal{U}_p$ which occur in Theorem 2 exist at the points of the coordinates planes. For example, $\frac{\partial Y}{\partial x}$ does not exist at the points $(0, y, z)$, because the partial derivative of $F_{1,p}\left(\frac{|y|^p}{|x|^p + |y|^p}\right)$ with respect to x does not exist at the points $(0, y, z)$.*

The expression of the inverse map of $\mathcal{U}_p$ is given in the next theorem.

Theorem 3. *The map $\mathcal{U}_p^{-1} : \mathcal{B}_1(r') \to \mathcal{B}_p(r)$ is defined by*

$$x = \frac{X + Y + Z}{c_p} G_{1,p}^{\frac{1}{p}}\left(\frac{Y}{X + Y}\right) G_{2,p}^{\frac{1}{p}}\left(\left(\frac{X + Y}{X + Y + Z}\right)^2\right), \tag{9}$$

$$y = \frac{X + Y + Z}{c_p}\left(1 - G_{1,p}\left(\frac{Y}{X + Y}\right)\right)^{\frac{1}{p}} G_{2,p}^{\frac{1}{p}}\left(\left(\frac{X + Y}{X + Y + Z}\right)^2\right), \tag{10}$$

$$z = \frac{X + Y + Z}{c_p}\left(1 - G_{2,p}\left(\left(\frac{X + Y}{X + Y + Z}\right)^2\right)\right)^{\frac{1}{p}}, \tag{11}$$

for every $(X, Y, Z) \in \mathcal{B}_1(r')$ and $X \geq 0$, $Y \geq 0$, $Z \geq 0$, $X + Y > 0$. If $X = Y = 0$, we have $\mathcal{U}_p^{-1}(0,0,Z) = (0,0,Z/c_p)$.

In the other seven octants, we define the inverse of the map $\mathcal{U}_p$ using the symmetry property (c) of $\mathcal{U}_p$.

Proof. Condition (4) is equivalent to

$$\frac{y^p}{x^p + y^p} = G_{1,p}\left(\frac{Y}{X + Y}\right).$$

Replacing (3) in (7) we obtain

$$\frac{X + Y}{X + Y + Z} = \sqrt{F_{2,p}\left(\frac{x^p + y^p}{x^p + y^p + z^p}\right)},$$

which is equivalent to

$$\frac{x^p + y^p}{x^p + y^p + z^p} = G_{2,p}\left(\left(\frac{X + Y}{X + Y + Z}\right)^2\right).$$

After some computations we can express x, y, z in terms of X, Y, Z to obtain (9)–(11). □

4. Particular Cases

4.1. The Cases $p = 1$ and $p = 2$

For $p = 1$ one has $c_1 = 1$, $F_{1,p}(t) = t$ and $F_{2,p}(t) = t^2$, therefore $\mathcal{U}_1$ is the identity.

For $p = 2$ one has $c_2 = \pi^{\frac{1}{3}}$, $F_{1,p}(t) = \frac{1}{\pi}\left(\arcsin(2t-1) + \frac{\pi}{2}\right) = \frac{2}{\pi}\arcsin\sqrt{t}$, $F_{2,p}(t) = 1 - \sqrt{1-t}$ and for $x, y, z > 0$, the map $\mathcal{U}_2$ is

$$X = 2\pi^{-2/3}\sqrt{x^2+y^2+z^2 - z\sqrt{x^2+y^2+z^2}}\arccos\frac{y}{\sqrt{x^2+y^2}},$$

$$Y = 2\pi^{-2/3}\sqrt{x^2+y^2+z^2 - z\sqrt{x^2+y^2+z^2}}\arcsin\frac{y}{\sqrt{x^2+y^2}},$$

$$Z = \pi^{1/3}\sqrt{x^2+y^2+z^2}\left(1 - \sqrt{1 - \frac{z}{\sqrt{x^2+y^2+z^2}}}\right).$$

If we use the spherical coordinates defined by $x = \rho\cos\theta\sin\varphi$, $y = \rho\sin\theta\sin\varphi$ and $z = \rho\cos\varphi$ we obtain relations (9), (10), (11) from [3], where we also gave the inverse, which has an explicit expression.

4.2. The Case $p = \infty$

In this case we will obtain a new map, different from the one constructed in [4].

We restrict again to the case $x, y, z > 0$ because of the symmetry property of the map.

First, a simple calculation shows that $c_\infty = 6^{1/3}$ and

$$\lim_{p\to\infty}(x^p + y^p + z^p)^{1/p} = \max(x, y, z).$$

In order to calculate the limits in (6)–(8) when $p \to \infty$ we use the following result.

Lemma 4. *For $\alpha, \beta > 0$ we have*

$$\lim_{p\to\infty}\frac{p}{B\left(\frac{\alpha}{p}, \frac{\beta}{p}\right)} = \frac{\alpha\beta}{\alpha+\beta}. \tag{12}$$

Proof. We use the equality $\Gamma(x) = \Gamma(x+1)/x$, which holds for $x > 0$. One has

$$\frac{p}{B\left(\frac{\alpha}{p}, \frac{\beta}{p}\right)} = \frac{p\,\Gamma\left(\frac{\alpha+\beta}{p}\right)}{\Gamma\left(\frac{\alpha}{p}\right)\cdot\Gamma\left(\frac{\beta}{p}\right)} = \frac{p\cdot\frac{\alpha}{p}\cdot\frac{\beta}{p}\cdot\Gamma\left(1+\frac{\alpha+\beta}{p}\right)}{\Gamma\left(1+\frac{\alpha}{p}\right)\cdot\Gamma\left(1+\frac{\beta}{p}\right)\cdot\frac{\alpha+\beta}{p}},$$

and now it is easy to see that the limit when $p \to \infty$ is the one in (12). □

Proposition 5. *For $x, y, z > 0$ we have*

$$\lim_{p\to\infty}F_{1,p}\left(\frac{y^p}{x^p+y^p}\right) = \begin{cases} \frac{y}{2x}, & x > y, \\ 1 - \frac{x}{2y}, & y \geq x. \end{cases}$$

Proof. We use the idea in [5].

Suppose $x > y$.

$$F_{1,p}\left(\frac{y^p}{x^p+y^p}\right) = \frac{1}{B\left(\frac{1}{p}, \frac{1}{p}\right)}\int_0^{y^p/(x^p+y^p)}(u(1-u))^{\frac{1}{p}-1}\,du.$$

With the change of variable $u = t^p$ we have

$$F_{1,p}\left(\frac{y^p}{x^p+y^p}\right) = \frac{p}{B\left(\frac{1}{p},\frac{1}{p}\right)} \int_0^{y/(x^p+y^p)^{1/p}} (1-t^p)^{\frac{1}{p}-1}\,\mathrm{d}t.$$

From $0 < t < y/(x^p+y^p)$ we further deduce that $x^p/(x^p+y^p) < 1-t^p < 1$, and therefore

$$\left(\frac{x^p}{x^p+y^p}\right)^{\frac{1}{p}-1} > (1-t^p)^{\frac{1}{p}-1} > 1.$$

After integration we obtain

$$\frac{y}{(x^p+y^p)^{\frac{1}{p}}}\left(\frac{x^p}{x^p+y^p}\right)^{\frac{1}{p}-1} \geq \int_0^{y/(x^p+y^p)^{1/p}} (1-t^p)^{\frac{1}{p}-1}\,\mathrm{d}t \geq \frac{y}{(x^p+y^p)^{\frac{1}{p}}},$$

and further,

$$\frac{p}{B\left(\frac{1}{p},\frac{1}{p}\right)}\frac{y}{(x^p+y^p)^{\frac{1}{p}}}\left(\frac{x^p}{x^p+y^p}\right)^{\frac{1}{p}-1} \geq F_{1,p}\left(\frac{y^p}{x^p+y^p}\right) \geq \frac{p}{B\left(\frac{1}{p},\frac{1}{p}\right)}\frac{y}{(x^p+y^p)^{\frac{1}{p}}}.$$

After applying Lemma 4 for $\alpha = \beta = 1$ and replacing the limits

$$\lim_{p\to\infty}(x^p+y^p)^{\frac{1}{p}} = \max(x,y) = x \text{ and } \lim_{p\to\infty}\frac{x^p}{x^p+y^p} = 1,$$

□

We finally obtain

$$\lim_{p\to\infty} F_{1,p}\left(\frac{y^p}{x^p+y^p}\right) = \frac{y}{2x}. \tag{13}$$

For the case $y \geq x$ we use the formula $F_{1,p}(1-t) = 1 - F_{1,p}(t)$ for $t = x^p/(x^p+y^p)$ and Formula (13), interchanging x and y.

Proposition 6. *For $x, y, z > 0$ we have*

$$\lim_{p\to\infty} F_{2,p}\left(\frac{x^p+y^p}{x^p+y^p+z^p}\right) = \begin{cases} \frac{1}{3z^2}\max(x,y)^2, & \text{if } z = \max(x,y,z), \\ 1 - \frac{2}{3}\frac{z}{\max(x,y)}, & \text{otherwise.} \end{cases}$$

Proof. *Case 1.* Suppose $\max(x,y,z) = z$.

With the change of variable $t = u^{2/p}$ we obtain

$$F_{2,p}\left(\frac{x^p+y^p}{x^p+y^p+z^p}\right) = \frac{p}{2B\left(\frac{2}{p},\frac{1}{p}\right)} \int_0^{\left(\frac{x^p+y^p}{x^p+y^p+z^p}\right)^{2/p}} \left(1-t^{\frac{p}{2}}\right)^{\frac{1}{p}-1}\,\mathrm{d}t.$$

Applying Lemma 4 for $\alpha = 2$, $\beta = 1$ we have

$$\lim_{p\to\infty}\frac{p}{2B\left(\frac{2}{p},\frac{1}{p}\right)} = \frac{1}{3}.$$

Further, from the condition that t belongs to the interval of integration we can write

$$\frac{z^p}{x^p + y^p + z^p} < 1 - t^{\frac{p}{2}} < 1,$$

and therefore

$$\left(\frac{z^p}{x^p + y^p + z^p}\right)^{\frac{1}{p}-1} > \left(1 - t^{\frac{p}{2}}\right)^{\frac{1}{p}-1} > 1.$$

After integration we obtain

$$\left(\frac{x^p + y^p}{x^p + y^p + z^p}\right)^{\frac{2}{p}} \left(\frac{z^p}{x^p + y^p + z^p}\right)^{\frac{1}{p}-1} \geq \int_0^{\left(\frac{x^p+y^p}{x^p+y^p+z^p}\right)^{\frac{2}{p}}} \left(1 - t^{\frac{p}{2}}\right)^{\frac{1}{p}-1} dt \geq \left(\frac{x^p + y^p}{x^p + y^p + z^p}\right)^{\frac{2}{p}}.$$

A simple calculation shows that

$$\lim_{p\to\infty} \left(\frac{x^p + y^p}{x^p + y^p + z^p}\right)^{2/p} = \frac{(\max(x,y))^2}{z^2} \quad \text{and} \quad \lim_{p\to\infty} \frac{z^p}{x^p + y^p + z^p} = 1,$$

which imply that

$$\lim_{p\to\infty} \int_0^{\left(\frac{x^p+y^p}{x^p+y^p+z^p}\right)^{\frac{2}{p}}} \left(1 - t^{\frac{p}{2}}\right)^{\frac{1}{p}-1} dt = \frac{(\max(x,y))^2}{z^2}.$$

Case 2. Suppose $\max(x, y, z) = x$ or y.
Using the equality

$$I_x(\alpha, \beta) = 1 - I_{1-x}(\beta, \alpha), \quad \alpha, \beta > 0, \quad x \in [0, 1],$$

we have

$$F_{2,p}\left(\frac{x^p + y^p}{x^p + y^p + z^p}\right) = 1 - \frac{1}{B\left(\frac{2}{p}, \frac{1}{p}\right)} \int_0^{\frac{z^p}{x^p+y^p+z^p}} u^{\frac{1}{p}-1}(1-u)^{\frac{2}{p}-1} du.$$

With the change of variable $u = t^p$ we get

$$F_{2,p}\left(\frac{x^p + y^p}{x^p + y^p + z^p}\right) = 1 - \frac{p}{B\left(\frac{2}{p}, \frac{1}{p}\right)} \int_0^{\left(\frac{z^p}{x^p+y^p+z^p}\right)^{1/p}} (1 - t^p)^{\frac{2}{p}-1} dt.$$

Similarly

$$\frac{z}{(x^p + y^p + z^p)^{1/p}} \left(\frac{x^p + y^p}{x^p + y^p + z^p}\right)^{\frac{2}{p}-1} \geq \int_0^{\left(\frac{z^p}{x^p+y^p+z^p}\right)^{1/p}} (1 - t^p)^{\frac{2}{p}-1} dt \geq \frac{z}{(x^p + y^p + z^p)^{1/p}}.$$

Using

$$\lim_{p\to\infty} \frac{z}{(x^p + y^p + z^p)^{1/p}} = \frac{z}{\max(x,y)} \quad \text{and} \quad \lim_{p\to\infty} \frac{x^p + y^p}{x^p + y^p + z^p} = 1,$$

the proof is complete. □

In conclusion, for $x, y, z > 0$, the map $\mathcal{U}_\infty$ has the values $(X, Y, Z) = \mathcal{U}_\infty(x, y, z)$ given by:

$$\begin{aligned}
&6^{1/3}\left(\frac{x}{2\sqrt{3}}, \frac{2y-x}{2\sqrt{3}}, z-\frac{y}{\sqrt{3}}\right), \; x \le y \le z,\\
&6^{1/3}\left(\frac{x}{2}\sqrt{1-\frac{2z}{3y}}, \left(y-\frac{x}{2}\right)\sqrt{1-\frac{2z}{3y}}, y\left(1-\sqrt{1-\frac{2z}{3y}}\right)\right), \; x \le z \le y,\\
&6^{1/3}\left(\frac{2x-y}{2\sqrt{3}}, \frac{y}{2\sqrt{3}}, z-\frac{x}{\sqrt{3}}\right), \; y \le x \le z,\\
&6^{1/3}\left(\left(x-\frac{y}{2}\right)\sqrt{1-\frac{2z}{3x}}, \frac{y}{2}\sqrt{1-\frac{2z}{3x}}, x\left(1-\sqrt{1-\frac{2z}{3x}}\right)\right), \; y \le z \le x,\\
&6^{1/3}\left(\frac{x}{2}\sqrt{1-\frac{2z}{3y}}, \left(y-\frac{x}{2}\right)\sqrt{1-\frac{2z}{3y}}, y\left(1-\sqrt{1-\frac{2z}{3y}}\right)\right), \; z \le x \le y,\\
&6^{1/3}\left(\left(x-\frac{y}{2}\right)\sqrt{1-\frac{2z}{3x}}, \frac{y}{2}\sqrt{1-\frac{2z}{3x}}, x\left(1-\sqrt{1-\frac{2z}{3x}}\right)\right), \; z \le y \le x,
\end{aligned}$$

and can be reduced to

$$\begin{aligned}
&6^{1/3}\left(\frac{x}{2\sqrt{3}}, \frac{2y-x}{2\sqrt{3}}, z-\frac{y}{\sqrt{3}}\right), \; x \le y \le z,\\
&6^{1/3}\left(\frac{x}{2}\sqrt{1-\frac{2z}{3y}}, \left(y-\frac{x}{2}\right)\sqrt{1-\frac{2z}{3y}}, y\left(1-\sqrt{1-\frac{2z}{3y}}\right)\right), \; x \le y,\ z \le y,\\
&6^{1/3}\left(\frac{2x-y}{2\sqrt{3}}, \frac{y}{2\sqrt{3}}, z-\frac{x}{\sqrt{3}}\right), \; y \le x \le z,\\
&6^{1/3}\left(\left(x-\frac{y}{2}\right)\sqrt{1-\frac{2z}{3x}}, \frac{y}{2}\sqrt{1-\frac{2z}{3x}}, x\left(1-\sqrt{1-\frac{2z}{3x}}\right)\right), \; y \le x,\ z \le x.
\end{aligned}$$

The above formulas can also be used in the case when $x = 0$ or $y = 0$ or $z = 0$, with the mention that the denominators cannot be zero, except the case when $x = y = z = 0$, when we take $\mathcal{U}_\infty(0, 0, 0) = (0, 0, 0)$.

After some calculations we get that, for $X, Y, Z > 0$ the inverse $\mathcal{U}_\infty^{-1}(X, Y, Z)$ is given by

$$\begin{aligned}
&6^{-1/3}\left(2\sqrt{3}X, \sqrt{3}(X+Y), X+Y+Z\right), \text{ on } D_1,\\
&6^{-1/3}\left(\frac{2X(X+Y+Z)}{X+Y}, X+Y+Z, \frac{3Z(2X+2Y+Z)}{2(X+Y+Z)}\right), \text{ on } D_2,\\
&6^{-1/3}\left(\sqrt{3}(X+Y), 2\sqrt{3}Y, X+Y+Z\right), \text{ on } D_3,\\
&6^{-1/3}\left(X+Y+Z, \frac{2Y(X+Y+Z)}{X+Y}, \frac{3Z(2X+2Y+Z)}{2(X+Y+Z)}\right), \text{ on } D_4,
\end{aligned}$$

where D_i, $i = 1, 2, 3, 4$ are the set of points (X, Y, Z) satisfying the following conditions, respectively:

$$\begin{aligned}
&X \le Y, \quad \sqrt{3}(X+Y) \le X+Y+Z,\\
&X \le Y, \quad \frac{3Z(2X+2Y+Z)}{2(X+Y+Z)} \le X+Y+Z,\\
&Y \le X, \quad (X+Y)\sqrt{3} \le X+Y+Z,\\
&Y \le X, \quad \frac{3Z(2X+2Y+Z)}{2(X+Y+Z)} \le X+Y+Z.
\end{aligned}$$

Condition

$$\frac{3Z(2X+2Y+Z)}{2(X+Y+Z)} \leq X+Y+Z$$

can be written as $3((X+Y+Z)^2 - (X+Y)^2) \leq 2(X+Y+Z)^2$, and is equivalent to $X+Y+Z \leq \sqrt{3}(X+Y)$, since $X, Y, Z > 0$.

Therefore,

$$\begin{aligned} D_1 &= \{X \leq Y, \quad \sqrt{3}(X+Y) \leq X+Y+Z\}, \\ D_2 &= \{X \leq Y, \quad X+Y+Z \leq \sqrt{3}(X+Y)\}, \\ D_3 &= \{Y \leq X, \quad (X+Y)\sqrt{3} \leq X+Y+Z\}, \\ D_4 &= \{Y \leq X, \quad X+Y+Z \leq \sqrt{3}(X+Y)\}. \end{aligned}$$

Finally, the expressions of $(x, y, z) = \mathcal{U}_\infty^{-1}(X, Y, Z)$ can be reduced to

$$\begin{aligned} x &= 6^{-1/3} \min\left(\sqrt{3}, 1 + \frac{Z}{X+Y}\right)(X + \min(X, Y)), \\ y &= 6^{-1/3} \min\left(\sqrt{3}, 1 + \frac{Z}{X+Y}\right)(Y + \min(X, Y)), \\ z &= 6^{-1/3} \min\left(X+Y+Z, 3Z\left(1 - \frac{Z}{2(X+Y+Z)}\right)\right). \end{aligned}$$

These formulas can also be used in the case when $Z = 0$ and in the case when $X = 0$ or $Y = 0$, but $X + Y > 0$. In the case when $X = Y = 0$ we take $\mathcal{U}_\infty^{-1}(0, 0, Z) = (0, 0, 6^{-1/3}Z)$.

If we take arbitrary numbers $p, \widetilde{p} > 0$, the application

$$\mathcal{U}_{\widetilde{p}}^{-1} \circ \mathcal{U}_p : \mathcal{B}_p(r) \to \mathcal{B}_{\widetilde{p}}(\widetilde{r}), \quad \text{with } \widetilde{r} = c_p c_{\widetilde{p}}^{-1} r,$$

is a volume preserving map, therefore we have defined a volume preserving map between arbitrary p-balls.

5. Possible Applications

A uniform grid of a 3D domain D is a grid in which all the cells have the same volume. This is required in statistical applications, in computer graphics in the theory of deformable bodies (see, for example, Ref. [6] and the references therein) and in construction of wavelet bases of the space $L^2(D)$. A refinement process is needed for multiresolution analysis or for multigrid methods, when a grid is not fine enough to solve a problem accurately. A refinement of a 3D grid is called uniform when each cell is divided into a given number of smaller cells having the same volume. To be efficient in practice, a refinement procedure should also be a simple one. One efficient way to construct a uniform and refinable (UR) grid on a domain D is to map on D an existing UR grid by a volume preserving map. In our case, we can construct (UR) grids on a ball $\mathcal{B}_{p'}$ by transporting from a ball $\mathcal{B}_p$ an already constructed (UR) grid. The simplest example of such a ball with (UR) grids is the cube $\mathcal{B}_\infty$, but we have also constructed such (UR) grids on the regular octahedron $\mathcal{B}_1$ (see [3,4]) and on the 3D Euclidean ball $\mathcal{B}_2$ (see [3,7]).

The technique used in [3] can be easily adapted to the p-ball $\mathcal{B}_p$ in order to construct multiresolution analysis of $L^2(\mathcal{B}_p)$ and orthonormal wavelet bases on the p-ball $\mathcal{B}_p$.

The centers of the cells in our (UR) grids in $\mathcal{B}_p$ can be taken as points in interpolation formulas, as Monte Carlo interpolation or adaptive interpolation formulas.

Another application of volume preserving maps is in the theory of partial differential equations on Lipschitz domains (see [8]).

Author Contributions: Writing—original draft preparation, A.H.; Writing—review and editing, A.H., D.R.; Visualization, D.R.

References

1. Barr, A.H. Superquadrics and Angle-Preserving Transformations. *IEEE-CGA* **1981**, *1*, 11–23. [CrossRef]
2. Barr, A.H. Rigid Physically Based Superquadrics. In *Graphics Gems III*; Kirk, D., Ed.; Academic Press: San Diego, CA, USA, 1992; pp. 137–159.
3. Holhoş, A.; Roşca, D. Orhonormal Wavelet Bases on the 3D Ball via Volume Preserving Map from the Regular Octahedron. Available online: https://arxiv.org/abs/1910.08067 (accessed on 10 October 2019).
4. Holhoş, A.; Roşca, D. Uniform refinable 3D grids of regular convex polyhedrons and balls. *Acta Math. Hung.* **2018**, *156*, 182–193. [CrossRef]
5. Holhoş, A. Two Area Preserving Maps from the Square to the p-Ball. *Math. Modell. Anal.* **2017**, *22*, 157–166. [CrossRef]
6. Savoye, Y. *Cage-Based Performance Capture*; Springer: Basel, Switzerland, 2014.
7. Roşca, D.; Morawiec, A.; De Graef, M. A new method of constructing a grid in the space of 3D rotations and its applications to texture analysis. *Model. Simul. Mater. Sci. Eng.* **2014**, *22*, 075013. [CrossRef]
8. Griepentrog, J.; Höpner, W.; Kaiser, H.C.; Rehberg, J. A bi-Lipschitz continuous, volume preserving map from the unit ball onto a cube. *Note Mat.* **2008**, *28*, 177–193.

11

Minimal Energy Configurations of Finite Molecular Arrays

Pablo V. Negrón-Marrero * and Melissa López-Serrano

Department of Mathematics, University of Puerto Rico, Humacao, PR 00791-4300, USA; melissa.lopez3@upr.edu
* Correspondence: pablo.negron1@upr.edu

Abstract: In this paper, we consider the problem of characterizing the minimum energy configurations of a finite system of particles interacting between them due to attractive or repulsive forces given by a certain intermolecular potential. We limit ourselves to the cases of three particles arranged in a triangular array and that of four particles in a tetrahedral array. The minimization is constrained to a fixed area in the case of the triangular array, and to a fixed volume in the tetrahedral case. For a general class of intermolecular potentials we give conditions for the homogeneous configuration (either an equilateral triangle or a regular tetrahedron) of the array to be stable that is, a minimizer of the potential energy of the system. To determine whether or not there exist other stable states, the system of first-order necessary conditions for a minimum is treated as a bifurcation problem with the area or volume variable as the bifurcation parameter. Because of the symmetries present in our problem, we can apply the techniques of equivariant bifurcation theory to show that there exist branches of non-homogeneous solutions bifurcating from the trivial branch of homogeneous solutions at precisely the values of the parameter of area or volume for which the homogeneous configuration changes stability. For the triangular array, we construct numerically the bifurcation diagrams for both a Lennard–Jones and Buckingham potentials. The numerics show that there exist non-homogeneous stable states, multiple stable states for intervals of values of the area parameter, and secondary bifurcations as well.

Keywords: molecular arrays; constrained optimization; equivariant bifurcation theory

---

1. Introduction

Consider a system of N molecules, modeled as identical spherical particles, enclosed in a bounded region $\mathcal{B}$ of $\mathbb{R}^3$. At any given instant (or in an equilibrium configuration), the total potential energy of the molecular array is given by:

$$E = \sum_{i<j} \phi(\|\vec{\mathbf{r}}_i - \vec{\mathbf{r}}_j\|), \tag{1}$$

where ϕ is the intermolecular potential energy with $\|\cdot\|$ the standard Euclidean or two-vector norm, and $\vec{\mathbf{r}}_1, \ldots, \vec{\mathbf{r}}_N \in \mathbb{R}^3$ are the positions of the particles. More general energy potentials have been considered of which (1) is a special case (cf. [1,2]), or those based on the eigenvalues of adjacency matrices like in [3]. The problem of minimizing (1) subject to certain type of global or local conditions have been studied extensively (see e.g., [4–6] and the references there in). In these models either the array is infinite, with some local repeating structure, or finite but with $N \to \infty$. In this paper, none of these conditions are required but we expect that our results can be extrapolated to such more general scenarios. Also, we do not commit to any particular intermolecular potential ϕ (but give examples for instance for Lennard–Jones type potentials) so that our results are applicable to any such smooth potential.

The particular problems that we consider in this paper are those of characterizing the minimum energy configurations of (1) in the case of three particles ($N = 3$) arranged in a triangle and that of four particles ($N = 4$) in a tetrahedral array. The minimization problem is subject to the constraint of fixed area for the triangular array and of fixed volume in the tetrahedral case. We are particularly interested on the dependence of the minimizing states on the parameter of area or volume in the constraint. Both problems have the particularity that they can be formulated in terms of the intermolecular distances only, that is, without specifying the coordinates corresponding to the positions of the particles, thus substantially reducing the number of unknowns in each problem.

The motivation to study these problems comes from the following phenomena observed both in laboratory experiments and molecular dynamics simulations (see e.g., [7,8]). As the density of a fluid is progressively lowered (keeping the temperature constant), there is a certain "critical" density such that if the density of the fluid is lower than this critical value, then bubbles or regions with very low density appear within the fluid. This phenomenon is usually called "cavitation" and it has been extensively studied as well in solids. (See for instance [9,10] for discussions and further references.) When using discrete models of materials like (1), distinguishing between regions of low vs. high density, or whether bubbles or holes have form within the array, is not obvious since one is dealing essentially with a set of points. (See for instance [11,12] for the use of Voronoi polyhedra to study such arrays.) Thus, to study this phenomenon within this discrete model, one is naturally led to study or characterize the stability of homogeneous energy minimizing configurations of such an array, as the density of the array changes. The problems considered in this paper are the simplest problems within such a model. Our main contribution is on the application of global bifurcation theory (as opposed to just local) to study the set of equilibrium configurations for (1) under the stated constraints. In particular, to give specific conditions in terms of the intermolecular potential ϕ for the existence of nontrivial states.

In Section 3 we consider the problem of three particles. By Heron's formula for the area of a triangle, any three numbers (representing the intermolecular distances) that yield a positive value for the area formula, represent a triangle. In this case, we show that the functional (1) subject to the constraint of fixed area A, has for any value of A, a critical point representing an equilateral triangle. Moreover, in Theorem 2 we give a necessary and sufficient condition (cf. (22)), in terms of the intermolecular potential, for this equilibrium point to be a (local) minimizer of the energy functional. This condition leads to a set of values $\mathcal{A}$ for the area parameter A for which the equilateral triangle is a stable configuration. We give examples of how this set looks for various intermolecular potentials including the classical Lennard–Jones [13] and Buckingham [14] potentials, and those that model hard and soft springs including the usual Hooke's law.

Next in Section 3.2 we turn to the question of whether there exists other (not equilateral) equilibrium configurations for those values of the area parameter A for which the equilateral triangle becomes unstable, that is, when it ceases to be a local minimizer. To answer this question, we treat the system of equations characterizing the equilibrium points (cf. (13)) as a bifurcation problem with the parameter A as a bifurcation parameter, and the set of equilateral equilibrium configurations as the trivial solution branch. We find that the necessary condition for bifurcation from the trivial branch for this system occurs exactly at the boundary points of the set $\mathcal{A}$ given by the stability condition (22). To check the sufficiency condition for bifurcation, one must consider the linearization of the system (13) about the trivial branch at a boundary point A_0 of $\mathcal{A}$. However, since the kernel of this linearization is two-dimensional we cannot immediately apply the usual or standard results from bifurcation theory (cf. [15–18]). Because of the symmetries present in this problem (cf (32)), we can apply bifurcation equivariant theory (cf. [16,19]) to construct a suitable reduced problem corresponding to isosceles triangular equilibrium configurations. The linearization of the reduced problem at the point where $A = A_0$ has now a one-dimensional kernel and provided that a certain transversality condition is satisfied (cf. (34)), we can show that there are three branches corresponding to isosceles triangles bifurcating from the trivial branch at the point where $A = A_0$. Since the stability of these bifurcating branches can only be determined numerically (because one must linearize about an unknown solution), in Section 5 we construct numerically the bifurcation diagrams, with their respective stability patterns, for instances of the Lennard–Jones and Buckingham potentials. These examples show that the primary bifurcations off the trivial branch are of trans-critical type, and that at least for the Lennard–Jones potential, there are secondary bifurcations corresponding to stable scalene triangles. Moreover, there are intervals of values of the area parameter, for which there exists multiple stable states of the system for each value of A in such an interval.

In Section 4 we consider an array of four molecules in a tetrahedron. The general treatment in this case is similar to the three particle case but with two main differences. First the characterization of when six numbers (representing the lengths of the sides of the tetrahedron) determine a tetrahedron, is given in terms of the Cayley-Menger determinant and the triangle inequalities of one of its faces (cf. [20]). The next complication arises from the fact that the tetrahedron has 24 symmetries as compared to only six for the triangle! To deal with this many possibilities, once again we make use of the basic techniques of equivariant theory to get suitable reduced problems to work with. As the Cayley–Menger determinant is proportional to the volume of the corresponding tetrahedron, the volume constraint in our problem is basically that of setting this determinant to a given value V for the volume. In this case we show in Section 4.1 that the functional (1) subject to the constraint of fixed volume V, has for any value of V a critical point representing a regular or equilateral tetrahedron. Moreover, in Theorem 4 we give necessary and sufficient conditions (cf. (43)), in terms of the intermolecular potential, for this equilibrium point to be a (local) minimizer of the energy functional. As for the triangular case, these conditions determine a set of values $\mathcal{V}$ for the volume parameter V for which the equilateral tetrahedron is a stable configuration.

In Section 4.2 we consider the question of the existence of non-equilateral equilibrium configurations. The equilibrium configurations in this case are given as solutions to a nonlinear system of seven equations in eight unknowns (cf. (40)). We treat this system as a bifurcation problem with the parameter V as a bifurcation parameter, and the set of equilateral tetrahedrons as the trivial solution branch. The necessary condition for bifurcation from the trivial branch for this system occurs exactly at the boundary points of the set $\mathcal{V}$ given by the stability conditions (43). At a boundary point V_0 of $\mathcal{V}$ there are two possibilities: the kernel of the linearization has dimension two or three. Using some of the machinery of equivariance theory as in [16], we can construct suitable reduced problems in each of these two cases, which enables us to establish the existence of non-equilateral equilibrium configurations and to get a full description of the symmetries of the bifurcating branches (cf. Theorems 5–7). As in the triangular case, the stability of this bifurcating branches can only be established numerically because one must linearize about the unknown bifurcating branch.

Notation: We let $\mathbb{R}^n$ denote the n dimensional space of column vectors with elements denoted by $\vec{\mathbf{x}}$, $\vec{\mathbf{y}}$,... The inner product of $\vec{\mathbf{x}}, \vec{\mathbf{y}} \in \mathbb{R}^n$ is denoted either by $\langle \vec{\mathbf{x}}, \vec{\mathbf{y}} \rangle$ or $\vec{\mathbf{x}}^t \vec{\mathbf{y}}$, where the superscript "t" denotes transpose. We denote the set of $n \times m$ matrices by $\mathbb{R}^{n \times m}$. For $L \in \mathbb{R}^{n \times m}$, we let $\ker(L) = \{\vec{\mathbf{x}} \in \mathbb{R}^n : L\vec{\mathbf{x}} = \vec{\mathbf{0}}\}$ and $\mathrm{Range}(L) = \{L\vec{\mathbf{x}} : \vec{\mathbf{x}} \in \mathbb{R}^n\}$. For a function $\vec{\mathbf{F}} : \mathbb{R}^n \to \mathbb{R}^m$, we denote its Fréchet derivative by $\mathrm{D}\vec{\mathbf{F}}$ which is given by the $m \times n$ matrix of partial derivatives of the components of $\vec{\mathbf{F}}$. If the variables in $\vec{\mathbf{F}}$ are given by $(\vec{\mathbf{x}}, \vec{\mathbf{y}})$, then $\mathrm{D}_{\vec{\mathbf{x}}}\vec{\mathbf{F}}$ denotes the derivative of $\vec{\mathbf{F}}$ with respect to the vector of variables $\vec{\mathbf{x}}$, i.e., the matrix of the partial derivatives of the components of $\vec{\mathbf{F}}$ with respect to the variables corresponding to $\vec{\mathbf{x}}$.

2. Equivariant Bifurcation from a Simple Eigenvalue

In this section, we provide an overview on some of the basic results on bifurcation theory from a simple eigenvalue for mappings between finite dimensional spaces, where the maps possess certain symmetries. The literature on this subject is extensive but we refer to [6,15,17] for details on the material presented in this section and further developments like for instance, the infinite dimensional case.

Let $\vec{\mathbf{F}} : \mathcal{U} \times \mathbb{R} \to \mathbb{R}^n$ where $\mathcal{U}$ is an open subset of $\mathbb{R}^n$, be a C^2 function, and consider the problem of characterizing the solution set of:

$$\vec{\mathbf{F}}(\vec{\mathbf{x}}, A) = \vec{\mathbf{0}}, \quad (\vec{\mathbf{x}}, A) \in \mathcal{U} \times \mathbb{R}. \tag{2}$$

We assume that there exists a (known) smooth function $\vec{\mathbf{g}}(\cdot)$ such that:

$$\vec{\mathbf{F}}(\vec{\mathbf{g}}(A), A) = \vec{\mathbf{0}}, \quad \forall\ A.$$

The set $\mathcal{T} = \{(\vec{\mathbf{g}}(A), A) : A \in \mathbb{R}\}$ is called the *trivial branch* of solutions of (2). We say that $(\vec{\mathbf{x}}_0, A_0) \in \mathcal{T}$ is a *bifurcation point* off the trivial branch $\mathcal{T}$, if every neighborhood of $(\vec{\mathbf{x}}_0, A_0)$ contains solutions of (2) not in $\mathcal{T}$. If we let

$$L(A) = \mathrm{D}_{\vec{\mathbf{x}}}\vec{\mathbf{F}}(\vec{\mathbf{g}}(A), A),$$

then by the Implicit Function Theorem, a necessary for $(\vec{\mathbf{x}}_0, A_0)$ to be a bifurcation point is that $L(A_0)$ must be singular, a condition well known to be not sufficient.

In many applications of bifurcation theory and for the problems considered in this paper, the mapping $\vec{\mathbf{F}}$ possesses symmetries due to the geometry of the underlying physical problem. The use of these symmetries in the analysis is useful for example to deal with problems in which $\dim\ker(L(A_0)) > 1$. Thus, we assume that for a proper subgroup $\mathcal{G}$ of $\mathbb{R}^{n\times n}$, characterizing the symmetries in the problem, the mapping $\vec{\mathbf{F}}$ satisfies:

$$\vec{\mathbf{F}}(P\vec{\mathbf{x}}, A) = P\vec{\mathbf{F}}(\vec{\mathbf{x}}, A), \quad \forall\, P \in \mathcal{G}. \tag{3}$$

Let $\vec{\mathbf{v}} \in \ker(L(A_0))$ and define the *isotropy subgroup* of $\mathcal{G}$ at $\vec{\mathbf{v}}$ by

$$\mathcal{H} = \{P \in \mathcal{G} : P\vec{\mathbf{v}} = \vec{\mathbf{v}}\}, \tag{4}$$

and the $\mathcal{H}$*–fixed point set* by

$$\mathbb{R}^n_{\mathcal{H}} = \{\vec{\mathbf{x}} \in \mathbb{R}^n : P\vec{\mathbf{x}} = \vec{\mathbf{x}},\ \forall P \in \mathcal{H}\}. \tag{5}$$

Clearly $\vec{\mathbf{v}} \in \mathbb{R}^n_{\mathcal{H}}$.

Let $\mathbb{P}_{\mathcal{H}} : \mathbb{R}^n \to \mathbb{R}^n$ be a linear map that projects onto $\mathbb{R}^n_{\mathcal{H}}$ that is $\mathrm{Range}(\mathbb{P}_{\mathcal{H}}) = \mathbb{R}^n_{\mathcal{H}}$ and $\mathbb{P}_{\mathcal{H}}(\mathbb{R}^n_{\mathcal{H}}) = \mathbb{R}^n_{\mathcal{H}}$. With $\mathcal{U}_{\mathcal{H}} = \mathbb{P}_{\mathcal{H}}(\mathcal{U}) = \mathcal{U} \cap \mathbb{R}^n_{\mathcal{H}}$, we define $\vec{\mathbf{F}}_{\mathcal{H}} : \mathcal{U}_{\mathcal{H}} \times \mathbb{R} \to \mathbb{R}^n_{\mathcal{H}}$ by:

$$\vec{\mathbf{F}}_{\mathcal{H}}(\vec{\mathbf{u}}, A) = \mathbb{P}_{\mathcal{H}}\vec{\mathbf{F}}(\vec{\mathbf{u}}, A), \quad (\vec{\mathbf{u}}, A) \in \mathcal{U}_{\mathcal{H}} \times \mathbb{R}. \tag{6}$$

An easy calculation now gives that

$$\mathrm{D}_{\vec{\mathbf{u}}}\vec{\mathbf{F}}_{\mathcal{H}}(\vec{\mathbf{u}}, A) = \mathbb{P}_{\mathcal{H}}\vec{\mathbf{F}}_{\vec{\mathbf{x}}}(\vec{\mathbf{u}}, A)\mathbb{P}_{\mathcal{H}}.$$

We assume that $\vec{\mathbf{g}}(A) \in \mathbb{R}^n_{\mathcal{H}}$ for all A, so that

$$\vec{\mathbf{F}}_{\mathcal{H}}(\vec{\mathbf{g}}(A), A) = \vec{\mathbf{0}}, \quad \forall\ A.$$

It follows now that $L_{\mathcal{H}}(A) : \mathbb{R}^n_{\mathcal{H}} \to \mathbb{R}^n_{\mathcal{H}}$ is given by:

$$L_{\mathcal{H}}(A) = \mathrm{D}_{\vec{\mathbf{u}}}\vec{\mathbf{F}}_{\mathcal{H}}(\vec{\mathbf{g}}(A), A) = \mathbb{P}_{\mathcal{H}}L(A)\mathbb{P}_{\mathcal{H}}.$$

Clearly $\vec{\mathbf{v}} \in \ker(L_{\mathcal{H}}(A_0))$. The $\mathcal{H}$*-reduced problem* is now given by:

$$\vec{\mathbf{F}}_{\mathcal{H}}(\vec{\mathbf{u}}, A) = \vec{\mathbf{0}}, \quad (\vec{\mathbf{u}}, A) \in \mathcal{U}_{\mathcal{H}} \times \mathbb{R}. \tag{7}$$

An important property relating (2) and (7) is that $(\vec{\mathbf{x}}, A) \in \mathcal{U}_{\mathcal{H}} \times \mathbb{R}$ is a solution of (2) if and only if $(\vec{\mathbf{x}}, A)$ is a solution of (7). The following result provides the required sufficient conditions for $(\vec{\mathbf{x}}_0, A_0)$ to be a bifurcation point of the $\mathcal{H}$-reduced problem.

Theorem 1 (Equivariant Bifurcation Theorem [19]). *Assume that for $A = A_0$ there exists $\vec{\mathbf{v}} \in \ker(L(A_0))$ that defines a proper isotropy subgroup $\mathcal{H}$ such that:*

$$\ker(L_{\mathcal{H}}(A_0)) = \text{span}\,\{\vec{\mathbf{v}}\}, \quad L'_{\mathcal{H}}(A_0)\vec{\mathbf{v}} \notin \text{Range}(L_{\mathcal{H}}(A_0)).$$

Then there exists a branch $\mathcal{C}_{\mathcal{H}}$ of nontrivial solutions of $\vec{\mathbf{F}}_{\mathcal{H}}(\vec{\mathbf{u}}, A) = \vec{\mathbf{0}}$ bifurcating from the trivial branch $\mathcal{T}$ at the point where $A = A_0$ and such that either:

1. *$\mathcal{C}_{\mathcal{H}}$ is unbounded in $\mathbb{R}^{n+1}$;*
2. *the closure of $\mathcal{C}_{\mathcal{H}}$ intersects the boundary $\partial\mathcal{U}$ of $\mathcal{U}$;*
3. *$\mathcal{C}_{\mathcal{H}}$ intersects $\mathcal{T}$ at a point $(\vec{\mathbf{x}}_*, A_*)$ where $A_* \neq A_0$.*

The proof of this theorem is basically an application of a result from Krasnoselski [18] that uses the homotopy invariance of the topological degree. The three alternatives in the statement of the theorem are usually referred to as the Crandall and Rabinowitz alternatives. The local version of this result (cf. [6]) that is, without the Crandall and Rabinowitz alternatives, can be obtained via the Lyapunov–Schmidt reduction method. A useful consequence of this reduction is an approximate formula for the bifurcating branch in a neighborhood of the bifurcation point. Let $\ker(L_{\mathcal{H}}(A_0)^t) = \text{span}\,\{\vec{\mathbf{v}}^*\}$, where $\langle \vec{\mathbf{v}}^*, \vec{\mathbf{v}} \rangle = 1$, so that $\text{Range}(L_{\mathcal{H}}(A_0)) = \{\vec{\mathbf{y}} \in \mathbb{R}^n_{\mathcal{H}} \, : \, \langle \vec{\mathbf{v}}^*, \vec{\mathbf{y}} \rangle = 0\}$. Now if we define

$$A^0 = \langle \vec{\mathbf{v}}^*, (\mathrm{D}_{\vec{\mathbf{u}}\vec{\mathbf{u}}}\vec{\mathbf{F}}^0_{\mathcal{H}}\vec{\mathbf{v}})\vec{\mathbf{v}} \rangle, \quad B^0 = \langle \vec{\mathbf{v}}^*, L'_{\mathcal{H}}(A_0)\vec{\mathbf{v}} \rangle. \tag{8}$$

(here the zero superscripts mean evaluated at $(\vec{\mathbf{x}}_0, A_0)$), then the bifurcating branch have the following asymptotic expansion (cf. [16]):

$$(\vec{\mathbf{x}}, A) = \left(\vec{\mathbf{g}}(A_0 + \varepsilon) + \varepsilon m \vec{\mathbf{v}} + O(\varepsilon^2), A_0 + \varepsilon\right), \tag{9}$$

where

$$m = -\frac{2B^0}{A^0}, \quad A^0 \neq 0. \tag{10}$$

3. The Three Particle Case

In this section, we consider the case in which the molecular array consists of three particles. The intermolecular energy is given by a smooth function $\phi \, : \, (0, \infty) \to \mathbb{R}$ called the *potential*. If (a, b, c) are the distances between the particles in the array, the total energy of the system is given by:

$$E(a, b, c) = \phi(a) + \phi(b) + \phi(c), \quad a, b, c > 0. \tag{11}$$

Also, the square of the area of the triangular array is given by Heron's formula:

$$g(a, b, c) \equiv s(s - a)(s - b)(s - c),$$

where $s = (a + b + c)/2$.

For any given number $A > 0$, we consider the following constrained minimization problem:

$$\begin{cases} \min\limits_{a,b,c>0} E(a, b, c) \\ \text{subject to } g(a, b, c) = A^2. \end{cases} \tag{12}$$

Thus, we are looking for minimizers of the energy functional E subject to the constraint that the area of the array is A. The first-order necessary conditions for a solution of this problem are given by:

$$\left\{ \begin{array}{rcl} g(a,b,c) - A^2 & = & 0, \\ \vec{\nabla} E(a,b,c) + \lambda \vec{\nabla} g(a,b,c) & = & \vec{\mathbf{0}}, \end{array} \right. \tag{13}$$

where $\lambda \in \mathbb{R}$ is the Lagrange multiplier corresponding to the restriction $g(a,b,c) = A^2$. For any given value of $A > 0$, this is a nonlinear system of equations for the unknowns (λ, a, b, c) in terms of A. In general this system can have multiple solutions depending on the characteristics of the potential ϕ and the value of A.

3.1. Existence and Stability of Trivial States

An easy calculation shows that:

$$\vec{\nabla} E = \begin{bmatrix} \phi'(a) \\ \phi'(b) \\ \phi'(c) \end{bmatrix}, \quad \vec{\nabla} g = \frac{1}{4} \begin{bmatrix} a(b^2 + c^2 - a^2) \\ b(a^2 + c^2 - b^2) \\ c(a^2 + b^2 - c^2) \end{bmatrix}. \tag{14}$$

Thus, the system (13) is equivalent to:

$$\left\{ \begin{array}{rcl} \frac{1}{8}(a^2b^2 + a^2c^2 + b^2c^2) - \frac{1}{16}(a^4 + b^4 + c^4) - A^2 & = & 0, \\ \phi'(a) + \frac{\lambda}{4} a(b^2 + c^2 - a^2) & = & 0, \\ \phi'(b) + \frac{\lambda}{4} b(a^2 + c^2 - b^2) & = & 0, \\ \phi'(c) + \frac{\lambda}{4} c(a^2 + b^2 - c^2) & = & 0. \end{array} \right. \tag{15}$$

This system always has a solution with $a = b = c$. In fact, upon setting $a = b = c$ in (15), this system reduces to:

$$\frac{3}{16} a^4 = A^2, \quad \phi'(a) = -\frac{\lambda a^3}{4}. \tag{16}$$

Thus, we have the following result:

Lemma 1. *For any value of $A > 0$, the system* (15) *has a solution of the form $(\lambda_A, a_A, a_A, a_A, A)$ where:*

$$a_A = \frac{2\sqrt{A}}{\sqrt[4]{3}}, \quad \lambda_A = -\frac{4\phi'(a_A)}{a_A^3}. \tag{17}$$

We now characterize for which values of A, the solution provided in Lemma 16 is actually a minimizer, i.e., a solution of (12). To do this we need to examine the matrix $\left[\nabla^2 E + \lambda \nabla^2 g\right](\vec{\mathbf{v}}_A)$ where $\vec{\mathbf{v}}_A = (\lambda_A, a_A, a_A, a_A)$, over the subspace:

$$\mathcal{M} = \left\{ (x,y,z) \; : \; \vec{\nabla} g(\vec{\mathbf{v}}_A) \cdot (x,y,z) = 0 \right\}. \tag{18}$$

A straightforward calculation gives that:

$$\nabla^2 E = \begin{bmatrix} \phi''(a) & 0 & 0 \\ 0 & \phi''(b) & 0 \\ 0 & 0 & \phi''(c) \end{bmatrix}, \tag{19}$$

$$\nabla^2 g = \frac{1}{4} \begin{bmatrix} b^2 + c^2 - 3a^2 & 2ab & 2ac \\ 2ab & a^2 + c^2 - 3b^2 & 2bc \\ 2ac & 2bc & a^2 + b^2 - 3c^2 \end{bmatrix}. \tag{20}$$

It follows now that

$$\left[\nabla^2 E + \lambda \nabla^2 g\right](\vec{\mathbf{v}}_A) = \begin{bmatrix} \phi''(a_A) - \frac{\lambda_A a_A^2}{4} & \frac{\lambda_A a_A^2}{2} & \frac{\lambda_A a_A^2}{2} \\ \frac{\lambda_A a_A^2}{2} & \phi''(a_A) - \frac{\lambda_A a_A^2}{4} & \frac{\lambda_A a_A^2}{2} \\ \frac{\lambda_A a_A^2}{2} & \frac{\lambda_A a_A^2}{2} & \phi''(a_A) - \frac{\lambda_A a_A^2}{4} \end{bmatrix}, \tag{21}$$

and that

$$\mathcal{M} = \{(x, y, z) \,:\, x + y + z = 0\}.$$

Using (17) one can show now that the matrix (21) is positive definite over $\mathcal{M}$ if and only if:

$$\phi''(a_A) + \frac{3}{a_A}\phi'(a_A) > 0. \tag{22}$$

Thus, we have the following result:

Theorem 2. *Let $\phi : (0, \infty) \to \mathbb{R}$ be twice continuously differentiable function. Then the uniform array $(a, b, c) = (a_A, a_A, a_A)$ in Lemma 1 is a relative minimizer for the problem* (12) *for those values of A for which* (22) *holds.*

Example 1. *Consider the case of a potential that has the following form:*

$$\phi(r) = \frac{c_1}{r^{\delta_1}} - \frac{c_2}{r^{\delta_2}}, \tag{23}$$

where c_1, c_2 are positive constants and $\delta_1 > \delta_2 > 2$. (These constants determine the physical properties of the particle or molecule in question. The classical Lennard–Jones [13] potential is obtained upon setting $\delta_1 = 12$ and $\delta_2 = 6$.) For this function:

$$\phi'(r) = -\frac{c_1\delta_1}{r^{\delta_1+1}} + \frac{c_2\delta_2}{r^{\delta_2+1}}, \quad \phi''(r) = \frac{c_1\delta_1(\delta_1+1)}{r^{\delta_1+2}} - \frac{c_2\delta_2(\delta_2+1)}{r^{\delta_2+2}},$$

so that:

$$\phi''(r) + \frac{3}{r}\phi'(r) = \frac{c_1\delta_1(\delta_1-2)}{r^{\delta_1+2}} - \frac{c_2\delta_2(\delta_2-2)}{r^{\delta_2+2}}.$$

Since a_A is directly proportional to $\sqrt{A}$ (see (17)*), we have that for* (23)*, the stability condition* (22) *holds if and only if $A < A_0$, where A_0 is determined from the condition:*

$$\frac{c_1\delta_1(\delta_1-2)}{a_A^{\delta_1+2}} - \frac{c_2\delta_2(\delta_2-2)}{a_A^{\delta_2+2}} = 0,$$

from which it follows that:

$$A_0 = \frac{\sqrt{3}}{4}\left[\frac{c_1\delta_1(\delta_1-2)}{c_2\delta_2(\delta_2-2)}\right]^{\frac{2}{\delta_1-\delta_2}}. \tag{24}$$

Thus, (17) *is a (local) solution of* (12) *if and only if $A < A_0$. We will show that for $A > A_0$ there exist solutions that break the symmetry $a = b = c$.*

Example 2. *A Buckingham potential has the form ([14,21]):*

$$\phi(r) = \alpha e^{-\beta r} - \frac{\Upsilon}{r^\eta}, \tag{25}$$

where $\alpha, \beta, \gamma, \eta$ are positive constants. Thus

$$\phi'(r) = -\alpha\beta e^{-\beta r} + \frac{\gamma\eta}{r^{\eta+1}}, \quad \phi''(r) = \alpha\beta^2 e^{-\beta r} - \frac{\gamma\eta(\eta+1)}{r^{\eta+2}},$$

from which it follows that:

$$\phi''(r) + \frac{3}{r}\phi'(r) = \alpha\beta\left[\beta - \frac{3}{r}\right]e^{-\beta r} - \frac{\gamma\eta(\eta-2)}{r^{\eta+2}}.$$

After rearrangement, the stability condition (22) is equivalent to:

$$F(a_A) > G(a_A), \tag{26}$$

where

$$F(r) = \alpha\beta(\beta r - 3)e^{-\beta r}, \quad G(r) = \frac{\gamma\eta(\eta-2)}{r^{\eta+1}}.$$

These functions generically look as in Figure 1 where for G we assumed that $\eta > 2$. Now clearly $F(r) < G(r)$ for r sufficiently large. Thus generically we expect the set of values of A for which (26) is satisfied to be of the form (A_0, A_1). Since F has a maximum at $r_m = \frac{4}{\beta}$, a sufficient condition for this is that $F(r_m) > G(r_m)$, or after rearrangement that the coefficients and exponents in (25) satisfy:

$$\alpha\beta e^{-4} > \gamma\eta(\eta-2)\left(\frac{\beta}{4}\right)^{\eta+1}. \tag{27}$$

To check this condition against the results in [21], we let $D, R, \xi > 0$ with $\xi > \eta$, and define

$$\alpha = \frac{D\eta e^{\xi}}{\xi - \eta}, \quad \beta = \frac{\xi}{R}, \quad \gamma = \frac{D\xi R^{\eta}}{\xi - \eta}. \tag{28}$$

It follows that (25) is now given in terms of D, R, ξ by:

$$\phi(r) = D\left[\frac{\eta}{\xi - \eta}e^{\xi(1-\frac{r}{R})} - \frac{\xi}{\xi - \eta}\left(\frac{R}{r}\right)^{\eta}\right].$$

It is easy to check now that provided $\xi > \eta + 1$, then ϕ has negative minimum value at $r = R$. The results in ([21], Table 1, Page 202) show that the best fit of a normalized Buckingham potential to a normalized Lennard–Jones (12-6) potential ($\delta_1 = 12$ and $\delta_2 = 6$ in (23)) is achieved for $\xi = 14.3863$ and $\eta = 5.6518$. For these values one can check that (28) satisfy the inequality (27) independent of the values of D and R.

Example 3. *Consider a potential of the form*

$$\phi(r) = \frac{1}{2}kr^2 + \frac{1}{4}\beta r^4, \tag{29}$$

with $k > 0$ and $\beta \in \mathbb{R}$. This potential corresponds to a Hook-type spring when $\beta = 0$, a hard spring if $\beta > 0$, and a soft spring if $\beta < 0$. (More general versions of (29) have been used in the study of the control of multi-agent systems, e.g., [22,23].) For this potential

$$\phi''(r) + \frac{3}{r}\phi'(r) = 4k + 6\beta r^2.$$

Please note that the stability condition (22) holds when $\beta \geq 0$ independent of the value of A! That is, the symmetric state (17) is a minimizer for all values of A. In the case $\beta = 0$ is easy to show that this state is a global minimizer.

On the other hand, if $\beta < 0$, the stability condition holds if and only if $A < A_0$ where

$$A_0 = -\frac{k}{2\sqrt{3}\,\beta}.$$

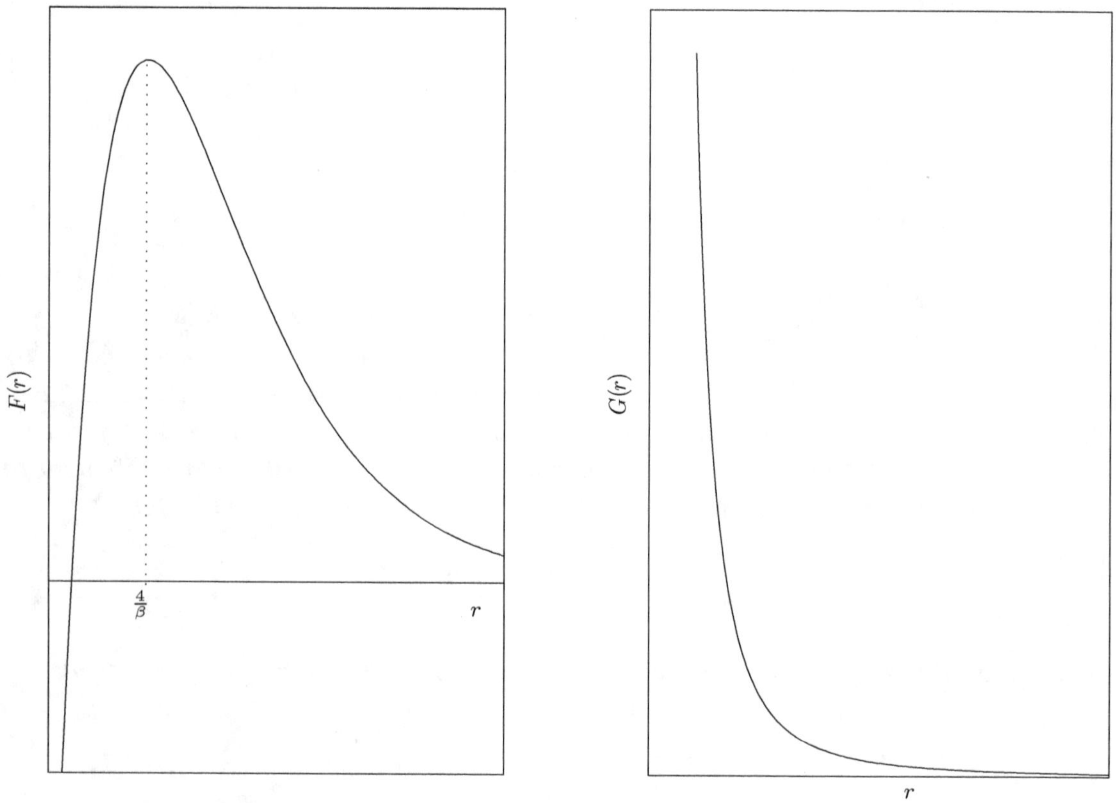

Figure 1. Generic graphs of the functions F and G appearing in the stability condition (26) for a Buckingham potential.

3.2. Existence and Stability of Nontrivial Solutions

We say that solutions of (13) are *trivial* if $a = b = c$ and call the set

$$\mathcal{T} = \{(\lambda_A, a_A, a_A, a_A, A) \,:\, \lambda_A, a_A \text{ given by (17)},\ A > 0\}, \tag{30}$$

the *trivial branch* parametrized by A. In this section, we show that there exist nontrivial solutions of (13) that bifurcate from the trivial branch.

If we let $\vec{\mathbf{x}} = (\lambda, a, b, c)$ and $\vec{\mathbf{G}} : \mathbb{R} \times (0, \infty)^4 \to \mathbb{R}^4$ be the left-hand side of (15), then this system is equivalent to $\vec{\mathbf{G}}(\vec{\mathbf{x}}, A) = \vec{\mathbf{0}}$. An easy calculation gives that

$$\mathrm{D}_{\vec{\mathbf{x}}}\vec{\mathbf{G}}(\vec{\mathbf{x}}, A) = \begin{bmatrix} 0 & (\vec{\nabla} g)^t \\ \vec{\nabla} g & \nabla^2 E + \lambda \nabla^2 g \end{bmatrix}. \tag{31}$$

If we evaluate now at the trivial branch (17), we get that

$$\mathrm{D}_{\vec{\mathbf{x}}}\vec{\mathbf{G}}(\vec{\mathbf{v}}_A, A) = \begin{bmatrix} 0 & \gamma & \gamma & \gamma \\ \gamma & \alpha & \beta & \beta \\ \gamma & \beta & \alpha & \beta \\ \gamma & \beta & \beta & \alpha \end{bmatrix},$$

where $\vec{\mathbf{v}}_A = (\lambda_A, a_A, a_A, a_A)$ and

$$\alpha = \phi''(a_A) - \frac{\lambda_A a_A^2}{4}, \quad \beta = \frac{\lambda_A a_A^2}{2}, \quad \gamma = \frac{a_A^3}{4}.$$

The eigenvalues of this matrix are $\alpha - \beta$, which is a double eigenvalue with geometric multiplicity two, and the simple eigenvalues

$$\frac{1}{2}\left[\alpha + 2\beta \pm \sqrt{(\alpha+2\beta)^2 + 12\gamma^2}\right],$$

which are always nonzero. A pair of linearly independent eigenvectors corresponding to $\alpha - \beta$ is $\{(0,-1,1,0)^t, (0,-1,0,1)^t\}$. Since

$$\alpha - \beta = \phi''(a_A) - \frac{3}{4}\lambda_A a_A^2 = \phi''(a_A) + \frac{3}{a_A}\phi'(a_A),$$

the double eigenvalue $\alpha - \beta$ becomes zero exactly at the value A_0 where the stability condition (22) fails by becoming zero. Thus according to standard theory of bifurcation theory, we can have either none, two or four branches of solutions of (15) bifurcating at the point where $A = A_0$. We now show that there are exactly four branches bifurcating from such a point: the trivial branch and three branches corresponding to isosceles triangles.

To avoid the complications of dealing with the two-dimensional kernel of (31) when evaluated at the trivial branch at $A = A_0$, we make use of the symmetries possessed by the mapping $\vec{\mathbf{G}}$. In particular, if we denote by $\mathcal{G}$ the subgroup of $\mathbb{R}^{4\times 4}$ of permutations that permute just the a, b, c components of any $\vec{\mathbf{x}} = (\lambda, a, b, c) \in \mathbb{R}^4$, then

$$\vec{\mathbf{G}}(P\vec{\mathbf{x}}, A) = P\vec{\mathbf{G}}(\vec{\mathbf{x}}, A), \quad \forall P \in \mathcal{G}. \tag{32}$$

Please note that every permutation in $\mathcal{G}$ changes the eigenvectors $\{(0,-1,1,0)^t, (0,-1,0,1)^t\}$ of $\alpha - \beta$. However, the eigenvector

$$\vec{\mathbf{v}} \equiv (0,-2,1,1)^t = (0,-1,1,0)^t + (0,-1,0,1)^t,$$

is unchanged by the proper subgroup of permutations $\mathcal{H}$ of $\mathcal{G}$ that permutes just the b, c components of any $\vec{\mathbf{x}} = (\lambda, a, b, c) \in \mathbb{R}^4$. Thus, $\mathcal{H}$ is the isotropy subgroup of $\mathcal{G}$ at $\vec{\mathbf{v}}$. The $\mathcal{H}$ fixed point set is given by:

$$\mathbb{R}^4_{\mathcal{H}} = \{(\lambda, a, b, b)^t \,:\, \lambda, a, b \in \mathbb{R}\}.$$

The projection $\mathbb{P}_{\mathcal{H}} : \mathbb{R}^4 \to \mathbb{R}^4_{\mathcal{H}}$ has matrix representation:

$$\mathbb{P}_{\mathcal{H}} = \begin{bmatrix} 1 & 0 & 0 & 0 \\ 0 & 1 & 0 & 0 \\ 0 & 0 & \frac{1}{2} & \frac{1}{2} \\ 0 & 0 & \frac{1}{2} & \frac{1}{2} \end{bmatrix},$$

and the $\mathcal{H}$ reduced problem is now:

$$\vec{\mathbf{G}}_{\mathcal{H}}(\vec{\mathbf{u}}, A) \equiv \mathbb{P}_{\mathcal{H}}\vec{\mathbf{G}}(\vec{\mathbf{u}}, A) = \vec{\mathbf{0}}, \quad (\vec{\mathbf{u}}, A) \in \mathbb{R}^4_{\mathcal{H}} \times (0, \infty).$$

Since $\mathcal{T} \subset \mathbb{R}^4_{\mathcal{H}} \times (0,\infty)$, it follows that $\mathcal{T}$ is a branch of solutions for the $\mathcal{H}$ reduced problem. Also, since

$$\mathrm{D}_{\vec{\mathbf{u}}}\vec{\mathbf{G}}_{\mathcal{H}}(\vec{\mathbf{u}}, A) = \mathbb{P}_{\mathcal{H}}\mathrm{D}_{\vec{\mathbf{x}}}\vec{\mathbf{G}}(\vec{\mathbf{u}}, A)\mathbb{P}_{\mathcal{H}}.$$

we have that $L_{\mathcal{H}}(A) : \mathbb{R}^4_{\mathcal{H}} \to \mathbb{R}^4_{\mathcal{H}}$ is given by:

$$L_{\mathcal{H}}(A) = \mathrm{D}_{\vec{\mathbf{u}}}\vec{\mathbf{G}}_{\mathcal{H}}(\vec{\mathbf{v}}_A, A) = \mathbb{P}_{\mathcal{H}}\mathrm{D}_{\vec{\mathbf{x}}}\vec{\mathbf{G}}(\vec{\mathbf{v}}_A, A)\mathbb{P}_{\mathcal{H}}.$$

Let

$$\mu(A) = \alpha - \beta = \phi''(a_A) + \frac{3}{a_A}\phi'(a_A). \tag{33}$$

We now establish a result for the existence of bifurcating branches for the reduced problem.

Theorem 3. *Let $\mu(A_0) = 0$ and assume that*

$$\frac{\mathrm{d}\mu}{\mathrm{d}A}(A_0) \neq 0. \tag{34}$$

Consider the system (13) *and its trivial branch of solutions* (30). *Then from the point $(\lambda_0, a_0, a_0, a_0) \in \mathcal{T}$ bifurcate three branches of nontrivial solutions of* (13) *each corresponding to isosceles triangles.*

Proof. With the definitions and notation as above, a lengthy but otherwise elementary calculation shows that for any $A > 0$, $\mu(A)$ is a simple eigenvalue of $L_{\mathcal{H}}(A)$ restricted to $\mathbb{R}^4_{\mathcal{H}}$ with corresponding eigenvector $\vec{\mathbf{v}} = (0, -2, 1, 1)^t$. Thus

$$L_{\mathcal{H}}(A)\vec{\mathbf{v}} = \mu(A)\vec{\mathbf{v}}, \quad \forall A > 0.$$

In particular $\ker(L_{\mathcal{H}}(A_0)) = \mathrm{span}\{\vec{\mathbf{v}}\}$. If we differentiate with respect to A in the equation above and set $A = A_0$, we get that

$$L'_{\mathcal{H}}(A_0)\vec{\mathbf{v}} = \mu'(A_0)\vec{\mathbf{v}}.$$

Since $L_{\mathcal{H}}(A_0)$ is symmetric, we have that $\mathrm{Range}(L_{\mathcal{H}}(A_0)) = \{\vec{\mathbf{y}} \in \mathbb{R}^n_{\mathcal{H}} : \langle \vec{\mathbf{v}}, \vec{\mathbf{y}} \rangle = 0\}$. Thus, the hypotheses in Theorem 1 are satisfied if and only if $\mu'(A_0) \neq 0$. Thus, we get a branch of solutions of the reduced problem, equivalently (15), bifurcating from the trivial branch at the point where $A = A_0$. Since this branch belongs to $\mathbb{R}^4_{\mathcal{H}} \times \mathbb{R}$, we can use (32) to get that there exist two additional branches of solutions, one belonging to $\left\{(\lambda, b, a, b, A)^t : \lambda, a, b, A \in \mathbb{R}\right\}$ and the other in $\left\{(\lambda, b, b, a, A)^t : \lambda, a, b, A \in \mathbb{R}\right\}$. □

4. Four Particles in a Tetrahedron

We now consider the case of four particles arranged in a tetrahedron T. Let a, b, c, A, B, C be the distances between the particles where a, b, c denote the lengths of the edges joining a vertex of T, A the length of the edge opposite to a, B the length of the edge opposite to b, and C the length of the edge opposite to c. The six-tuple $\vec{\mathbf{a}} = (a, b, c, A, B, C)^t$ generates a tetrahedron ([20]) if and only if

$$g(\vec{\mathbf{a}}) > 0, \quad A < B + C, \quad B < A + C, \quad C < A + B, \tag{35}$$

where $g(\vec{\mathbf{a}})$ is given by the Cayley–Menger determinant:

$$g(\vec{\mathbf{a}}) = \begin{vmatrix} 0 & a^2 & b^2 & c^2 & 1 \\ a^2 & 0 & C^2 & B^2 & 1 \\ b^2 & C^2 & 0 & A^2 & 1 \\ c^2 & B^2 & A^2 & 0 & 1 \\ 1 & 1 & 1 & 1 & 0 \end{vmatrix}. \tag{36}$$

If we let $\mathbb{R}_+$ denote the set of positive real numbers, then we define

$$\mathcal{S} = \left\{ \vec{\mathbf{a}} = (a, b, c, A, B, C)^t \in \mathbb{R}^6_+ \ : \ (35) \text{ holds} \right\}.$$

Please note that $\mathcal{S}$ is open in $\mathbb{R}^6$. Moreover, any *regular* tetrahedron in which $a = b = c = A = B = C > 0$, is contained in $\mathcal{S}$.

If $\vec{\mathbf{a}} = (a, b, c, A, B, C)^t$ generates a tetrahedron, then so does $P\vec{\mathbf{a}}$ where $P = RQ$ and

1. R permutes (a, b, c) and (A, B, C) with the same permutation of three elements;
2. Q is any permutation of (a, b, c, A, B, C) in which the base of the tetrahedron is changed to another face. For example, (c, A, B, C, a, b) corresponds to reorienting the tetrahedron so that the base is given by (C, a, b).

Since there are six permutations of the type R and four of the type Q, we get that there are 24 permutations of the form $P = RQ$. These 24 permutations form a subgroup $\mathcal{R}$ of the group of permutations of six letters. Also, it is easy to show that

$$g(P\vec{\mathbf{a}}) = g(\vec{\mathbf{a}}), \quad \forall P \in \mathcal{R}. \tag{37}$$

As the Cayley–Menger determinant is directly proportional to the square of the volume of the tetrahedron (cf. (39)), this identity simply states that the volume of the tetrahedron remains the same after rotations of the base and independent of which face we use as the base.

The total energy of the system of four particles is given now by:

$$E(\vec{\mathbf{a}}) = \phi(a) + \phi(b) + \phi(c) + \phi(A) + \phi(B) + \phi(C), \tag{38}$$

where the intermolecular potential ϕ is as before. For any $V > 0$ we consider the constrained minimization problem:

$$\begin{cases} \min\limits_{\mathcal{S}} E(\vec{\mathbf{a}}) \\ \text{subject to } g(\vec{\mathbf{a}}) = 288V^2. \end{cases} \tag{39}$$

The constraint here specifies that the tetrahedron determined by $\vec{\mathbf{a}}$ has volume V (cf. [20]). The first-order necessary conditions for a solution of this problem are given by (Since the inequality constraints in the definition of the set $\mathcal{S}$ are strict (non-active), the multipliers corresponding to these constraints are zero.): confirm.

$$\begin{cases} g(\vec{\mathbf{a}}) - 288V^2 &= 0, \\ \vec{\nabla}E(\vec{\mathbf{a}}) + \lambda\vec{\nabla}g(\vec{\mathbf{a}}) &= \vec{\mathbf{0}}, \end{cases} \tag{40}$$

which is now a nonlinear system for the seven unknowns $(\lambda, \vec{\mathbf{a}})$ in terms of the parameter V.

4.1. *Existence and Stability of Trivial States*

Expanding the determinant in (36) and computing its partial derivatives, we get that

$$\begin{aligned} \vec{\nabla}g(\vec{\mathbf{a}}) = 4\big[& a(A^2(b^2 + c^2 + B^2 + C^2 - 2a^2 - A^2) + (b^2 - c^2)(B^2 - C^2)), \\ & b(B^2(a^2 + c^2 + A^2 + C^2 - 2b^2 - B^2) + (a^2 - c^2)(A^2 - C^2)), \\ & c(C^2(a^2 + b^2 + A^2 + B^2 - 2c^2 - C^2) + (a^2 - b^2)(A^2 - B^2)), \\ & A(a^2(b^2 + c^2 + B^2 + C^2 - 2A^2 - a^2) - (b^2 - C^2)(c^2 - B^2)), \\ & B(b^2(a^2 + c^2 + A^2 + C^2 - 2B^2 - b^2) - (a^2 - C^2)(c^2 - A^2)), \\ & C(c^2(a^2 + b^2 + A^2 + B^2 - 2C^2 - c^2) - (a^2 - B^2)(b^2 - A^2))\big]. \end{aligned}$$

Since $\vec{\nabla}E(\vec{\mathbf{a}}) = (\phi'(a), \phi'(b), \phi'(c), \phi'(A), \phi'(B), \phi'(C))^t$, the system (40) when evaluated at the regular tetrahedron $\vec{\mathbf{a}} = (a, a, a, a, a, a)$, reduces to

$$4a^6 = 288V^2, \quad \phi'(a) + 4\lambda a^5 = 0.$$

Thus, we have the following result.

Lemma 2. *For any $V > 0$ the system (40) has the solution $(\lambda_V, \vec{\mathbf{a}}_V, V)$ where $\vec{\mathbf{a}}_V = (a_V, a_V, a_V, a_V, a_V, a_V)$ and*

$$a_V^3 = 6\sqrt{2}\,V, \quad \lambda_V = -\frac{\phi'(a_V)}{4a_V^5}. \tag{41}$$

We now examine the stability of the trivial state (41). A lengthy but otherwise elementary calculation shows that

$$H_V \equiv [\nabla^2 E + \lambda_V \nabla^2 g](\vec{\mathbf{a}}_V) = \begin{bmatrix} \alpha & \beta & \beta & 0 & \beta & \beta \\ \beta & \alpha & \beta & \beta & 0 & \beta \\ \beta & \beta & \alpha & \beta & \beta & 0 \\ 0 & \beta & \beta & \alpha & \beta & \beta \\ \beta & 0 & \beta & \beta & \alpha & \beta \\ \beta & \beta & 0 & \beta & \beta & \alpha \end{bmatrix}, \tag{42}$$

where

$$\alpha = \phi''(a_V) + \frac{3}{a_V}\phi'(a_V), \quad \beta = -\frac{2}{a_V}\phi'(a_V).$$

Since $\vec{\nabla} g(\vec{\mathbf{a}}_V) = 4a_V^5(1,1,1,1,1,1)^t$, we need examine the structure of H_V on the subspace of $\mathbb{R}^6$ given by

$$\mathcal{M} = \left\{ \vec{\mathbf{y}} \in \mathbb{R}^6 \,:\, y_1 + y_2 + y_3 + y_4 + y_5 + y_6 = 0 \right\}.$$

We have now the following result:

Theorem 4. *Let $\phi : (0,\infty) \to \mathbb{R}$ be twice continuously differentiable function. Then the matrix (42) is positive definite over $\mathcal{M}$ if and only if $\alpha > 0$ and $\alpha - 2\beta > 0$, which in turn are equivalent to*

$$\phi''(a_V) + \frac{3}{a_V}\phi'(a_V) > 0, \quad \phi''(a_V) + \frac{7}{a_V}\phi'(a_V) > 0. \tag{43}$$

Thus, the regular tetrahedron $\vec{\mathbf{a}}_V$ is a relative minimizer for the problem (39) for those values of V for which conditions (43) hold.

Proof. It is easy to check that $\mathcal{M} = \text{range}(M)$ where

$$M = \begin{bmatrix} 1 & 0 & 0 & 0 & 0 \\ 0 & 1 & 0 & 0 & 0 \\ 0 & 0 & 1 & 0 & 0 \\ -1 & -1 & -1 & -1 & -1 \\ 0 & 0 & 0 & 1 & 0 \\ 0 & 0 & 0 & 0 & 1 \end{bmatrix}.$$

The matrix corresponding to the quadratic form of H_V restricted to $\mathcal{M}$ is now given by $U_V = M^t H_V M$. The eigenvalues of U_V are:

$$\alpha \text{ (double)}, \quad \alpha - 2\beta, \quad \frac{1}{2}\left[7\alpha - 6\beta \pm \sqrt{16\alpha^2 + 9(\alpha - 2\beta)^2}\right].$$

Since the product of the last two of these eigenvalues is $6\alpha(\alpha - 2\beta)$, and $7\alpha - 6\beta = 3(\alpha - 2\beta) + 4\alpha$, we can conclude now that they are all positive if and only if $\alpha > 0$ and $\alpha - 2\beta > 0$. Thus, H_V restricted to $\mathcal{M}$ is positive definite provided these two conditions hold, which in turn implies that $\vec{\mathbf{a}}_V$ is a relative minimizer for problem (39). That $\alpha > 0$ and $\alpha - 2\beta > 0$ are equivalent to (43) follows from the definitions of α and β. □

Example 4. *For the Lennard–Jones potential* (23) *we have that:*

$$\begin{aligned}\phi''(r)+\frac{3}{r}\phi'(r) &= \frac{c_1\delta_1(\delta_1-2)}{r^{\delta_1+2}}-\frac{c_2\delta_2(\delta_2-2)}{r^{\delta_2+2}},\\ \phi''(r)+\frac{7}{r}\phi'(r) &= \frac{c_1\delta_1(\delta_1-6)}{r^{\delta_1+2}}-\frac{c_2\delta_2(\delta_2-6)}{r^{\delta_2+2}}.\end{aligned}$$

For simplicity, we assume $\delta_1 > 6$. *We now have two cases:*

1. *Assume that* $\delta_2 \in (2,6]$. *Then the second condition in* (43) *is automatically satisfied and the first condition holds if and only if* $V < V_0$, *where* V_0 *is determined from the condition (cf.* (41)*):*

$$\frac{c_1\delta_1(\delta_1-2)}{r_0^{\delta_1+2}}-\frac{c_2\delta_2(\delta_2-2)}{r_0^{\delta_2+2}}=0,\quad r_0=a_{V_0},$$

   from which it follows that:

$$V_0=\frac{\sqrt{2}}{12}\left[\frac{c_1\delta_1(\delta_1-2)}{c_2\delta_2(\delta_2-2)}\right]^{\frac{3}{\delta_1-\delta_2}}.$$

   Thus, in this case the regular tetrahedron $\vec{\mathbf{a}}_V$ *is a (local) solution of* (39) *if and only if* $V < V_0$.
2. *If* $\delta_2 > 6$, *then the second condition in* (43) *holds if and only if* $V < V_1$, *where* V_1 *is determined from the condition:*

$$\frac{c_1\delta_1(\delta_1-6)}{r_1^{\delta_1+2}}-\frac{c_2\delta_2(\delta_2-6)}{r_1^{\delta_2+2}}=0,\quad r_1=a_{V_1},$$

   from which it follows that:

$$V_1=\frac{\sqrt{2}}{12}\left[\frac{c_1\delta_1(\delta_1-6)}{c_2\delta_2(\delta_2-6)}\right]^{\frac{3}{\delta_1-\delta_2}}.$$

   Since $\delta_1 > \delta_2 > 6$, *it follows that* $V_1 < V_0$. *Thus, in this case the regular tetrahedron* $\vec{\mathbf{a}}_V$ *is a (local) solution of* (39) *if and only if* $V < V_1$.

Example 5. *For the Buckingham* (25), *we have that*

$$\begin{aligned}\phi''(r)+\frac{3}{r}\phi'(r) &= \alpha\beta\left[\beta-\frac{3}{r}\right]\mathrm{e}^{-\beta r}-\frac{\gamma\eta(\eta-2)}{r^{\eta+2}},\\ \phi''(r)+\frac{7}{r}\phi'(r) &= \alpha\beta\left[\beta-\frac{7}{r}\right]\mathrm{e}^{-\beta r}-\frac{\gamma\eta(\eta-6)}{r^{\eta+2}}.\end{aligned}$$

Please note that the first condition in (43) *holds for an interval* (V_0, V_1) *of volume values under the conditions* (27) *in Example 2. The analysis now becomes rather complicated and we just describe it qualitatively. If* $\eta > 6$, *then the second condition in* (43) *would hold as well for values of* V *in an interval of the form* (V_2, V_3) *provided some condition similar to* (27) *holds. Depending as to whether or not the intersection* $(V_0, V_1) \cap (V_2, V_3)$ *is non-empty, we might get stable regular tetrahedrons. On the other hand, if* $\eta \in (2,6]$, *then the second condition in* (43) *would hold as well for values of* V *in an interval of the form* (V_4, ∞) *and again the existence of trivial states will depend on whether the corresponding intersection is non-empty.*

Example 6. *For the potential* (29)

$$\phi''(r)+\frac{3}{r}\phi'(r)=4k+6\beta r^2,\quad \phi''(r)+\frac{7}{r}\phi'(r)=8k+10\beta r^2.$$

Please note that the stability conditions (43) *holds when* $\beta \geq 0$ *independent of the value of* V! *That is, the regular tetrahedron* $\vec{\mathbf{a}}_V$ *is a relative minimizer for the problem* (39) *for all values of* V. *In the case* $\beta = 0$, *since the functional* (38) *is convex, this state is a global minimizer.*

On the other hand, if $\beta < 0$, the first condition in (43) holds if $V < V_1$ and the second condition if $V < V_2$ where

$$V_1^2 = -\frac{k^3}{243\,\beta^3}, \quad V_2^2 = -\frac{8k^3}{1125\,\beta^3}.$$

Since $V_1 < V_2$ we get that conditions (43) hold both for $V < V_1$. For $V > V_1$ either one or both conditions fail.

4.2. Existence and Stability of Nontrivial States

Let $\vec{\mathbf{G}} : \mathbb{R} \times \mathcal{S} \times (0,\infty) \to \mathbb{R}^7$ be given by the left-hand side of (40):

$$\vec{\mathbf{G}}(\vec{\mathbf{x}}, V) = \begin{bmatrix} g(\vec{\mathbf{a}}) - 288V^2 \\ \vec{\nabla}E(\vec{\mathbf{a}}) + \lambda\vec{\nabla}g(\vec{\mathbf{a}}) \end{bmatrix},$$

where $\vec{\mathbf{x}} = (\lambda, \vec{\mathbf{a}})$. We have now that

$$D_{\vec{\mathbf{x}}}\vec{\mathbf{G}}(\vec{\mathbf{x}}, V) = \begin{bmatrix} 0 & (\vec{\nabla}g(\vec{\mathbf{a}}))^t \\ \vec{\nabla}g(\vec{\mathbf{a}}) & \nabla^2 E(\vec{\mathbf{a}}) + \lambda\nabla^2 g(\vec{\mathbf{a}}) \end{bmatrix}. \tag{44}$$

It follows from (37) that

$$\vec{\nabla}g(P\vec{\mathbf{a}}) = P\vec{\nabla}g(\vec{\mathbf{a}}), \tag{45a}$$

$$\nabla^2 g(P\vec{\mathbf{a}}) = P\nabla^2 g(\vec{\mathbf{a}})P^t, \quad P \in \mathcal{R}, \tag{45b}$$

with similar relations for the total energy E. It follows now from (45a) that

$$\vec{\mathbf{G}}(Q\vec{\mathbf{x}}, V) = Q\vec{\mathbf{G}}(\vec{\mathbf{x}}, V), \quad Q \in \mathcal{G}, \tag{46}$$

where

$$\mathcal{G} = \left\{ Q = \begin{bmatrix} 1 & \vec{\mathbf{0}}^t \\ \vec{\mathbf{0}} & P \end{bmatrix} : P \in \mathcal{R} \right\}.$$

Thus, the system (40) remains the same, up to reordering of the equations, when $\vec{\mathbf{x}} = (\lambda, \vec{\mathbf{a}})$ is replaced by $Q\vec{\mathbf{x}}$.

We now begin the analysis of the existence of solutions of the system (40) bifurcating from the trivial branch:

$$\mathcal{T} = \{(\lambda_V, \vec{\mathbf{a}}_V, V) : \lambda_V, \vec{\mathbf{a}}_V \text{ given by (41)}, V > 0\}.$$

If we evaluate (44) at the trivial state $(\lambda_V, \vec{\mathbf{a}}_V, V)$, then this matrix reduces to:

$$D_{\vec{\mathbf{x}}}\vec{\mathbf{G}}(\lambda_V, \vec{\mathbf{a}}_V, V) = \begin{bmatrix} 0 & (\vec{\nabla}g(\vec{\mathbf{a}}_V))^t \\ \vec{\nabla}g(\vec{\mathbf{a}}_V) & H_V \end{bmatrix}, \tag{47}$$

where $\vec{\nabla}g(\vec{\mathbf{a}}_V) = 4a_V^5(1,1,1,1,1,1)^t$ and H_V is given by (42). The matrix (47) has two eigenvalues which are nonzero for every value of $V > 0$, with the remaining eigenvalues given by:

1. $\mu_1(V) = \phi''(a_V) + \frac{3}{a_V}\phi'(a_V)$ with algebraic and geometric multiplicity three, and corresponding eigenvectors:

$$(0,-1,0,0,1,0,0)^t, \quad (0,0,-1,0,0,1,0)^t, \quad (0,0,0,-1,0,0,1)^t. \tag{48}$$

2. $\mu_2(V) = \phi''(a_V) + \frac{7}{a_V}\phi'(a_V)$ with algebraic and geometric multiplicity two, and corresponding eigenvectors:

$$(0,-1,1,0,-1,1,0)^t, \quad (0,-1,0,1,-1,0,1)^t. \tag{49}$$

Remark 1. *Please note that the expressions for these eigenvalues are the ones that appear in Theorem 4 characterizing the stability of the trivial state $(\lambda_V, \vec{\mathbf{a}}_V, V)$. Thus, the trivial state can change stability exactly when one of these two eigenvalues becomes zero.*

To deal with these kernels with dimension greater than one, we proceed as in the previous section by considering a suitable reduced problem in each case. These reductions are determined by the symmetries present in this problem which are embodied in (46).

4.2.1. The Eigenvalue $\mu_1(V)$

Let us take the eigenvector $\vec{\mathbf{g}} = (0,0,0,-1,0,0,1)^t$ of the eigenvalue $\mu_1(V)$ above. (The analysis for the other two eigenvectors is similar.) By inspection it is easy to get that the isotropy subgroup $\mathcal{H}$ of $\mathcal{G}$ at $\vec{\mathbf{g}}$ is given by:

$$\mathcal{H} = \left\{ \begin{pmatrix} \lambda & a & b & c & A & B & C \\ \lambda & a & b & c & A & B & C \end{pmatrix}, \begin{pmatrix} \lambda & a & b & c & A & B & C \\ \lambda & b & a & c & B & A & C \end{pmatrix}, \right.$$
$$\left. \begin{pmatrix} \lambda & a & b & c & A & B & C \\ \lambda & B & A & c & b & a & C \end{pmatrix}, \begin{pmatrix} \lambda & a & b & c & A & B & C \\ \lambda & A & B & c & a & b & C \end{pmatrix} \right\}.$$

The $\mathcal{H}$-*fixed point set* is now given by:

$$\mathbb{R}^7_{\mathcal{H}} = \left\{ (\lambda, a, a, c, a, a, C)^t \ : \ \lambda, a, c, C \in \mathbb{R} \right\}. \tag{50}$$

The projection $\mathbb{P}_{\mathcal{H}} : \mathbb{R}^7 \to \mathbb{R}^7_{\mathcal{H}}$ has matrix representation:

$$\mathbb{P}_{\mathcal{H}} = \begin{bmatrix} 1 & 0 & 0 & 0 & 0 & 0 & 0 \\ 0 & \frac{1}{4} & \frac{1}{4} & 0 & \frac{1}{4} & \frac{1}{4} & 0 \\ 0 & \frac{1}{4} & \frac{1}{4} & 0 & \frac{1}{4} & \frac{1}{4} & 0 \\ 0 & 0 & 0 & 1 & 0 & 0 & 0 \\ 0 & \frac{1}{4} & \frac{1}{4} & 0 & \frac{1}{4} & \frac{1}{4} & 0 \\ 0 & \frac{1}{4} & \frac{1}{4} & 0 & \frac{1}{4} & \frac{1}{4} & 0 \\ 0 & 0 & 0 & 0 & 0 & 0 & 1 \end{bmatrix},$$

and the $\mathcal{H}$ reduced problem is now:

$$\vec{\mathbf{G}}_{\mathcal{H}}(\vec{\mathbf{u}}, V) \equiv \mathbb{P}_{\mathcal{H}} \vec{\mathbf{G}}(\vec{\mathbf{u}}, V) = \vec{\mathbf{0}}, \quad (\vec{\mathbf{u}}, V) \in \mathbb{R}^7_{\mathcal{H}} \times (0, \infty).$$

Since $\mathcal{T} \subset \mathbb{R}^7_{\mathcal{H}} \times (0, \infty)$, it follows that $\mathcal{T}$ is a branch of solutions for the $\mathcal{H}$ reduced problem. Also, since

$$\mathrm{D}_{\vec{\mathbf{u}}} \vec{\mathbf{G}}_{\mathcal{H}}(\vec{\mathbf{u}}, V) = \mathbb{P}_{\mathcal{H}} \mathrm{D}_{\vec{\mathbf{x}}} \vec{\mathbf{G}}(\vec{\mathbf{u}}, V) \mathbb{P}_{\mathcal{H}}.$$

we have that $L_{\mathcal{H}}(V) \ : \ \mathbb{R}^7_{\mathcal{H}} \to \mathbb{R}^7_{\mathcal{H}}$ is given by:

$$L_{\mathcal{H}}(V) = \mathrm{D}_{\vec{\mathbf{u}}} \vec{\mathbf{G}}_{\mathcal{H}}(\lambda_V, \vec{\mathbf{a}}_V, V) = \mathbb{P}_{\mathcal{H}} \mathrm{D}_{\vec{\mathbf{x}}} \vec{\mathbf{G}}(\lambda_V, \vec{\mathbf{a}}_V, V) \mathbb{P}_{\mathcal{H}}.$$

It easy to check now that $\mu_1(V)$ is a simple eigenvalue of $L_{\mathcal{H}}(V)$ restricted to $\mathbb{R}^7_{\mathcal{H}}$ with corresponding eigenvector $\vec{\mathbf{g}}$ that is

$$L_{\mathcal{H}}(V) \vec{\mathbf{g}} = \mu_1(V) \vec{\mathbf{g}}, \quad \forall \, V > 0.$$

We now have the result for the existence of bifurcating branches for the reduced problem. We omit the proof as it is similar to that of Theorem 3.

Theorem 5. *Let $\mu_1(V_1) = 0$ and assume that $\frac{d\mu_1}{dV}(V_1) \neq 0$. Then the system* (40) *has a branch of nontrivial solutions in $\mathbb{R}^7_{\mathcal{H}} \times (0, \infty)$ bifurcating from the trivial branch $\mathcal{T}$ at the point where $V = V_1$, where $\mathbb{R}^7_{\mathcal{H}}$ is given by* (50).

Remark 2. *It follows from* (46) *that there are two additional branches of nontrivial solutions of the system* (40) *of the forms:*

$$\{(\lambda, a, c, a, a, C, a) \,:\, \lambda \in \mathbb{R},\, a, c, C > 0\},$$
$$\{(\lambda, c, a, a, C, a, a) \,:\, \lambda \in \mathbb{R},\, a, c, C > 0\}.$$

We now consider the case of the eigenvector $\vec{\mathbf{g}} = (0, -1, -1, -1, 1, 1, 1)^T$ of $\mu_1(V)$. This eigenvector is obtained by adding the three eigenvectors in (48). By inspection, the isotropy subgroup $\mathcal{H}$ of $\mathcal{G}$ at $\vec{\mathbf{g}}$ is given by those permutations in $\mathcal{G}$ that permute the symbols (a, b, c) and (A, B, C) in $(\lambda, a, b, c, A, B, C)$ with the same permutation. (Thus, $\mathcal{H}$ has six elements.) The $\mathcal{H}$–fixed point set is now given by:

$$\mathbb{R}^7_{\mathcal{H}} = \{(\lambda, a, a, a, A, A, A)^t \,:\, \lambda, a, A \in \mathbb{R}\}. \tag{51}$$

The projection $\mathbb{P}_{\mathcal{H}} : \mathbb{R}^7 \to \mathbb{R}^7_{\mathcal{H}}$ has matrix representation:

$$\mathbb{P}_{\mathcal{H}} = \begin{bmatrix} 1 & 0 & 0 & 0 & 0 & 0 & 0 \\ 0 & \frac{1}{3} & \frac{1}{3} & \frac{1}{3} & 0 & 0 & 0 \\ 0 & \frac{1}{3} & \frac{1}{3} & \frac{1}{3} & 0 & 0 & 0 \\ 0 & \frac{1}{3} & \frac{1}{3} & \frac{1}{3} & 0 & 0 & 0 \\ 0 & 0 & 0 & 0 & \frac{1}{3} & \frac{1}{3} & \frac{1}{3} \\ 0 & 0 & 0 & 0 & \frac{1}{3} & \frac{1}{3} & \frac{1}{3} \\ 0 & 0 & 0 & 0 & \frac{1}{3} & \frac{1}{3} & \frac{1}{3} \end{bmatrix},$$

It follows now that for the $\mathcal{H}$ reduced problem, $\mu_1(V)$ is a simple eigenvalue with corresponding eigenvector $\vec{\mathbf{g}}$. The proof of the following result is as that of Theorem 3.

Theorem 6. *Let $\mu_1(V_1) = 0$ and assume that $\frac{d\mu_1}{dV}(V_1) \neq 0$. Then the system* (40) *has a branch of nontrivial solutions in $\mathbb{R}^7_{\mathcal{H}} \times (0, \infty)$ bifurcating from the trivial branch $\mathcal{T}$ at the point where $V = V_1$, where $\mathbb{R}^7_{\mathcal{H}}$ is given by* (51).

Remark 3. *By applying all the transformations in $\mathcal{G}$, it follows from* (46) *that there are three additional branches of solutions of the system* (40) *of the forms:*

$$\{(\lambda, A, A, a, a, a, A) \,:\, \lambda \in \mathbb{R},\, a, A > 0\},$$
$$\{(\lambda, a, A, A, A, a, a) \,:\, \lambda \in \mathbb{R},\, a, A > 0\},$$
$$\{(\lambda, A, a, A, a, A, a) \,:\, \lambda \in \mathbb{R},\, a, A > 0\}.$$

Thus, combining both theorems, we get that there are seven branches of nontrivial solutions bifurcating from the trivial branch $\{(\lambda_V, \vec{\mathbf{a}}_V, V) \,:\, V > 0\}$ at the value of $V = V_1$ where $\mu_1(V_1) = 0$ and $\mu_1'(V_1) \neq 0$.

4.2.2. The Eigenvalue $\mu_2(V)$

We now consider the case of the eigenvector $\vec{\mathbf{g}} = (0, -2, 1, 1, -2, 1, 1)^T$ of $\mu_2(V)$. This eigenvector is obtained by adding the eigenvectors in (49). By inspection, the isotropy subgroup $\mathcal{H}$ of $\mathcal{G}$ at $\vec{\mathbf{g}}$ is given by

$$\mathcal{H} = \left\{ \begin{pmatrix} \lambda & a & b & c & A & B & C \\ \lambda & a & b & c & A & B & C \end{pmatrix}, \begin{pmatrix} \lambda & a & b & c & A & B & C \\ \lambda & a & c & b & A & C & B \end{pmatrix}, \right.$$
$$\begin{pmatrix} \lambda & a & b & c & A & B & C \\ \lambda & A & C & b & a & c & B \end{pmatrix}, \begin{pmatrix} \lambda & a & b & c & A & B & C \\ \lambda & A & b & C & a & B & c \end{pmatrix},$$
$$\left. \begin{pmatrix} \lambda & a & b & c & A & B & C \\ \lambda & A & B & c & a & b & C \end{pmatrix}, \begin{pmatrix} \lambda & a & b & c & A & B & C \\ \lambda & A & c & B & a & C & b \end{pmatrix} \right\},$$

with $\mathcal{H}$–fixed point set given by:

$$\mathbb{R}^7_{\mathcal{H}} = \{(\lambda, a, b, b, a, b, b) \,:\, \lambda, a, b \in \mathbb{R}\}. \tag{52}$$

The projection $\mathbb{P}_{\mathcal{H}} : \mathbb{R}^7 \to \mathbb{R}^7_{\mathcal{H}}$ has matrix representation:

$$\mathbb{P}_{\mathcal{H}} = \begin{bmatrix} 1 & 0 & 0 & 0 & 0 & 0 & 0 \\ 0 & \frac{1}{2} & 0 & 0 & \frac{1}{2} & 0 & 0 \\ 0 & 0 & \frac{1}{4} & \frac{1}{4} & 0 & \frac{1}{4} & \frac{1}{4} \\ 0 & 0 & \frac{1}{4} & \frac{1}{4} & 0 & \frac{1}{4} & \frac{1}{4} \\ 0 & \frac{1}{2} & 0 & 0 & \frac{1}{2} & 0 & 0 \\ 0 & 0 & \frac{1}{4} & \frac{1}{4} & 0 & \frac{1}{4} & \frac{1}{4} \\ 0 & 0 & \frac{1}{4} & \frac{1}{4} & 0 & \frac{1}{4} & \frac{1}{4} \end{bmatrix},$$

It follows now that for the $\mathcal{H}$-reduced problem, $\mu_2(V)$ is a simple eigenvalue with corresponding eigenvector $\vec{\mathbf{g}}$. The proof of the following result is as that of Theorem 3.

Theorem 7. *Let $\mu_2(V_2) = 0$ and assume that $\frac{d\mu_2}{dV}(V_2) \neq 0$. Then the system* (40) *has a branch of nontrivial solutions in $\mathbb{R}^7_{\mathcal{H}} \times (0, \infty)$ bifurcating from the trivial branch $\mathcal{T}$ at the point where $V = V_2$, where $\mathbb{R}^7_{\mathcal{H}}$ is given by* (52).

Remark 4. *By applying all the transformations in $\mathcal{G}$, it follows from* (46) *that there are two additional branches of solutions of the system* (40) *of the forms:*

$$\{(\lambda, b, a, b, b, a, b) \,:\, \lambda \in \mathbb{R},\, a, b > 0\},$$
$$\{(\lambda, b, b, a, b, b, a) \,:\, \lambda \in \mathbb{R},\, a, b > 0\}.$$

5. Numerical Examples

In this section, we present some numerical examples illustrating the results of the previous sections. For simplicity we limit ourselves to the three particle problem. The examples show that the structure of the bifurcation diagrams is quite rich and complex. To construct the pictures in this section, we use the results of Theorem 3, in particular the symmetries given by (32), together with various numerical techniques to get full or detailed descriptions of the corresponding bifurcation diagram.

To compute approximations of the bifurcating solutions predicted by Theorem 3, one employs a predictor-corrector continuation method (cf. [24,25]). The bifurcation points off the trivial branch can be determined, by Theorem 3, from the solutions of the equation $\mu(A) = 0$ (cf. (17), (33)). Secondary bifurcation points off nontrivial branches can be detected by monitoring the sign of a certain determinant. Once a sign change in this determinant is detected, the bifurcation point can be computed by a bisection or secant type iteration. After detection and computation of a bifurcation point, then one can use formulas (8)–(10) to get an approximate point on the solution curve from which the continuation of the bifurcating branch can proceed.

Any trivial or nontrivial computed solution $(\vec{\mathbf{x}}^*, A^*)$ will be called *stable*, if the matrix $(\nabla^2 E + \lambda \nabla^2 g)(\vec{\mathbf{x}}^*, A^*)$ (cf. (31)) is positive definite when restricted to the tangent space at $\vec{\mathbf{x}}^*$ of the constraint of fixed area. Otherwise the point $(\vec{\mathbf{x}}^*, A^*)$ will be called *unstable*. We recall that the tangent space at $\vec{\mathbf{x}}^*$ of the constraint of fixed area is given by

$$\mathcal{M} = \left\{(x,y,z) \, : \, \vec{\nabla} g(\vec{\mathbf{x}}^*) \cdot (x,y,z) = 0\right\}.$$

Please note that since this space depends on the point $\vec{\mathbf{x}}^*$, then except for the trivial branch where the solution is known explicitly, the stability of a solution can only be determined numerically.

In our first example we consider the Lennard–Jones potential (23) with $c_1 = 1$, $c_2 = 2$, $\delta_1 = 12$, and $\delta_2 = 6$. (We obtained similar results for other values of c_1, c_2 like those for argon in which $c_1/c_2 = 3.4^6$ Å^6.) From equation (24) we get that the bifurcation point off the trivial branch is given approximately by $A_0 = 0.5877$. For the case of Theorem 3 in which $a = b$, we show in Figure 2 a close-up of the bifurcating branch near this bifurcation point, for the projection onto the A–a plane. In this figure and the others, the color red indicates unstable solutions while the stable ones are marked in green. Please note that the bifurcation is of the trans-critical type. It is interesting to note that for an interval of values of the parameter A to the left of A_0 in the figure (approximately $(0.5855, 0.5877)$), there are multiple states (trivial and nontrivial) which are stable, the trivial one with an energy less than the nontrivial state in this case. In Figure 3 we look at the same branches of solutions, again the projection onto the A–a plane, but for a larger interval of values of A. We now discover that there are two secondary bifurcation points (In Figure 3 there are bifurcations only corresponding to the values of $A_0 = 0.5877$, $A_1 = 0.6251$ and $A_2 = 0.6670$. The apparent crossing of a branch of scalene triangles and the trivial branch is just an artifact of the projection onto the A–a plane.) at approximately $A_1 = 0.6251$ and $A_2 = 0.6670$, and once again we have multiple stable states (with different symmetries) existing for an interval of values of the parameter A. The branches of solutions bifurcating at these values of A correspond to stable scalene triangles. Once a branch of solutions is computed, we can use the symmetries (32) to generate other branches of solutions. In Figure 4 we show all the solutions obtained via this process, projected to the *abc* space (no A dependence). Figure 5 show the same set of solution but with the branch or axis of trivial solutions coming out of the page. The figures clearly show the variety of solutions (stable and unstable) for the problem (12) as well as the rather complexity of the corresponding solution set.

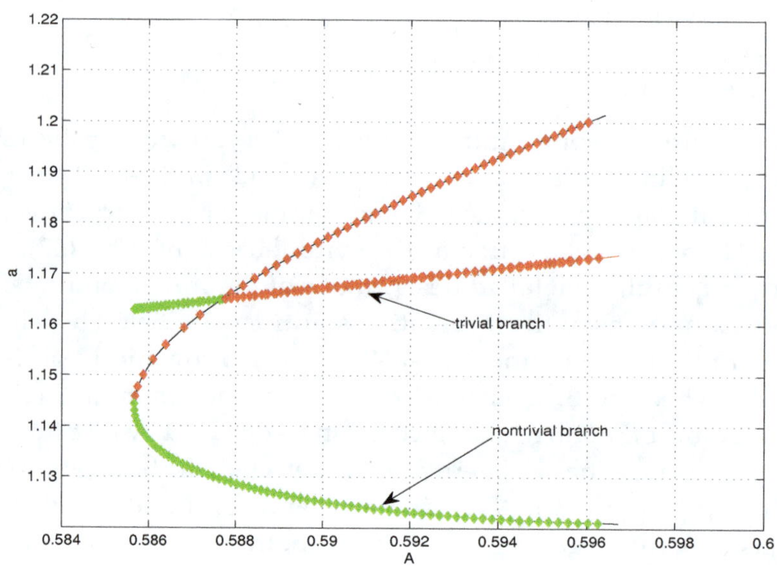

Figure 2. Bifurcation diagram for the a component vs. A for the system (13) in the case $a = b$ and a Lennard–Jones potential. The points in green represent local minima of (12) while those in red are either maxima or none.

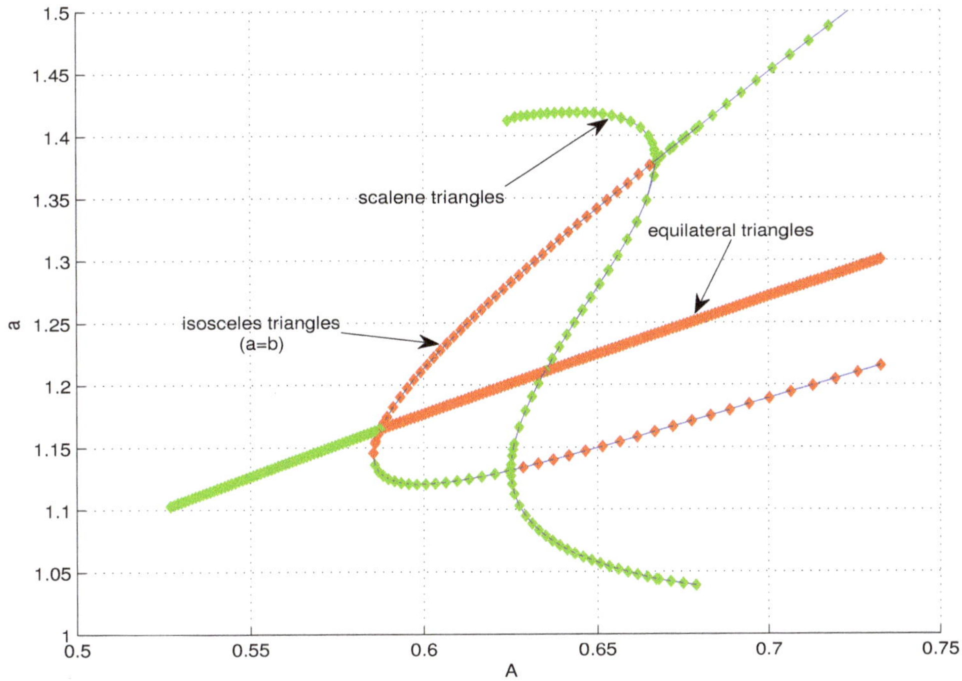

Figure 3. Bifurcation diagram for the system (13) in the case of a Lennard–Jones potential for a larger interval of values of A. There are secondary bifurcations into stable scalene triangles.

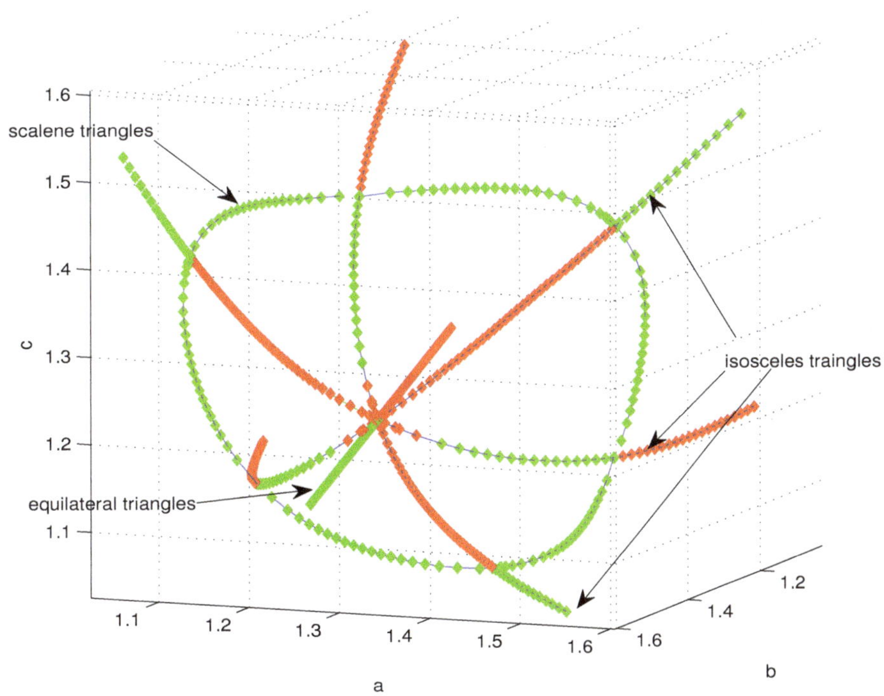

Figure 4. Solution set for the system (13) in the case of a Lennard–Jones potential without the A dependence.

For our next numerical example, we consider the Buckingham potential (25) with parameter values $\alpha = \beta = \gamma = 1$ and $\eta = 4$, which satisfy (27). In this case, we have two bifurcation points off the trivial branch (which correspond to solutions of $\mu(A) = 0$) at approximately $A_0 = 5.3154$ and $A_1 = 74.2253$. The trivial branch is stable for $A \in (A_0, A_1)$ and unstable otherwise. Both bifurcations are into isosceles triangles, and both are of trans-critical type but with different stability patterns. In Figure 6 we show the solution set for the case $a = b$. The plot shows the dependence of the a and c components on the area parameter A. In Figure 7 we show the projection of this set onto the c vs.

A plane where one can appreciate somewhat better the stability patterns at each bifurcation point, and that there is a turning point for $A \approx 46$ on the branch bifurcating from A_0. Please note that once again we have multiple stable states existing for an interval of values of the parameter A. No secondary bifurcations were detected in this case.

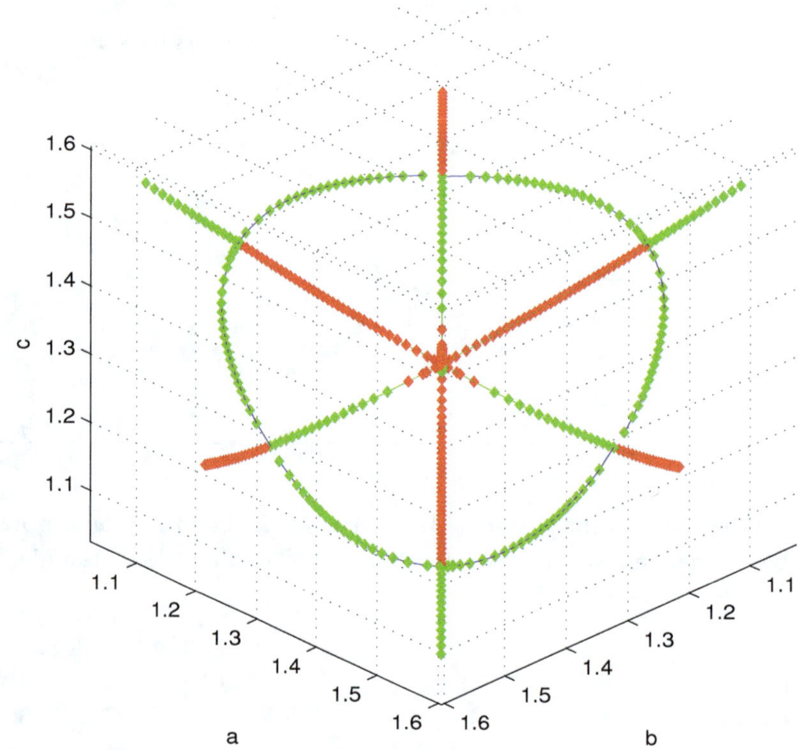

Figure 5. Solution set for the system (13) in the case of a Lennard–Jones potential without the A dependence with the branch of trivial solutions coming out of the page.

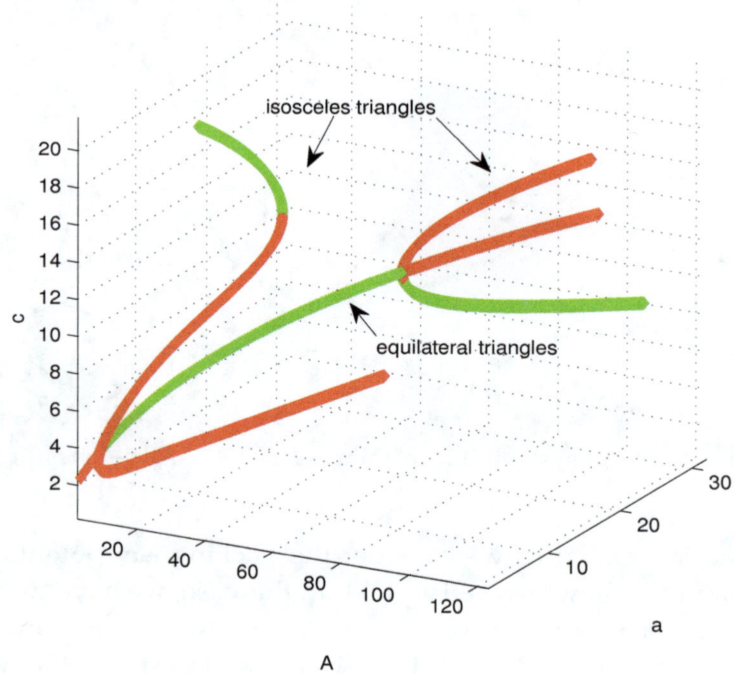

Figure 6. Dependence of the a and c components on the parameter A for the system (13) in the case $a = b$ and for a Buckingham potential.

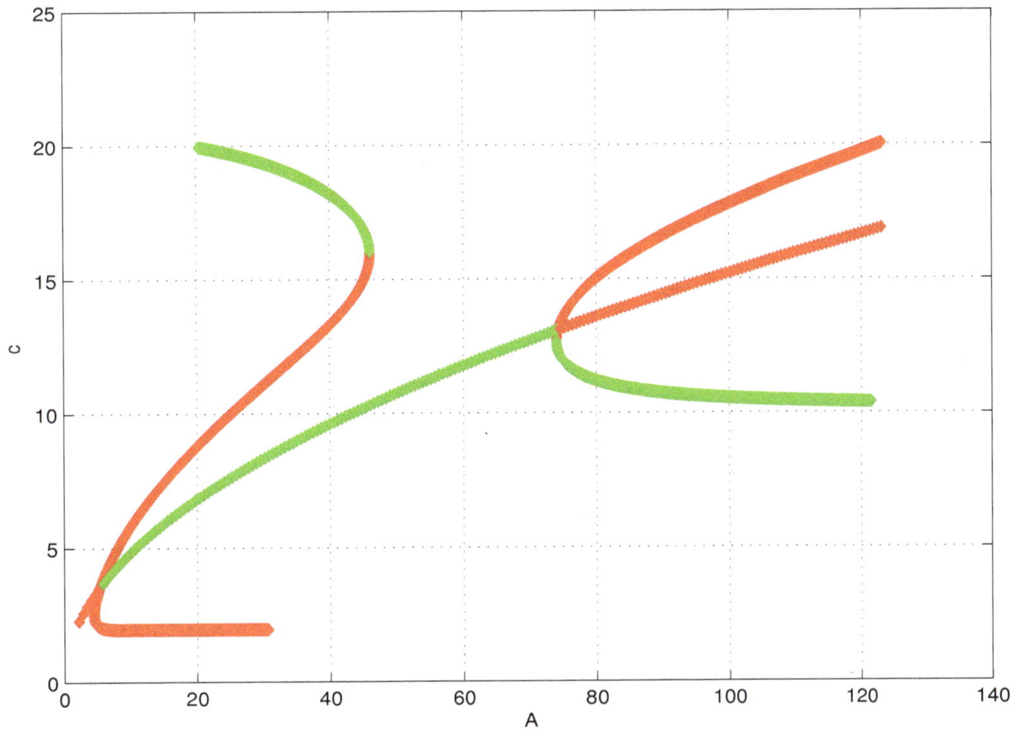

Figure 7. Projection onto the c vs. A plane of the set in Figure 6.

6. Final Comments

The variety or type of solutions obtained from Theorems 3, 5–7, could be predicted generically from an analysis of the symmetries present in our problem as given by (32) and (46). However, such an analysis does not guaranty the existence of solutions with the predicted symmetries. It is the application of the Equivariant Bifurcation Theorem 1 that actually yields the result that such solutions exist. The generic analysis however is a preliminary step in identifying the spaces in which Theorem 1 can be applied. We should also point out that the results on the bifurcating branches in Theorems 3, 5–7 are global in the sense that the so-called Crandall and Rabinowitz alternatives in Theorem 1 hold. That is, any bifurcating branch is either unbounded, or it intersects the boundary of the domain of definition of the operator in the equilibrium conditions, or it intersects the trivial branch at another eigenvalue.

The results of this paper might be useful in the analysis of the more general and complex problem of arrays with many particles. As the total area or volume of such an array is increased, thus reducing its density, one might expect that locally situations similar to the ones discussed in this paper might be taking place in different parts of the array. It is interesting to note here that the existence of multiple stable configurations detected in the numerical examples of Section 5, opens the possibility for the existence of multiple equilibrium (local) states in a large molecular array, reminiscent of the bubble formation phenomena mentioned in the introduction. Thus, as further analysis either via molecular dynamics simulations or theoretically would be the questions as to whether the local results in this paper are related or can predict the initiation of cavitation bubbles of different sizes in actual fluid or gases.

Author Contributions: Conceptualization, P.N.M., M.L.S.; methodology, P.N.M., M.L.S.; validation, M.L.S.; simulation, M.L.S.; formal analysis and investigation, P.N.M., M.L.S.; writing original draft preparation, writing and editing, P.N.M.; funding acquisition, P.N.M., M.L.S.

Acknowledgments: The author would like to thank the academic editor and the three anonymous referees for their constructive comments that have improved the presentation of this manuscript.

References

1. Lim, T.-C. Mathematical relationships for development of a molecular potential function converter. *Commun. Math. Comput. Chem.* **2003**, *49*, 155–169.
2. Tersoff, J. Modeling solid-state chemistry: Interatomic potentials for multicomponent systems. *Phys. Rev. B* **1989**, *39*, 5566–5568. [CrossRef]
3. Shang, Y. Lower Bounds for Gaussian Estrada Index of Graphs. *Symmetry* **2018**, *10*, 325. [CrossRef]
4. Collevecchio, A.; Konig, W.; Morters, P.; Sidorova, N. Phase transitions for dilute particle systems with Lennard-Jones potential. *Commun. Math. Phys.* **2010**, *299*, 603–630. [CrossRef]
5. Discher, D.E.; Boal, D.H.; Boey, S.K. Phase transitions and anisotropic responses of planar triangular nets under large deformation. *Phys. Rev. E* **1997**, *55*, 4762. [CrossRef]
6. Golubitsky, M. The Bénard problem, symmetry and the lattice of isotropy subgroups. In *Bifurcation Theory, Mechanics and Physics*; Bruter, C.P., Aragnol, A., Lichnorowicz, A., Eds.; Reidel: Dordrecht, The Netherlands, 1983; pp. 225–256.
7. Bazhirov, T.T.; Norman, G.E.; Stegailov, V.V. Cavitation in liquid metals under negative pressures. Molecular dynamics modeling and simulation. *J. Phys. Condens. Matter* **2008**, *20*, 114113. [CrossRef]
8. Blander, M.; Katz, J. Bubble Nucleation in Liquids. *AlChE J.* **1975**, *21*, 833–848. [CrossRef]
9. Fond, C. Cavitation Criterion for Rubber Materials: A Review of Void-Growth Models. *J. Polym. Sci. Part B Polym. Phys.* **2001**, *39*, 2081–2096. [CrossRef]
10. Horgan, C.O.; Polignone, D.A. Cavitation in nonlinearly elastic solids: A review. *Appl. Mech. Rev.* **1995**, *48*, 471–485. [CrossRef]
11. Chakraborty, D.; Chandra, A. An analysis of voids and necks in supercritical water. *J. Mol. Liquids* **2011**, *163*, 1–6. [CrossRef]
12. David, E.E.; David, C.W. Voronoi polyhedra as a tool for studying solvation structure. *J. Chem. Phys.* **1982**, *76*, 4611. [CrossRef]
13. Lennard-Jones, J.E. On the Determination of Molecular Fields. *Proc. R. Soc. Lond. A* **1924**, *106*, 463–477.
14. Buckingham, R.A. The classical equation of state of gaseous helium, neon and argon. *Proc. R. Soc. Lond. Ser. A Math. Phys. Sci.* **1938**, *168*, 264–283.
15. Ambrosetti, A. *A Premier on Nonlinear Analysis*; Cambridge University Press: Cambridge, UK, 1993.
16. Golubitsky, M.; Stewart, I.; Schaeffer, D. Singularities and Groups in Bifurcation Theory: Volume II. In *Applied Mathematical Sciences (Book 69)*; Springer: New York, NY, USA, 2000.
17. Kielhöfer, H. *Bifurcation Theory: An Introduction with Applications to Pdes*; Springer: New York, NY, USA, 2004.
18. Krasnoselski, M.A. *Topological Methods in the Theory of Nonlinear Integral Equations*; Pergamon: Oxford, UK, 1965.
19. Healey, T.H. Global bifurcation and continuation in the presence of symmetry with an application to solid mechanics. *SIAM J. Math Anal.* **1988**, *19*, 824–840. [CrossRef]
20. Wirth, K.; Dreiding, A.S. Edge lengths determining tetrahedrons. *Elem. Math.* **2009**, *64*, 160–170. [CrossRef]
21. Lim, T.-C. Alignment of Buckingham parameters to generalized Lennard-Jones potential functions. *Z. Naturforsch.* **2009**, *64*, 200–204. [CrossRef]
22. Shang, Y. Average consensus in multi-agent systems with uncertain topologies and multiple time-varying delays. *Linear Algebra Appl.* **2014**, *459*, 411–429. [CrossRef]
23. Shang, Y. Couple-group consensus of continuous-time multi-agent systems under Markovian switching topologies. *J. Frankl. Inst.* **2015**, *352*, 4826–4844, [CrossRef]
24. Keller, H.B. *Lectures on Numerical Methods in Bifurcation Problems;* Tata Institute of Fundamental Research, Springer: New York, NY, USA, 1986.
25. Allgower, E.L.; Georg, K. Introduction to numerical continuation methods. In *Classics in Applied Mathematics (Book 45)*; Society for Industrial and Applied Mathematics: Philadelphia, PA, USA, 2003.

12

Facility Location Problem Approach for Distributed Drones

Jared Lynskey, Kyi Thar, Thant Zin Oo and Choong Seon Hong *

Department of Computer Science and Engineering, Kyung Hee University, Yongin-si, Gyeonggi-do 17104, Korea; jared@khu.ac.kr (J.L.); kyithar@khu.ac.kr (K.T.); tzoo@khu.ac.kr (T.Z.O.)

* Correspondence: cshong@khu.ac.kr

Abstract: Currently, industry and academia are undergoing an evolution in developing the next generation of drone applications. Including the development of autonomous drones that can carry out tasks without the assistance of a human operator. In spite of this, there are still problems left unanswered related to the placement of drone take-off, landing and charging areas. Future policies by governments and aviation agencies are inevitably going to restrict the operational area where drones can take-off and land. Hence, there is a need to develop a system to manage landing and take-off areas for drones. Additionally, we proposed this approach due to the lack of justification for the initial location of drones in current research. Therefore, to provide a foundation for future research, we give a justified reason that allows predetermined location of drones with the use of drone ports. Furthermore, we propose an algorithm to optimally place these drone ports to minimize the average distance drones must travel based on a set of potential drone port locations and tasks generated in a given area. Our approach is derived from the Facility Location problem which produces an efficient near optimal solution to place drone ports that reduces the overall drone energy consumption. Secondly, we apply various traveling salesman algorithms to determine the shortest route the drone must travel to visit all the tasks.

Keywords: drone deployment; drone port; traveling salesman; facility location problem

---

1. Introduction

Recent reports from the Federal Aviation Agency state that there will be an increase from 2.75 to 4.47 million small drones operating in the United States by 2021. Since the end of May 2017, more than 772,000 owners have already registered with the Federal Aviation Administration (FAA) [1]. The main reason for the sudden increase in drone ownership is due to consumers purchasing drones for their high mobility and applications in the field of computer vision [2]. Thus, a shift in vision related jobs (building inspection, traffic monitoring and temporary cellular coverage extension) slowly being taken over by drones to perform these tasks. This is because drones can provide the required perspective for jobs such as bird's eye view. Additionally, by using machines to take pictures in hazardous areas, we can minimize the risk to human safety. However, there is still no proposal on the initial deployment of drones which do not include random placement. Furthermore, the cost of privately owning drones can be far too expensive for companies who may only require drones for a single task, in comparison to renting drones [3]. In the case of a rental system, companies and users are not required to purchase a drone allowing the cost to be fairly distributed amongst them. Although the cost to rent can reduce the overall cost when compared to owning a drone. Typically, users must visit a rental center to collect the drone. Otherwise, the drone must fly from the shop to the task, reducing the total energy available for completing tasks. Therefore, to overcome these foreseen issues, an unmanned drone rental service that utilizes drones placed at distributed drone ports is necessary. By providing a public service that allows the rental of distributed autonomous drones waiting at drone ports, this can reduce the total number of drones in the sky and the total cost of utilizing drones to complete tasks requested by the user. Thus, we propose an algorithm that can be applied to a shared drone service to reduce the excessive utilization to increase the efficiency by intelligently placing drone ports in respect to the demand and limitations of drones.

1.1. Motivation for Distributed Drone Ports

In this paper, we assume that drones will return to the drone port and charge after each cycle since there are already companies creating landing pads, including Skysense (San Francisco, CA, USA) [4]. We imagine that drones must land, take off and be stored at the base of the drone port to charge drones while they are not completing tasks in the air to remove the risk of users being injured by rotating blades and electrical components that charge the drone. Multiple charging pads can be distributed over an area to reduce the average distance drones must travel to recharge. By doing so, we believe the cost of operating a fleet of drones located at distributed drone ports will be less than the cost of operating a central drone port because miniature drones don't require a large drone and can be located from small areas such as building tops to open fields closer to tasks. Finally, the introduction of drone ports will give future research a foundation to justify their initial placement of drones.

1.2. Related Work

This section encompasses the work done by researchers who look at the problem of optimally placing drones in the sky. However, from our research, we noticed that none of their work justified the reason for the initial and final location of drones in their work. Thus, we proposed an idea to justify further work's reasoning of the initial and final location of drones. The authors in paper [5] propose a discrete and continuous environment to determine the location of drones based on the users. However, they fail to consider where these drones will begin and end their mission. Related work in the field of deploying drones to provide coverage has been growing in interest. In Ref. [6], authors proposed an optimal transport approach to minimize the energy consumption to gather data from moving clusters of IoT (Internet of Things) devices. Works related to a rental system outline the challenges of providing enough assets to satisfy the number of users in the area.

1.3. Challenges

One challenge faced with proposing a rental system is the finite number of combinations possible between the number of drone ports, the location of drone ports and the association between tasks and drones located at drone ports. Thus, we require an efficient mixed integer problem approach to solve our facility location framework. To complement our facility location framework, we apply a subfield of machine learning, clustering, to quickly identify the most efficient central points as drone ports.

Using the original facility location framework to solve the problem is a complex procedure that does not guarantee an optimal solution because it only considers the average distance between the drone ports and facilities. We also need to find the round-trip path to ensure that the drone can complete all the tasks in its cluster.

Drones come in a variety of classes as outlined in Table 1. However, previous research shows, among the available class of drones, the outlook looks the most promising for mini drones. This is due to the fact that mini drones are expected to reach a level of autonomous control within the next five years that meets FAA requirements. Despite the current legal roadblocks imposed by the FAA, there is likely still going to be an increasing demand for mini drones in civilian. Mostly due to their small size compared with larger drones, mini drones have a much greater intrinsic safety rating [7]. Furthermore, the reason for choosing mini drones to complete tasks is due to their high mobility and much lower cost compared with larger drones. Academics are already working to implement vision systems to allow drones to land [8].

Table 1. The various class of drones available.

Category Name	Mass [kg]	Range [km]	Flight Altitude [m]	Endurance [Hours]
Micro	<5	<10	250	1
Mini	<20/30/150	<10	150/250/300	<2
Close Range	25–150	10–30	3000	2–4
Medium Range	50–250	30–70	3000	3–6
High Alt. Long Endurance	>250	>70	>3000	>6

1.4. Facility Location Problem

The Facility Location problem consists of a set of potential drone port locations L. We use this set of potential drone locations to discretize our solution space. Secondly, there is also a set of task locations D that must be serviced, as seen in Figure 1. The objective is to pick a subset l of drone ports to open that minimizes the average distance between each customer and facility. The Uncapacitated Facility Location (UFLP) and Capacitated Facility Location (CFLP) constitute the basic discrete facility location formulation with an abundance of papers based on their extensions by relaxing one or more of the underlying assumptions. The current state-of-the-art algorithm for solving Facility Location problems is proposed by [9]. They provide a close approximation to the global optimum. We can apply the facility location problem formulation since a drone port is considered to be a facility with limited output and a task can be considered as a customer, with a required demand. However, because our problem is not related to delivery trucks, we must make some adjustments to the original problem to ensure that the drones can perform the task without running out of energy. For the situation where there is delay, a congested facility location problem can be applied to minimize the waiting times for customers [10]. Next, we apply the shortest path algorithm to confirm if a generated cluster of tasks can be served by the drone port facility. In Figure 1, we illustrate the decision-making process. The light shaded gray drone ports are the drone ports that were not constructed. Otherwise, the drone ports selected to optimally serve tasks based on information such the population density and task arrival rate. Each cell in Figure 1 represents the discrete space of potential drone port places.

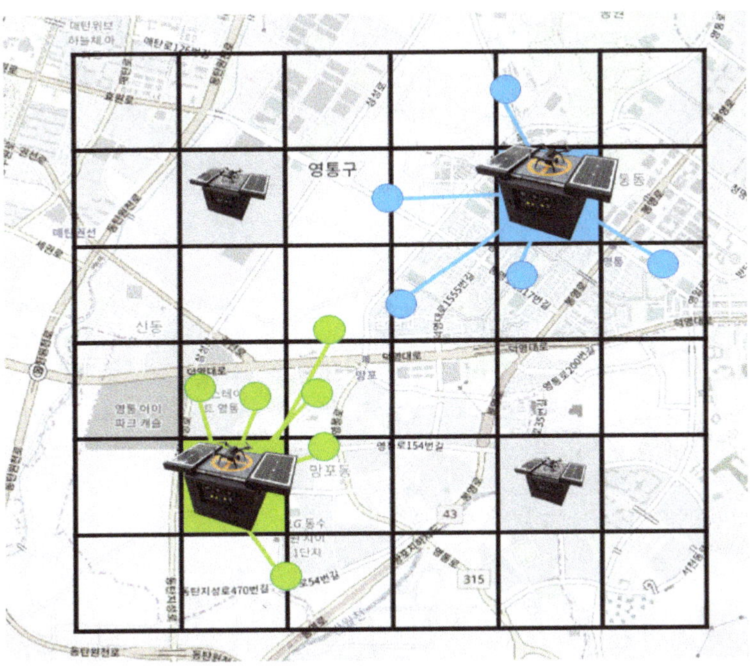

Figure 1. An example of selecting drone ports based on a set of potential 'facility' locations.

1.5. K-Means Clustering Algorithms

Papers that utilize drones to provide coverage to a cluster of users such as [11] attempt to maximize the ground users' battery life by deploying drones in areas where existing macro base stations can not provide sufficient coverage. Their numerical analysis showed promising results due to the fact that their algorithm was able to position drones so that over 60% of users fell within a drone's coverage. However, they did not benchmark their algorithm with other clustering methods, therefore making it difficult to truly understand if their approach was any better than other methods including exhaustive search. One interesting point was their data was based on real Beijing downtown trajectory data. The authors who proposed a k-means stochastic approach gave promising results with a level of complexity $O(\frac{1}{t})$ under general conditions [12] with mini batches improving the speed of convergence even more.

2. Materials and Methods

We envision that multiple drone ports will be managed by a central controller that is responsible for assigning autonomous grounded drones waiting at a drone port to tasks located within the drones operational coverage, removing the need for users to privately own a drone. A drone port is designed to be a small designated take-off, landing and charging area for autonomous drones. Our method to solve this is based on a facility location problem. Furthermore, our tasks only include those related to computers, which just requires the drone to visit the task location, take a photo, and then return to the base or move on to the next task. The tasks considered in this paper assumed that the drone is capable of completing the digital task at the requested location without the need to adjust its path to deliver a physical object, such as a delivery service. We apply k-means to reduce the complexity by implementing a clusters subject to our framework based on the facility problem. Once the model is matured, it allows us to determine the optimal location to install drone ports based on task demand. Then, for additional tasks, we can immediately assign tasks to their respective drone port, thus reducing the computational time to optimally associate drone port and tasks. Lastly, we distributed the drone ports over an area to increase the average number of tasks a drone can complete in one charge compared to a centralized drone port approach. In this section, we explain our system model and justify our approach to solve the issue of placing drones efficiently in an area. Finding the optimal solution for a Facility location problem is NP-Hard, it is not therefore advisable to use an exhaustive method to determine the optimal number of drones. Our algorithm utilizes clustering to reduce the initial search space of possible paths, thus improving the efficiency of our algorithm while giving a near optimal route for drones. We also perform heuristic analysis to find the point at which each drone can complete all of its tasks assigned to it during the cluster phase.

2.1. Central Controller

In our model, we consider a central controller that is responsible for a set of drone ports that each have one drone. By using a central controller, we can assume that information containing the drone state, drone port and task state is available. Therefore, giving a large advantage over a completely distributed systems since information is accessible in a single place and can be used to increase the co-operation between drones waiting at drone ports.

2.2. Drone Port

We propose the idea of a drone port that is connected via existing infrastructure to a central controller then controls the drone and dispatches it to a job. Our model will consider a drone's initial position (x, y) at a drone port as well as the energy consumption e to fly to and from the task, and the energy consumed while completing a task, and the distance flown before it can perform a task τ.

2.3. Objective

Drones are considered to be resourced constrained devices; therefore, it is important that our cost function puts an emphasis on maximizing the utility of a drone in one charge cycle, while ensuring that it can safely return back to the drone to charge. Thus, the objective function is to minimize the number of drone ports in a given area to serve as many tasks as possible. Then, we analyze the result by applying various traveling salesman algorithms to see which algorithm maximizes the number of tasks that can be completed subject to task delay and energy constraints.

2.4. Tasks

Tasks are generated by actual users or artificial intelligent agents that require data or images that can only be collected by drones due to the location or danger to human safety. Since we can assume that our tasks will be periodically requested, it is feasible to estimate an arrival rate based on factors around the area, such as population density. The tasks in terms of a facility location problem are known as the customers that must be served by some facility. In terms of drone demand, we can apply different degrees of energy requirements to complete a task. For example, a task may require v units of energy to perform a task, so, therefore, there must be v units of energy available in the drone's battery to complete the task.

In Figure 2, we illustrate that a drone can only land and take off at its own drone port, in the case of a single drone at a drone port.

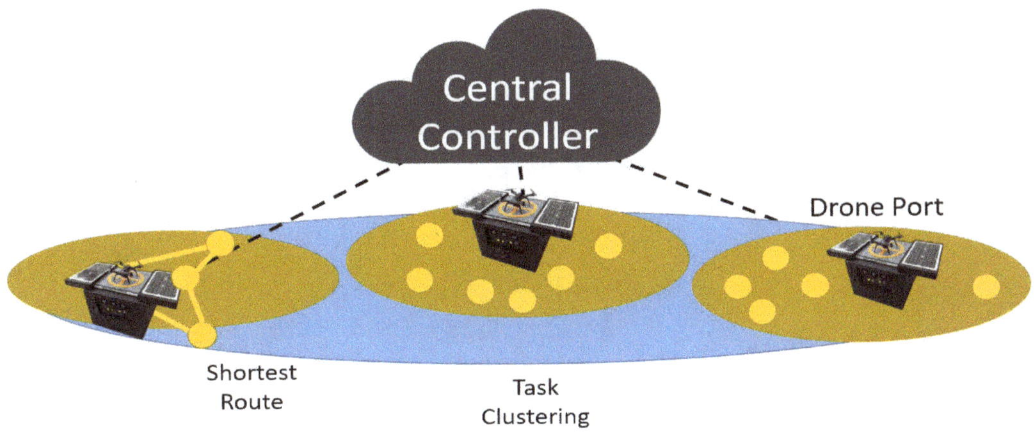

Figure 2. An example of a distributed drone port system.

2.5. Energy Consumption

We estimate the energy consumption to be the following:

$$e_{c+1} = e_c - \sum_{T \in d_i} t_p + \sum_{T \in d_i} ||v * ||\gamma. \quad (1)$$

The energy consumption is calculated by the measuring the distance the drone must travel multiplied by its energy consumption rate and the total energy required by its respective tasks. The approximate energy function is vital because it calculates the energy remaining in a drone and allows us to determine if the path generated is feasible. The energy level after a cycle is denoted as e_{c+1}. This energy level is calculated from the current energy state minus the task energy t_p and distance traveled $||v * ||$ multiplied with an energy consumption constant γ [13]. Equation (1) shows that the drone's energy in the next cycle is its current energy minus the power requirements of the task and the distance traveled to complete those tasks and return to its station.

2.6. Facility Location Problem

The objective of an capacitated facility location is to minimize the distance between the tasks and central drone port. We use this technique in combination with a shortest route algorithm to ensure that the coverage size and the number of drones is suitable to satisfy all the tasks generated near the drone port. In our initial formulation, introduce variable x_i, x_j containing the co-ordinates of potential drone ports and tasks, respectively, where $z_i = 1, \ldots, n$ where $x_i = 1$ if drone port i is built, and $x_i = 0$, otherwise. We can treat the drone port acts as a facility because it contains the service subject to the energy constraint, known as the supply.

2.7. Capital Expenditure

In respect to capital expenditure, our objective is to minimize the total monetary cost of building new infrastructure. In the below cost function, our goal is to reduce the total distance between the drone port x_i and tasks x_j. Since we are limited by the budget F, it is impossible to build every potential drone port i:

$$\begin{aligned}
\text{minimize} \quad & \sum_{d_i} d_{ij}, \\
\text{subject to} \quad & \sum_{i \in A} c_i z_i \leq F, && i = 1, \ldots, n, \\
& x_j \in \{0, 1\}, && j = 1, \ldots, m, \\
& d_{ij} = ||x_i - x_j||_2.
\end{aligned} \tag{2}$$

2.8. Operational Expenditure

In respect to operational expenditure, our objective is to maximize the number of jobs denoted as y_{ij} the index of a job while satisfying constraints such as τ_j delay for every job. Furthermore, each drone must return back to its drone port. Lastly, the path generated must not exceed the drone's energy capacity denoted as the function $f(d_{ij})$

$$\begin{aligned}
\text{maximize} \quad & \sum_i \sum_j y_{ij}, \\
\text{subject to} \quad & \sum_j y_{ij} f(d_{ij}) \leq B_i && i = 1, \ldots, n, \\
& x_0 = x_{fac}, \\
& x_{final} = x_{fac}, \\
& \tau_j \leq \hat{\tau}_j, \\
& f(d_{ij}) = P * d_{ij} + \eta, \\
& \tau_j = t_{complete} - t_{arrival}.
\end{aligned} \tag{3}$$

2.9. Drone Port Placement Algorithm

Algorithm 1 is an overview of our proposed algorithm to efficiently place drones in a distributed area. We develop a heuristic approach by implementing a clustering and traveling salesman problem ensemble to determine the feasibility our algorithm's output data. We continue to increase the number of drone ports until each drone has enough energy to perform all of the tasks in its coverage.

Algorithm 1 Facility Location Problem for Drones

```
procedure DRONE PORT PLACEMENT AND SHORTEST ROUTE
    i ← max number of droneports
    T ← Task Locations array
    P ← Set of drone ports array
    P_max ← max number of droneports
    p ← Initial number of droneports
    while p ≤ P_max do
        clusters ← GenerateCluster(Tasks, p)
        shortestroute ← TSP(tasks, droneport)
        e ← f(d_ij)
        if e ≥ θ then
            Break
        else
            p ← p + 1
    return Drone Port location
    return Shortest Routes
```

2.10. Traveling Salesman Problem

After creating the clusters, we apply a shortest route algorithm to ensure that the drone can perform all of the tasks allocated to it while completing a round-trip back to its original drone port. It is denoted as the Traveling Salesman Problem (TSP) below.

3. Performance Evaluation

Our simulation was conducted on a PC with the operating system Ubuntu 16.04. We applied the algorithms in Table 2 written in Python 3.5 to conduct our simulations. The area size for our simulation is 1000×1000 units.

Table 2. Algorithms we compared.

Algorithm	Type	Author
2-opt	Shortest Path	[14]
Genetic Algorithm	Shortest Path	[15]
Exhaustive Search	Shortest Path	
Ant Colony	Shortest Path	[16]
k-means	Clustering	[17]

3.1. Coverage Size effect on the Combinatorial Search Space

In Figure 3, we analyzed the combination space size in log units if clustering was not applied. Increasing the potential coverage size and the number of potential drone ports exponentially increases the search space. The average complexity is given by $O(kn^T)$, where n is the number of combinations and T is the maximum number of iterations. In practice, the k-means algorithm is fast, but it tends to return a local minima. To avoid this, we used a bottom up approach to avoid falling in to any local minima [18]. This means we can effectively reduce the number of combinations greatly by applying k-means instead of using a naive approach such as a fixed coverage area. The constraints in our proposal ensure that no task appears outside a drone port's range due to improper clustering.

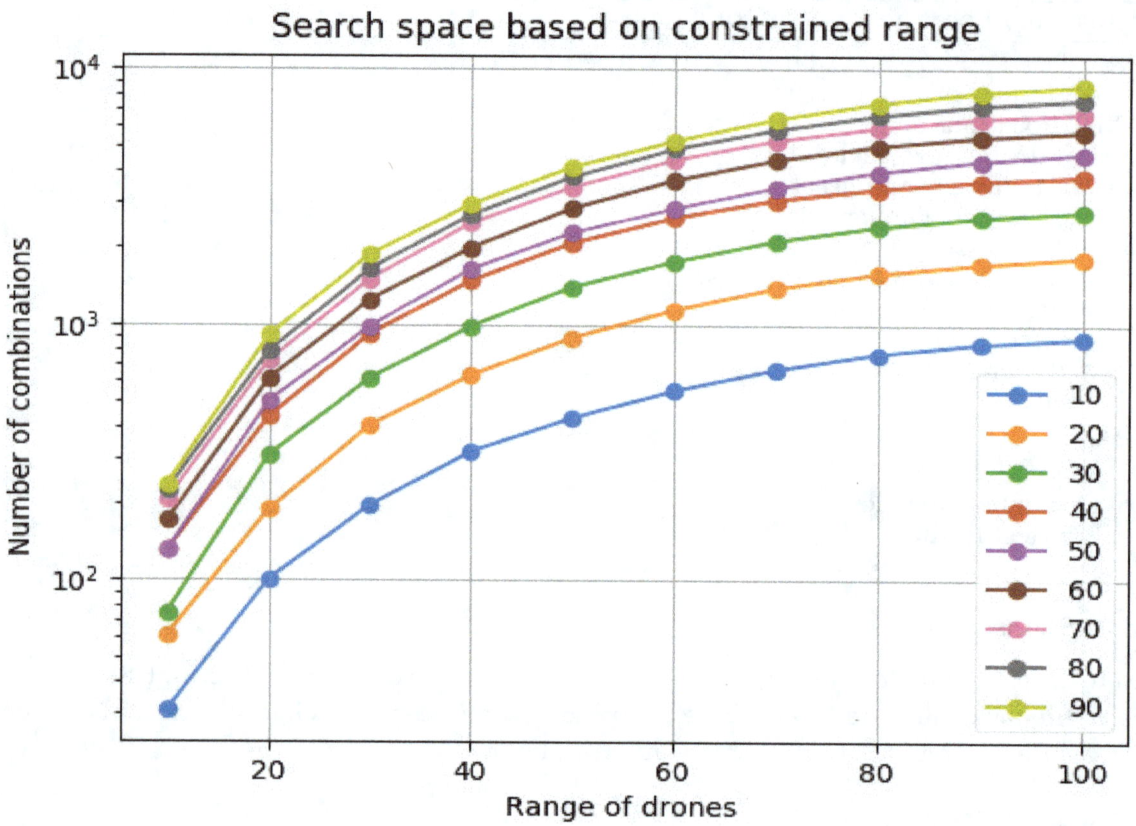

Figure 3. The worst case to determine the optimal solution for possible drone port coverage sizes.

3.2. Average Round Trip Distance

In Figure 4, we determined the coverage and round trip distance. The result comes from our problem formulation constraint, which is that the drone must visit all the tasks assigned within its cluster. Our k-means selects locations based on shortest average distance between the potential drone port point and task locations. The red line shows the optimal solution. It may provide the shortest path and least cost, but it is computationally expensive by almost 50% on average compared with the approximation algorithms. The issue with the genetic algorithm shown in Figure 4 is due to the random approach it takes to converge to the optimal solution. This may have been due to not allowing enough generations to fine the combination that gives the shortest path. Secondly, the 2-opt algorithm performs slight worse, due to being a single thread and with a limited number of iterations to find the optimal solution. Each time the number of drone ports increases, the number of combinations between drone ports and tasks is restricted—thus allowing the solver to find the optimal solution with multiple solution sub-spaces to solve producing a near-optimal solution. The genetic algorithm was able to find a near-optimal solution once the sub-spaces were small enough so its random choices had a larger impact on the distance traveled.

3.3. Infrastructure and Energy Cost

In Figure 5, we compared the cost of having drones fly further versus the cost of installing a drone port. For this experiment, we set the cost of a drone flying per unit distance γ to 0.5. In addition, the cost of a drone port to c 2. If our cost to install a drone port was a lot less than the operating cost of a drone, then there would likely be a lot more drone ports since more ports would be possible without increasing the budget. However, this cost does not consider the on-going costs of maintenance for the drones and drone ports. These values can be configured later to reflect actual prices of drone ports and electricity.

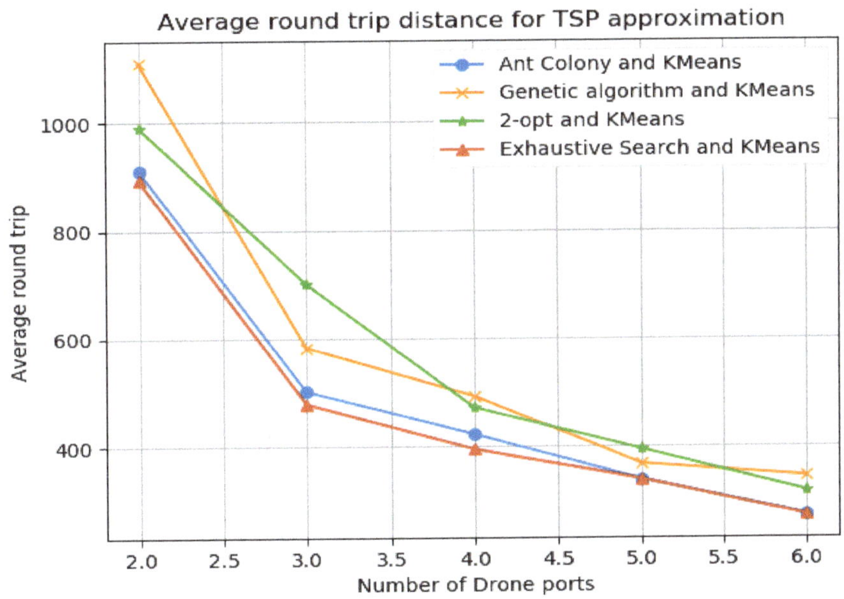

Figure 4. Average round trip for drones based on k-means clustering and Traveling Salesman Approximation.

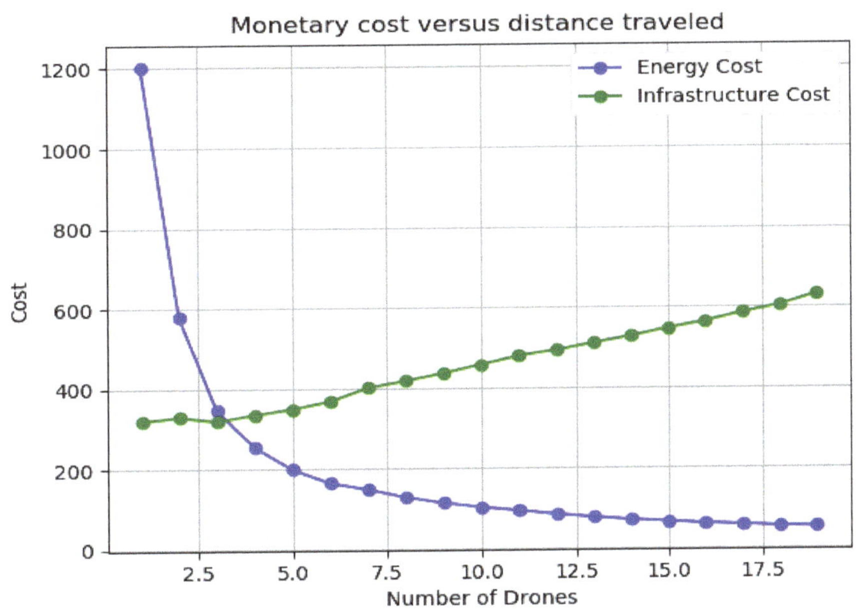

Figure 5. Infrastructure vs. monetary cost.

3.4. Discussion

Further studies that are concerned with the efficient placement of drone ports can now consider our model to further reduce their model's complexity and reduce the distance each drone requires to visit each task. The algorithm is also robust to environments with dynamic arrival rates since we are not randomly placing drone port areas. Instead, by using the k-means cluster, we can efficiently determine areas to install drone ports, centroids where there is a high density of tasks appearing. Although our research was able to show a reduction in the distance covered by drones, our model does not consider factors such as external costs such as maintaining a distributed system. We imagine this cost will be negligible since there is little to no moving parts associated with a drone port, and only the drone. Furthermore, the charging speed and properties of a drone port are still undecided. This system can also be further improved by including terrain data to minimize the difference in the drone port and flight altitude height to which the drone must fly. We did not cover the implications of security or privacy in this paper, but it is something to realize since our framework could be used to spy as noted in [19], therefore identifying rouge tasks is important. Several additional avenues for drone port include the following—firstly, considering information such as buildings or landmarks to include in the final decision-making to create a path that the drone can follow to complete the task; secondly, adding more capacity at drone ports to allow multiple drones to land and take off; thirdly, optimizing the coverage in a way that maximizes co-operation between drones such as a chain link deliver system; and, fourthly, creating boundaries for partitions between two central controllers would allow for a hybrid solution with each sub-space of an area having its own controller. Lastly, we also wish to work on improving the computational efficiency of our proposed algorithm.

4. Conclusions

We proposed a novel algorithm to manage distributed drone ports with a centralized controller to ensure maximum co-operation between drones and fair allocation of tasks. The main goal of this system is to efficiently assign grounded drones at drone ports with their respective tasks. Our combination of approximation algorithms ensures that the cluster of tasks belonging to each drone port are within the drone's coverage. Second, the drone can perform the maximum number before returning to the drone port to recharge. We show that utilizing the Ant algorithm for our cluster round trip for drones minimizes the distance traveled for each drone. By utilizing this approach, further tasks assigned to an area can be immediately be assigned without the need to recalculate the entire environment.

Author Contributions: Conceptualization, J.L.; Methodology, J.K., K.T., T.Z.O.; Software, J.L.; Validation, J.L., Y.Y. and Z.Z.; Formal Analysis, J.L., K.T., T.Z.O.; Investigation, J.K., K.T.; Resources, X.X.; Data Curation, J.K.; Writing—Original Draft Preparation, J.K., K.T.; Writing—Review and Editing, J.L.; Visualization, J.K.; Supervision, C.S.H.

Abbreviations

Notation used in this article

Notation	Explanation
e_c	Set of drone ports
t_p	Set of drones
d_i	Drone i
$v*$	Vector
γ	Drone energy level
d_{ij}	Distance between task j and drone port i
c_i	Cost to build drone port
z_i	Decision variable to build drone port
x_j	Decision variable for drone i to complete task j
y_{ij}	Task completion state $\{0,1\}$
x_0	Drone initial position
x_0	Drone final position
τ_j	Calculated task completion delay
$\hat{\tau}_j$	Earliest deadline first constraint
$f(d_{ij})$	Power Function
$\tau_{complete}$	Drone energy function
$\tau_{arrival}$	Minimum drone energy
η	Task's drone energy consumption
η	Task's drone energy consumption

References

1. FAA. Unmanned Aircraft System. *FAA Aerosp. Forecast.* **2018**. Available online: https://www.faa.gov/data_research/aviation/aerospace_forecasts/media/Unmanned_Aircraft_Systems.pdf (accessed on 12 November 2018).
2. Kanellakis, C.; Nikolakopoulos, G. Survey on Computer Vision for UAVs: Current Developments and Trends. *J. Intell. Robot. Syst.* **2017**, *87*, 141–168. [CrossRef]
3. Purwanda, I.G.; Adiono, T.; Situmorang, S.; Dawani, F.; Samhany, H.A.; Fuada, S. Prototyping design of a low-cost bike sharing system for smart city application. In Proceedings of the 2017 International Conference on ICT For Smart Society (ICISS), Tangerang, Indonesia, 18–19 September 2017; pp. 1–6. [CrossRef]
4. Puiatti, A. Dronesense: Drone Charging Pad. 2014. Available online: https://www.skysense.co/ (accessed on 3 April 2018).
5. Sharafeddine, S.; Islambouli, R. On-Demand Deployment of Multiple Aerial Base Stations for Traffic Offloading and Network Recovery. *arXiv* **2018**, arXiv:1807.02009.
6. Chen, M.; Mozaffari, M.; Saad, W.; Yin, C.; Debbah, M.; Hong, C.S. Caching in the sky: Proactive deployment of cache-enabled unmanned aerial vehicles for optimized quality-of-experience. *IEEE J. Sel. Areas Commun.* **2017**, *35*, 1046–1061. [CrossRef]
7. Floreano, D.; Wood, R.J. Science, technology and the future of small autonomous drones. *Nature* **2015**, *521*, 460. [CrossRef] [PubMed]
8. Cesetti, A.; Frontoni, E.; Mancini, A.; Zingaretti, P.; Longhi, S. A Vision-Based Guidance System for UAV Navigation and Safe Landing Using Natural Landmarks. *J. Intell. Robot. Syst.* **2010**, *57*, 233–257. [CrossRef]
9. Ahmadian, S.; Swamy, C. Improved approximation guarantees for lower-bounded facility location. In Proceedings of the International Workshop on Approximation and Online Algorithms, Ljubljana, Slovenia, 13–14 September 2012; pp. 257–271.
10. Desrochers, M.; Marcotte, P.; Stan, M. The congested facility location problem. *Locat. Sci.* **1995**, *3*, 9–23. [CrossRef]
11. Iellamo, S.; Lehtomaki, J.J.; Khan, Z. Placement of 5G Drone Base Stations by Data Field Clustering. In Proceedings of the 2017 IEEE 85th Vehicular Technology Conference (VTC Spring), Sydney, Australia, 4–7 June 2017; pp. 1–5. [CrossRef]
12. Tang, C.; Monteleoni, C. Convergence rate of stochastic k-means. *arXiv* **2016**, arXiv:1610.04900.
13. Geng, Q.; Zhao, Z. A kind of route planning method for UAV based on improved PSO algorithm. In Proceedings of the 2013 25th Chinese Control and Decision Conference (CCDC), Guiyang, China, 25–27 May 2013; pp. 2328–2331. [CrossRef]
14. Croes, G.A. A method for solving traveling-salesman problems. *Oper. Res.* **1958**, *6*, 791–812. [CrossRef]
15. Mitchell, M. *An Introduction to Genetic Algorithms*; MIT Press: Cambridge, MA, USA, 1998.
16. Dorigo, M.; Birattari, M.; Blum, C.; Clerc, M.; Stützle, T.; Winfield, A. In *Ant Colony Optimization and Swarm Intelligence, Proceedings of the 6th International Conference, ANTS 2008, Brussels, Belgium, 22–24 September 2008*; Springer: Berlin/Heidelberg, Germany, 2008; Volume 5217.
17. Kanungo, T.; Mount, D.M.; Netanyahu, N.S.; Piatko, C.D.; Silverman, R.; Wu, A.Y. An efficient k-means clustering algorithm: Analysis and implementation. *IEEE Trans. Pattern Anal. Mach. Intell.* **2002**, *27*, 881–892. [CrossRef]
18. Arthur, D.; Vassilvitskii, S. k-means++: The advantages of careful seeding. In Proceedings of the Eighteenth Annual ACM-SIAM Symposium on Discrete Algorithms. Society for Industrial and Applied Mathematics, Philadelphia, PA, USA, 7–9 January 2007; pp. 1027–1035.
19. Gonzalez, G. Autonomous Vehicles, Drones Offer New Insurer Risks and Opportunities. Available online: https://www.businessinsurance.com/article/20171207/NEWS06/912317799/Autonomous-vehicles,-drones-offer-new-insurer-risks-and-opportunities(accessed on 12 June 2018).

13

A Test Detecting the Outliers for Continuous Distributions Based on the Cumulative Distribution Function of the Data Being Tested

Lorentz Jäntschi [1,2]

[1] Department of Physics and Chemistry, Technical University of Cluj-Napoca, 400641 Cluj, Romania; lorentz.jantschi@chem.utcluj.ro or lorentz.jantschi@gmail.com

[2] Chemical Doctoral School, Babeș-Bolyai University, 400084 Cluj-Napoca, Romania

Abstract: One of the pillars of experimental science is sampling. Based on the analysis of samples, estimations for populations are made. There is an entire science based on sampling. Distribution of the population, of the sample, and the connection among those two (including sampling distribution) provides rich information for any estimation to be made. Distributions are split into two main groups: continuous and discrete. The present study applies to continuous distributions. One of the challenges of sampling is its accuracy, or, in other words, how representative the sample is of the population from which it was drawn. To answer this question, a series of statistics have been developed to measure the agreement between the theoretical (the population) and observed (the sample) distributions. Another challenge, connected to this, is the presence of outliers - regarded here as observations wrongly collected, that is, not belonging to the population subjected to study. To detect outliers, a series of tests have been proposed, but mainly for normal (Gauss) distributions—the most frequently encountered distribution. The present study proposes a statistic (and a test) intended to be used for any continuous distribution to detect outliers by constructing the confidence interval for the extreme value in the sample, at a certain (preselected) risk of being in error, and depending on the sample size. The proposed statistic is operational for known distributions (with a known probability density function) and is also dependent on the statistical parameters of the population—here it is discussed in connection with estimating those parameters by the maximum likelihood estimation method operating on a uniform U(0,1) continuous symmetrical distribution.

Keywords: test for outliers; order statistics; extreme values; confidence intervals; Monte-Carlo simulation

---

1. Introduction

Many statistical techniques are sensitive to the presence of outliers and all calculations, including the mean and standard deviation can be distorted by a single grossly inaccurate data point. Therefore, checking for outliers should be a routine part of any data analysis.

To date, several tests have been developed for the purpose of identifying outliers of certain distributions. Most of the studies are connected with the Normal (or Gauss) distribution [1]. The first paper that attracted attention on this matter is [2] and this was followed by studies that identified the derivation of the distribution of the extreme values in samples taken from Normal distributions [3]. Then, a series of tests were developed by Thompson in 1935 [4], these were subjected to evaluation [5], and revised [6,7].

For other distributions such as the Gamma distribution, procedures for detecting outliers were proposed [8], revised [9], and unfortunately proved to be inefficient [10].

The first attempt to generalize the criterion for detecting outliers for any distribution can be found in [11], but further research on this subject is scarce apart from a notable recent attempt by Bardet and Dimby [12].

The Grubbs test is a frequently used test for detecting the outliers of a Normal distribution [7]. For a sample (x), the Grubbs' test statistic takes the largest absolute deviation from the sample mean ($\overline{x}$) in units of the sample standard deviation (s) in order to calculate the risk of being in error (α_G) when stating that the most departed values from the mean (min(x), max(x) or both) are not outliers (see Table 1). The associated probabilities of the observed (p_G) are obtained from the Student t distribution [13].

Table 1. The Grubbs statistic.

Sample statistic (G)	Associated probability (p_G = 1-α_G)	Equation
$G_{\text{"min"}} = \frac{\overline{x}-\min(x)}{s}$ $G_{\text{"max"}} = \frac{\max(x)-\overline{x}}{s}$	$\alpha_G = n \cdot \text{CDF}_{\text{"Student t"}}\left(-\sqrt{\frac{n(n-2)}{\left(\frac{n-1}{G}\right)^2 - n}}, n-2\right)$	(1)
$G_{\text{"all"}} = \max(G_{\text{"min"}}, G_{\text{"max"}})$	$\alpha_G = 2n \cdot \text{CDF}_{\text{"Student t"}}\left(-\sqrt{\frac{n(n-2)}{\left(\frac{n-1}{G}\right)^2 - n}}, n-2\right)$	(2)

One should note that the Grubbs test statistic produces a symmetrical confidence interval (see Equations (1) and (2)). The Grubbs statistic as given in Table 1, is intended to be used with the parameters of the population (μ and σ), which are determined using the central moments (CM) method ($\hat{\mu} = \overline{x} = \sum x/n$; $\hat{\sigma} = s = \left(\sum (x - \overline{x})^2\right)^{1/2}/n$).

Here, a method is proposed for constructing the confidence intervals for the extreme values of any continuous distribution for which the cumulative distribution function is also obtainable. The method involves the direct application of a simple test for detecting the outliers. The proposed method is based on deriving the statistic for the extreme values for the uniform distribution. Also, the proposed method provides a symmetrical confidence interval in the probability space.

2. Materials and Methods

The Grubbs test (Table 1) is based on the fact that if outliers exist, then these are "localized" as the maximum value and/or the minimum value in the dataset. Thus, the Grubbs test is essentially a sort of order statistic [14].

Some introductory elements are required for describing the proposed procedure. When a sample of data is tested under the null hypothesis that it follows a certain distribution, it is intrinsically assumed that the distribution is known. The usual assumption is that we possess its probability density function (PDF, for a continuous distribution) or its probability distribution function (PDF for a discrete distribution). The discussion below relates to continuous distributions, although the treatment of discrete distributions are similar to certain degree. Nevertheless, a major distinction between continuous and discrete distributions in the treatment of data is made here; that is, a continuous distribution is "dense", e.g., between any two distinct observations it is possible to observe another while in the case of a discrete distribution, this is generally not true.

Even when the PDF is known (possibly intrinsically), its (statistical) parameters may not necessarily be known, and this raises the complex problem of estimating the parameters of the (population) distribution from the sample; however, this issue is outside the scope of this paper. In general, the estimation of the parameters of the distribution of the data is biased by the presence of the outliers in the data, and thus, identifying the outliers along with the estimation of the parameters of the distribution is a difficult task because two statistical hypotheses are operating. Assuming that the parameters ("parameters") of the distribution (of the PDF) are obtained using the maximum likelihood estimation method (MLE, Equation (3); see [15]), there is some suggestion that the uncertainty accompanying this estimation is transmitted to the process of detecting the outliers.

$$\prod \text{PDF(X; "parameters")} \rightarrow \text{max.} \Rightarrow \sum \ln\left(\text{PDF(X; "parameters")}\right) \rightarrow \text{min.} \quad (3)$$

It should be noted that Equation (3) is a simplified version of the MLE method, since the real use of it requires and involves partial derivatives of the parameters; see Source code (MathCad language) for the MLE estimations in the Supplementary Materials available online.

Either way (whether the uncertainty accompanying this estimation is transmitted to the process of detecting the outliers or not), once an estimate for the parameters of the distribution is available, a test (most desirably, a test based on a statistic) for detecting the presence of an outlier must provide the probability of observing that (assumed) "outlier" as a randomly drawn value from the distribution. What to do next with the probability is another statistical "trick": to observe a value with a probability less than an imposed "level" (usually 5%) is defined as an unlikely event, and therefore, the suspicion regarding the presence of the outlier is justified. With regard to the statistical "trick" mentioned above, the opinion of the author of this manuscript is that one "observation" is not enough. Actually, there should be a series of observations, that come from a series of statistics, each providing a probability. Then, the unlikeliness of the event can be safely ascertained by using Fisher's "combining probability from independent tests" method (FCS, Equation (4); see [16–18]:

$$-\sum_{i=1}^{\tau} \ln(p_i) \sim \chi^2(\tau) \rightarrow \alpha_{FCS} = 1 - CDF_{\chi^2}(-\sum_{i=1}^{\tau} \ln(p_i); \tau) \tag{4}$$

where $p_1, \ldots, p_\tau$ are probabilities from τ independent tests, $CDF\chi^2$ is the χ^2 cumulative distribution function (see also up until Equation (6) below), and p_{FCS} is the combined probability from independent tests.

Taking the general case, for $(x_1, \ldots, x_n)$ as n independent draws (or observations) from a (assumed known) continuous distribution defined by its probability density function, PDF $(x; (\pi_j)_{1\le j\le m})$ where $(\pi_j)_{1\le j\le m}$ are the (assumed unknown) m statistical parameters of the distribution, by way of integration for a (assumed known) domain (D) of the distribution, we may have access to the associated cumulative density function (CDF) $CDF(x; (\pi_j)_{1\le j\le m}; PDF)$, simply expressed as (Equation (5)):

$$CDF(x; (\pi_j)_{1\le j\le m}) = \int_{\inf(D)}^{x} PDF(x; (\pi_j)_{1\le j\le m}) \tag{5}$$

where inf(D) was used instead of min(D) to include unbounded domains (e.g., when inf(D) = -∞; "inf" stands for infimum, "min" stands for minimum). Please note that having the PDF and CDF does not necessarily imply that we have an explicit formula (or expression) for any of them. However, with access to numerical integration methods [19], it is enough to have the possibility of evaluating them at any point (x).

Unlike $PDF(x; (\pi_j)_{1\le j\le m})$, $CDF(x; (\pi_j)_{1\le j\le m})$ is a bijective function and therefore, it is always invertible (even if we do not have an explicit formula; let "InvCDF" be its inverse, Equation (6)):

$$\text{if } p = CDF(x; (\pi_j)_{1\le j\le m}), \text{ then } x = InvCDF(p; (\pi_j)_{1\le j\le m}), \text{ and vice-versa} \tag{6}$$

$CDF(x; (\pi_j)_{1\le j\le m}; \text{"PDF"})$ is a strong tool that greatly simplifies the problem at hand: the problems of analyzing any distribution function (PDF) are translated such that only one needs to be analyzed (the continuous uniform distribution). That is, a series of observed data $(x_i)_{1\le i\le n}$ is expressed through their associated probabilities $p_i = CDF(x_i; (\pi_j)_{1\le j\le m})$ (for $1\le i\le n$) and the analysis can be conducted on the $(p_i)_{1\le i\le n}$ series instead.

Since the analysis of the $(p_i)_{1\le i\le n}$ series of probabilities is a native case of order statistics, the discussion now turns to order statistics. The first studies in this area were by the fathers of modern statistics, Karl Pearson [20] and Ronald A. Fisher [3] while the first order statistic applicable to any distribution (not only the normal distribution) was first studied by Cramér and Von Mises (see [21,22]).

An order statistic operating on probabilities ($(p_i)_{1\le i\le n}$) will sort the values (let $(q_i)_{1\le i\le n}$ be the series of sorted $(p_i)_{1\le i\le n}$ values, Equation (7)) and will assess its departure from the continuous uniform distribution (where it is assumed that SORT is a procedure that sorts ascending the values).

$$(q_i)_{1\le i\le n} \leftarrow \mathrm{SORT}((p_i)_{1\le i\le n}) \tag{7}$$

Since the assessment of the departure from the continuous uniform distribution cannot be made directly, the use of a series of order statistics was proposed by several authors including: Cramér and Von Mises [21,22], Kolmogorov-Smirnov [23–25], Anderson-Darling [26,27], Kuiper V [28], Watson U^2 [29], and the H1 Statistic [18]; see Equation (8). They remain in use today.

For instance, the Kolmogorov-Smirnov (KS) method (see Equation (8); the Kolmogorov-Smirnov statistic) calculates the $KS_{Statistic}$ and later tests the value (from a sample) against the threshold of a chosen significance level (usually 5%).

In order to have certain thresholds for a series of significance levels, these statistics can be derived from Monte-Carlo ("MC") simulations [30], and deployed for a large number of samples in order to reflect, as best as possible, the state of the population.

$$\begin{aligned}
KS_{Statistic} &= \sqrt{n}\cdot\max_{1\le i\le n}\left(q_i - \tfrac{i-1}{n}, \tfrac{i}{n} - q_i\right)\\
KV_{Statistic} &= \sqrt{n}\cdot\left(\max_{1\le i\le n}\left(q_i - \tfrac{i-1}{n}\right) + \max_{1\le i\le n}\left(\tfrac{i}{n} - q_i\right)\right)\\
AD_{Statistic} &= -n - \tfrac{1}{n}\cdot\sum_{i=1}^{n}(2i-1)\cdot\ln(q_i\cdot(1-q_{n-i}))\\
CM_{Statistic} &= \tfrac{1}{12n} + \sum_{i=1}^{n}\left(\tfrac{2\cdot i-1}{2\cdot n} - q_i\right)^2\\
WU_{Statistic} &= CM_{Statistic} + \left(\tfrac{1}{2} - \tfrac{1}{n}\sum_{i=1}^{n} q_i\right)^2\\
H1_{Statistic} &= -\sum_{i=1}^{n} q_i\cdot\ln(q_i) - \sum_{i=1}^{n}(1-q_i)\cdot\ln(1-q_i)
\end{aligned} \tag{8}$$

3. Proposed Outlier Detection Statistic

A statistic was developed to be applicable to any distribution. For a series of probabilities ($(p_i)_{1\le i\le n}$) or (sorted probabilities, $(q_i)_{1\le i\le n}$) associated with a series of (repeated drawing) observations ($(x_i)_{1\le i\le n}$), the $(r_i)_{1\le i\le n}$ differences are calculated as Equation (9):

$$r_i = \left|p_i - 0.5\right|, \text{ for } 1 \le i \le n \tag{9}$$

The statistic called "g1" (see below) was generated based on the formula given in Equation (9) (given as Equation (10)).

$$g1 = \max_{1\le i\le n} r_i \tag{10}$$

It should be noted that Equations (9) and (10) provide the same result regardless of whether the calculation is made on a sorted series of probabilities ($(q_i)_{1\le i\le n}$) or not (then it is made on $(p_i)_{1\le i\le n}$).

Regarding the name of this new proposed statistic ("g1"), when Equations (1) and (2) ($G_{\text{"min"}}$, $G_{\text{"max"}}$, $G_{\text{"all"}}$) and Equation (9) are compared, for a standard normal distribution $N(x; \mu=0,\sigma=1)$ the equation defining $G_{\text{"all"}}$ becomes much more like Equation (9), with the difference being that in Equation (2) the sample mean ($\overline{x}$) is used as an estimate for the mean of the population (μ) and the sample standard deviation (s) is used as an estimate for the standard deviation of the population (σ) while Equation (9) basically expresses the same in terms of associated probabilities ($p_i = P(X \le x_i) = CDF_{\text{"Normal"}}(x_i; \mu,\sigma)$, $0.5 = P(X \le \mu) = CDF_{\text{"Normal"}}(\mu; \mu,\sigma)$).

Therefore, the proposed statistic very much resembles the Grubbs test for normality (and hence its name). One difference is that in the Grubbs test sample statistics are used to calculate the sample $G_{\text{"all"}}$ value ($\overline{x}$ and s), thereby reducing the degrees of freedom associated with the value (from n to n-2) while for the g1 value (and statistic) the degrees of freedom remain unchanged (n). The major difference is actually the one that makes the proposed statistic generalizable to any distribution—the mean used in the Grubbs test is replaced by the median—the beauty of this change is that for symmetrical distributions (including a Normal distribution) these two coincide.

A further connection with other statistics must also be noted. If any sample is resampled by extracting only the smallest and the largest of its values, then the Kolmogorov-Smirnov statistic for those subsamples almost perfectly resembles (by setting n = 2 in Equations (8)–(1)) the proposed "g1" statistic.

Since CDF is a bijective function (see Equation (6)), the proposed generalization of the Grubbs test for detecting the outliers for Normal distribution into the "g1" statistic for detecting the outliers for any distribution is a natural extension of it. The "g1" test associated with the "g1" statistic will be able to operate in the probability space ($(p_i)_{1\leq i\leq n}$ or $(q_i)_{1\leq i\leq n}$) instead of the observed space ($(x_i)_{1\leq i\leq n}$), the calculation formula (Equations (9) and (10)) is slightly different (to those given in Equations (1) and (2)), and the probability associated with the departure will no longer be extracted from the Student t distribution (as in Equations (1) and (2)). The change from mean (μ for $G_{\text{"all"}}$) to median (0.5 in Equation (9)) is a safe extension for any distribution type, since Equation (9) measures (or accounts for) the extreme departures from the equiprobable point—having an observation y ($y \leftarrow X$) with $y \leq \text{InvCDF}_{\text{"Any distribution"}}(0.5; \text{"parameters"})$ and an observation z ($z \leftarrow X$) with $z \geq \text{InvCDF}_{\text{"Any distribution"}}(0.5; \text{"parameters"})$ is equiprobable.

One way to associate a probability with the "g1" statistic is to do a Monte-Carlo (MC) simulation.

4. Simulation Study

A MC study was conducted. Two different strategies were developed in order to deal efficiently with a very large amount of data, and specifically, to solve the order statistics problem (that is, first sampling from the uniform distribution, and later using Equations (7)–(10). One of those alternatives has been described in [14] and the other is described below. Table 2 shows the details of the conducted MC study.

Table 2. Details of the MC simulation on "g1" outlier detection statistic.

Parameter	Meaning	Setting
n	sample size of the observed	from 2 to 12
m	sample size of the MC simulation	10^8
p	control points for the probability	999
resa	internal resamples (repetitions)	10
repe	external repetitions	7

For each sample size of the observed *n* in each run *m* samples (see Table 2) were generated from the standard uniform continuous distribution (e.g., from the [0, 1] interval). The outlier detection statistic "g1" was calculated (Equations (9) and (10)). From a large pool of sampled and resampled data ($m \cdot resa \cdot repe = 7 \cdot 10^9$ in Table 2, repetitions were joined *(n, p, g1)* as pairs from the *p·n* control points, that is, where the probability was from 0.001 to 0.999 with a step of 0.001 for each n (from 2 to 12). The external repetitions (resa = 7 in Table 2) were joined together by taking the median (since the median is a sufficiency statistic [31] for any order statistic such as in the extraction of *(n, p, g1)* pairs from the *p·n* control points). The MC simulation was conducted with the configuration set as defined in Table 2. The obtained data were recorded in separate files by sample size and analyzed as such.

The objective associated (with any statistic) is to obtain the cumulative distribution function (CDF, Equation (5)), and thus by evaluating the CDF for the value of the statistic obtained from the sample

(Equations (9) and (10)) to obtain a probability for the sampling. Please note that only in the lucky cases are we able to do this; Generally only the critical values (values corresponding to certain risks of being in error) or approximation formulas are available (see for instance [21,24,26,28,29]). Here, the analytical CDF formula was obtained for the "g1" outlier detection statistic.

5. The Analytical Formula of CDF for g1

The "g1" statistic have a very simple calculation formula (see Equation (9)) and, as expected, its CDF formula is also very simple (see Equation (11)). Thus, for a calculated sample statistic g1 (x ← g1 in Equation (11)), the significance level (α ← 1-p) is immediate (Equation (11), where P represents the probability that the random variable X takes on a value less than or equal to x).

$$p = \mathrm{CDF}_{\text{“}g1\text{”}}(x;n) = \mathrm{P}(\mathrm{X} \le \mathrm{x}) = (2{\cdot}x)^n,\ \alpha = 1 - p = 1 - (2{\cdot}x)^n \quad (11)$$

6. Simulation Results for the Distribution of the "g1" Statistic

The results of the simulation for n varying from 2 to 10 were sufficient to provide a clear indication of the analytical formula for the CDF of "g1". Descriptive statistics including Standard Error (SE, the standard error formula is given as Equation (12)) between the expected probability (from MC simulation) and the calculated probability (from Equation (11), $\hat{p}_i \leftarrow (2{\cdot}x_i)^n$) and the highest positive and highest negative departures are given in Table 3.

$$SE = \sqrt{\frac{1}{999}\sum_{i=1}^{999}(p_i - \hat{p}_i)^2},\ p_i = \frac{i}{1000} \quad (12)$$

Table 3. Descriptive statistics for the agreement in the calculation of the "g1" statistic (Equation (10) vs. Equation (11)).

n	SE	$\min(p_i - \hat{p}_i)$		$\max(p_i - \hat{p}_i)$	
2	**2.9×10^{-6}**	-7.9×10^{-6}	at p = 0.694	5.7×10^{-6}	at p = 0.427
3	5.6×10^{-6}	-1.2×10^{-5}	at p = 0.787	1.6×10^{-6}	at p = 0.118
4	2.2×10^{-6}	-5.6×10^{-6}	at p = 0.234	3.7×10^{-6}	at p = 0.613
5	6.0×10^{-6}	-1.2×10^{-5}	at p = 0.546	2.3×10^{-6}	at p = 0.080
6	3.5×10^{-6}	-5.8×10^{-6}	at p = 0.797	9.2×10^{-6}	at p = 0.196
7	5.0×10^{-6}	-9.6×10^{-6}	at p = 0.777	3.8×10^{-6}	at p = 0.035
8	4.2×10^{-6}	-8.4×10^{-6}	at p = 0.675	3.9×10^{-6}	at p = 0.948
9	3.3×10^{-6}	-9.1×10^{-6}	at p = 0.269	7.9×10^{-6}	at p = 0.689
10	2.8×10^{-6}	-6.4×10^{-6}	at p = 0.443	6.6×10^{-6}	at p = 0.652

As can be observed in Table 3 the standard error (SE) slowly decreases beginning with n = 7, being two orders of magnitude smaller (actually it is about 200 times smaller) than the step from the MC experiment. Since the standard error alone is not proof that Equation (11) is the true CDF formula for providing the probability for the g1 statistic, the smallest and the highest difference between the observed and the expected probabilities are also given in Table 3. They substantiate that Equation (11) is indeed the right estimate for the CDF of g1. For convenience, Figure 1 shows the value of the error in each observation point (999 points corresponding to p = 0.001 up to p = 0.999 for each n from 2 to 12).

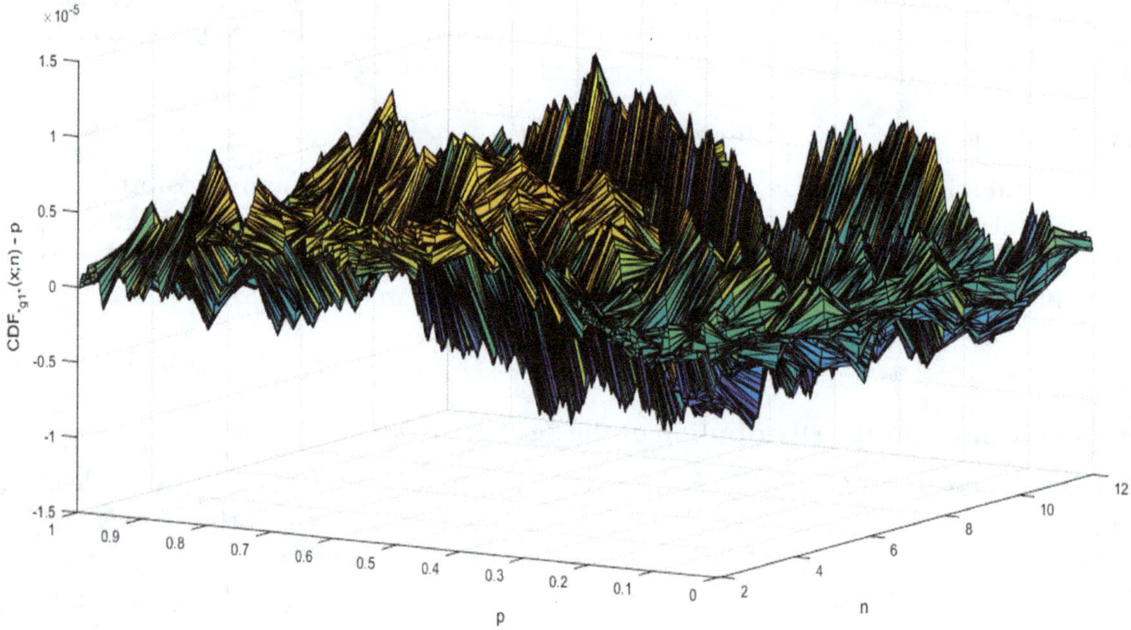

Figure 1. Departures between expected and observed probabilities for g1 statistic (Equation (10) vs. Equation (11)).

Regarding the estimation error (of the "g1" statistic) depicted in Figure 1, the "g1" statistic is rarely bigger than 10^{-5}, never bigger than $1.5 \cdot \times 10^{-5}$ and tends to become smaller with the increase in sample size (n). Using Equation (11), Figure 2 depicts the shape of the $CDF_{\text{"g1"}}(x;n)$.

With regard to the "g1" statistic (depicted in Figure 2), the domain for a variable distributed by the "g1" statistic (see Equation (11)) has values between 0 and 0.5 with the mode at p = 0 (a vertical asymptote at p = 0), a median of $n^{-1} \cdot 2^{-1/n}$ (and having a left asymmetry decreasing with the increasing of n and converging (for $n \to \infty$) to symmetry) and mean of 1/2(n+1).

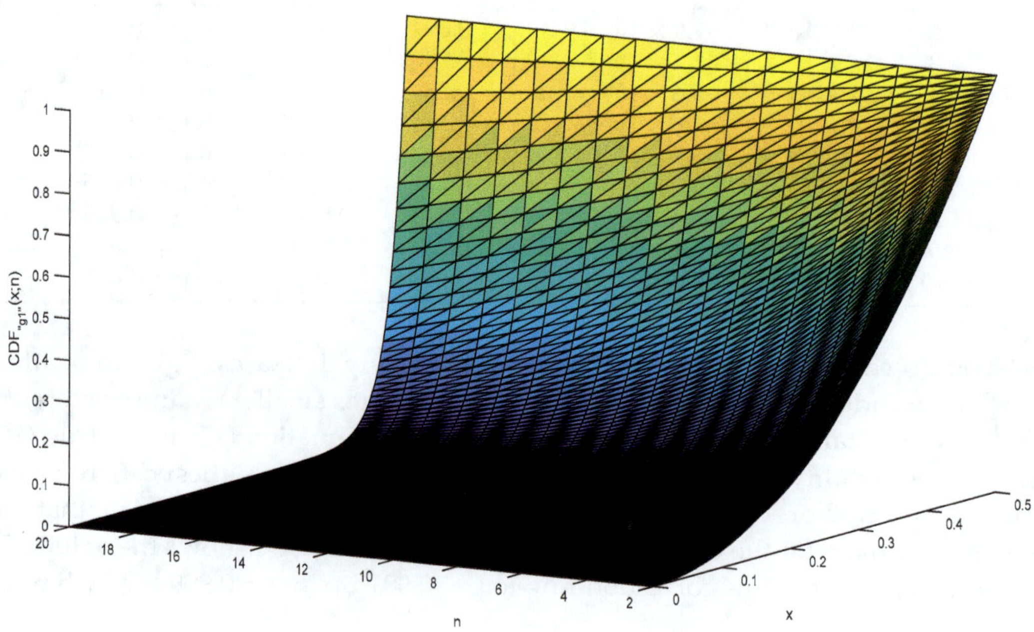

Figure 2. $CDF_{\text{"g1"}}(x;n)$ for n = 2 to n = 20.

The expression of CDF“g1” is easily inverted (see Equation (13)).

$$\mathrm{CDF}_{\text{“g1”}}(x;\, n) \;=\; (2x)^{n} \;\rightarrow\; \mathrm{InvCDF}_{\text{“g1”}}(p;\, n) \;=\; \sqrt[n]{p}/2 \tag{13}$$

7. from “g1” Statistic to “g1” Confidence Intervals for the Extreme Values

Equation (13) can be used to calculate the critical values of the “g1” statistic for any values of α ($\alpha \leftarrow$ 1-p) and n. The critical values of the “g1” statistic acts as the boundaries of the confidence intervals.

By setting the risk of being in error α (usually at 5%), then p = 1-α and Equation (13) can be used to calculate the statistic associated with it ($\mathrm{InvCDF}_{\text{“g1”}}(1\text{-}\alpha;n) = \sqrt[n]{1-\alpha}/2$). By placing this value into Equations (9) and (10), the (extreme) probabilities can be extracted (Equation (14)).

$$\max_{1 \le i \le n} \left| p_i - 0.5 \right| = \sqrt[n]{1-\alpha}/2. \;\rightarrow\; p_{\text{extreme}}(\alpha) \;=\; 0.5 \pm \sqrt[n]{1-\alpha}/2 \tag{14}$$

One should note that the confidence interval defined by Equation (14) is symmetric.

In order to arrive at the confidence intervals for the extreme values in the sampled data (Equation (15)) it is necessary to use the inverse of the CDF (again), and for the distribution of the sampled data.

$$x_{\text{extreme}}(\alpha) = \mathrm{InvCDF}_{\text{“Distribution”}}(0.5 \pm \sqrt[n]{1-\alpha}/2; \text{“parameters”}) \tag{15}$$

To illustrate the calculation of the confidence intervals for the extreme values in the sampled data, a series of 206 data was chosen from [32]. The data were tested against the assumption that it follows a generalized Gauss-Laplace distribution (Equation (16), a symmetrical distribution), and later if there were some observations suspected to be outliers. The steps of this analysis and the obtained results are given in Table 4.

$$\mathrm{PDF}_{\text{“GL”}}(x; \mu, \sigma, k) = c_1 \sigma^{-1} e^{-|c_0 z|^k},\; c_0 = \left(\frac{\Gamma(3/k)}{\Gamma(1/k)}\right)^{1/2},\; c_1 = \frac{k c_0}{2\Gamma(1/k)},\; z = \frac{x-\mu}{\sigma} \tag{16}$$

The greatest departure from the median (0.5) for the 206 PCB dataset (Table 4) was 9.603 ($\mathrm{CDF}_{\text{“GL”}}(9.603;\ \mu = 6.47938,\ \sigma = 0.82828,\ k = 1.79106) = 0.9998$). Due to the force of this deviation from the median, 9.603 was suspected as being an outlier and was removed (it should be noted that in a broader context, an outlier can be also seen as an atypical observation, correctly collected from the population observation, as part of the data generation process and thus it may be maintained in the sample but probably with a less weight). The same procedure (as in Table 4) can be applied to the remaining data (205 observations). Then, $\mathrm{InvCDF}_{\text{“g1”}}(1\text{-}0.05;\ 205) = 0.499875$, $p_{min}(n{=}205) = 0.0001251$; and $p_{max}(n{=}205) = 0.9998749$. The MLE estimates for the parameters of the Gauss-Laplace distribution remain unchanged ($\mu = 6.47938$, $\sigma = 0.82828$, $k = 1.79106$) and the removed observation (9.603) is still not an outlier ($x_{max} = \mathrm{InvCDF}_{\text{“GL”}}(0.9998749;\ \mu = 6.47938,\ \sigma = 0.82828,\ k = 1.79106) = 9.7166 > 9.603$).

Table 4. Distribution analysis for a series of 206 measurements for the octanol water partition coefficient (K_{ow}) of polychlorinated biphenyls expressed in logarithmic scale ($\log_{10}(K_{ow})$)

Step	Results
Dataset (given for convenience)	4.151; 4.401; 4.421; 4.601; 4.941; 5.021; 5.023; 5.150; 5.180; 5.295; 5.301; 5.311; 5.311; 5.335; 5.343; 5.404; 5.421; 5.447; 5.452; 5.452; 5.481; 5.504; 5.517; 5.537; 5.537; 5.551; 5.561; 5.572; 5.577; 5.577; 5.627; 5.637; 5.637; 5.667; 5.667; 5.671; 5.677; 5.677; 5.691; 5.717; 5.743; 5.751; 5.757; 5.761; 5.767; 5.767; 5.787; 5.811; 5.817; 5.827; 5.867; 5.897; 5.897; 5.904; 5.943; 5.957; 5.957; 5.987; 6.041; 6.047; 6.047; 6.047; 6.057; 6.077; 6.091; 6.111; 6.117; 6.117; 6.137; 6.137; 6.137; 6.137; 6.137; 6.142; 6.167; 6.177; 6.177; 6.177; 6.204; 6.207; 6.221; 6.227; 6.227; 6.231; 6.237; 6.257; 6.267; 6.267; 6.267; 6.291; 6.304; 6.327; 6.357; 6.357; 6.367; 6.367; 6.371; 6.427; 6.457; 6.467; 6.487; 6.497; 6.511; 6.517; 6.517; 6.523; 6.532; 6.547; 6.583; 6.587; 6.587; 6.587; 6.607; 6.611; 6.647; 6.647; 6.647; 6.647; 6.647; 6.657; 6.657; 6.671; 6.671; 6.677; 6.677; 6.677; 6.697; 6.704; 6.717; 6.717; 6.737; 6.737; 6.737; 6.747; 6.767; 6.767; 6.767; 6.797; 6.827; 6.857; 6.867; 6.897; 6.897; 6.937; 6.937; 6.957; 6.961; 6.997; 7.027; 7.027; 7.027; 7.057; 7.071; 7.087; 7.087; 7.117; 7.117; 7.117; 7.121; 7.123; 7.147; 7.151; 7.177; 7.177; 7.187; 7.187; 7.207; 7.207; 7.207; 7.211; 7.247; 7.247; 7.277; 7.277; 7.277; 7.281; 7.304; 7.307; 7.307; 7.321; 7.337; 7.367; 7.391; 7.427; 7.441; 7.467; 7.516; 7.527; 7.527; 7.557; 7.567; 7.592; 7.627; 7.627; 7.657; 7.657; 7.717; 7.747; 7.751; 7.933; 8.007; 8.164; 8.423; 8.683; 9.143; 9.603
For n = 206 calculate the probability that the extreme values contain an outlier by using Equation (13)	At α = 5% risk being in error $\mathrm{InvCDF}_{\text{“g1”}}(1\text{-}0.05;\ 206) = 0.498755$

Table 4. *Cont.*

Step	Results
Calculate the critical probabilities for the extreme values by using Equations (9) and (10)	$g1 = 0.498755 \rightarrow \lvert 0.5 - p_{min/max} \rvert = 0.498755 \rightarrow 1 - 2p_{min/max} = \pm 0.99751 \rightarrow$ $p_{min} = 0.0001245$; $p_{max} = 0.9998755$
Estimate the parameters of the distribution fitting the dataset (distribution: Gauss-Laplace; μ - location parameter; σ - scale parameter; k - shape parameter)	Initial estimates (from a hybrid CM & MLE method): $\mu = 6.4806$; $\sigma = 0.83076$; $k = 1.4645$; MLE estimates (by applying eq.3): $\mu = 6.47938$; $\sigma = 0.82828$; $k = 1.79106$;
Calculate the lower and the upper bound for the extreme values by using InvCDF of the distribution fitting the data (Equation (15))	InvCDF"GL"(0.0001245; $\mu = 6.47938$, $\sigma = 0.82828$, $k = 1.79106$) = 3.2409 InvCDF"GL"(0.9998755; $\mu = 6.47938$, $\sigma = 0.82828$, $k = 1.79106$) = 9.7178
Make the conclusion regarding the outliers	Since the smallest value in the dataset is 4.151 (> 3.24) and the largest value is 9.603 (< 9.71), at 5% risk being in error there are no outliers in the dataset on the assumption that data follows the Gauss-Laplace distribution

8. Proposed Procedure for Detecting the Outliers

The procedure for detecting the outliers should start with measuring the agreement between the observed and estimated (Figure 3).

Figure 3 contains a statistical "trick", namely, when there are no outliers the statistics measuring the gap between the observation and the model (order statistics, Equation (6)) are in agreement (their associated probabilities are not too far from each other). When outliers exist, the order statistics are also sensitive to their presence. Since this is a separate subject, for further discussion please see the series of papers beginning with [32–34].

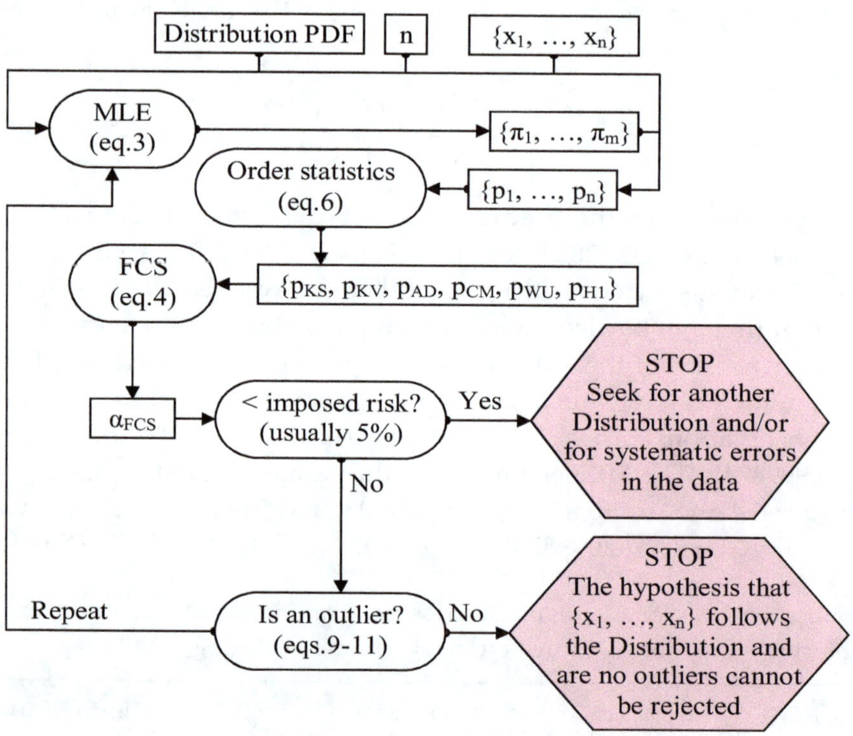

Figure 3. The procedure for detecting outliers.

9. Second Simulation Assessing "Grubbs" and "g1" Outlier Detection Alternatives

Another MC study was designed to test the claim that the proposed method provides consistent results. This second MC simulation is much simpler than the one used to derive the data for constructing the outlier statistics (Figure 4).

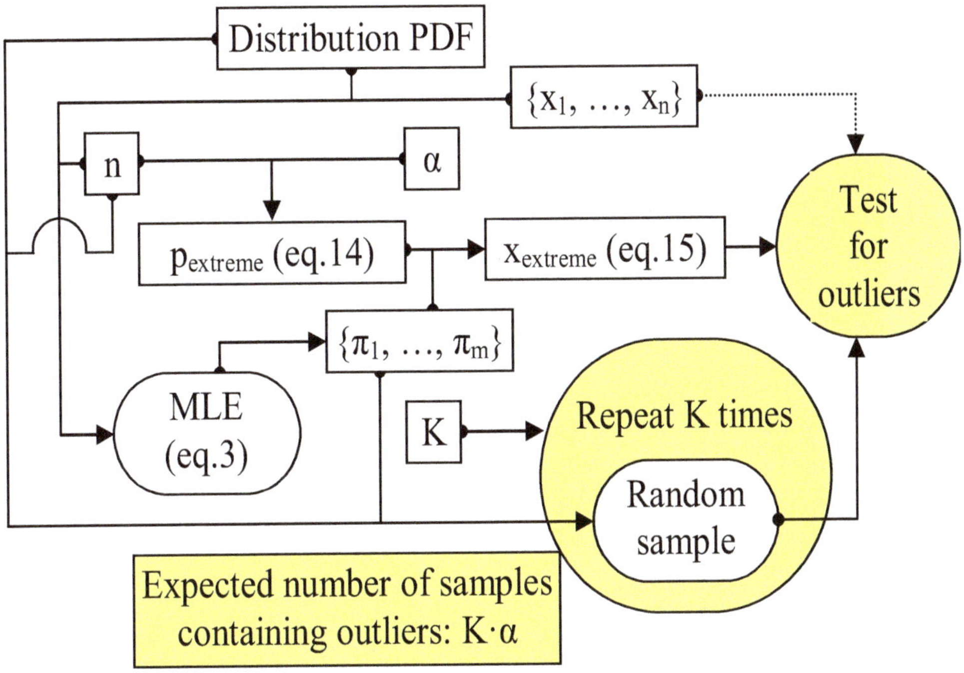

Figure 4. The procedure for testing the outlier statistics.

The data used here as a proof of the facts are from [7] and all cases involve a Normal distribution (Distribution = Normal in Equation (15); PDF and CDF for Normal distribution in Equation (18); a symmetrical distribution) with $\alpha = 5\%$ risk being in error. The parameters of the Normal distribution (μ and σ) are determined for each case, as well as the sample size (Equation (17)).

$$x_{\text{extreme}}(\alpha) = \text{InvCDF}_{\text{"Normal"}}(0.5 \pm 0.5 \cdot \sqrt[n]{1-\alpha}; \mu, \sigma) \tag{17}$$

$$\text{PDF}_{\text{"Normal"}}(x; \mu, \sigma) = \frac{e^{-\frac{(x-\mu)^2}{\sigma^2}}}{\sigma\sqrt{2\pi}}, \ \text{CDF}_{\text{"Normal"}}(x; \mu, \sigma) = \int_{-\infty}^{x} \text{PDF}_{\text{"Normal"}}(t; \mu, \sigma)dt \tag{18}$$

For comparison, the same strategy for calculating the confidence intervals of the extreme values for the Normal distribution with the Grubbs test statistic (Equation (2)) was used to provide an alternate result (Equation (19)).

$$x_{\text{crit}}(\alpha) = \bar{x} \pm G_{\text{crit}}(\alpha) \cdot s, G_{\text{crit}}(\alpha) = \frac{n-1}{\sqrt{n}}\sqrt{\frac{t_G^2(\alpha)}{n-2+t_G^2(\alpha)}}, \ t_G = \text{InvCDF}_{\text{"Student t"}}(\frac{\alpha}{2n}, n-2) \tag{19}$$

The steps followed in this analysis are given in the Table 5.

Table 5. Comparison of the steps of the analysis and simulation for extreme values confidence intervals (proposed method vs. Grubbs test)

Step	Action (step 0 is setting the dataset; $\alpha \leftarrow 0.05$)
1	Estimate (with MLE, Equation (3)) parameters (μ, σ) of the Normal distribution; calculate the associated CDFs (Equation (18))
2	Calculate the order statistics, their associated risks being in error, FCS and p_{FCS} (Equations (6) and (4))
3	For n and α calculate the confidence intervals for the extreme values by using (a) Equation (6) and (17) and (b) Equation (19)
4	Run the MC experiment (Figure 4) for K = 10000 (and then the expected number of outliers is 500) samples and count the samples containing outliers for the existing method (Grubbs, Equation (19); with μ and σ from CM method) and for the proposed method (g1, Equations (13)–(15) and (17); with μ and σ from the MLE method)

Results of the analysis using the steps given in Table 5 for the first dataset are given in Table 6.

Table 6. Outlier analysis results for {568, 570, 570, 570, 572, 572, 572, 578, 584, 596} dataset.

Step	Results (for $\alpha = 5\%$)							
1	$\mu = 575.2$; $\sigma = 8.256$ (MLE) $\rightarrow$ CPs = {0.1916, 0.2644, 0.2644, 0.2644, 0.3492, 0.3492, 0.3492, 0.6328, 0.8568, 0.9941}							
2	**Statistic**	**AD**	**KS**	**CM**	**KV**	**WU**	**H1**	**FCS**
	Value	1.137	1.110	0.206	1.715	0.182	5.266	12.293
	$\alpha_{Statistic}$	0.288	0.132	0.259	0.028	0.049	0.343	0.056
3	$x_{crit}(5\%) = 575.2 \pm 2.29 \cdot 8.7025$; $p_{extreme}(5\%) = 0.5 \pm \text{InvCDF}_{\text{"g1"}}(1\text{-}0.05; 10)$; $x_{extreme}(5\%) = \{552.086, 598.314\}$							
4	**Number of samples containing outliers**	**Existing method (Grubbs)**	**Proposed method (g1)**					
	First run	1977 (19.77%)	510 (5.1%)					
	Second run	2009 (20.09%)	526 (5.26%)					

In regard to the results given in Table 6:

At step 1, CPs are the cumulative probabilities ($\{p_1, \ldots, p_{10}\}$ in Figure 3) associated with the series of the observations from the sample ($\{x_1, \ldots, x_{10}\}$ in Figure 3).

At step 2, the data passes the normality test ($\alpha_{FCS} = 7\% > 5\% = \alpha$, see Figure 3).

Step 3 was made for n = 10 (see Figure 4). (a) The proposed method does not detect outliers in the sample (552.086 < 568, 596 < 598.314); (b) Grubbs test detect 596 as being an outlier (596 > 595.13).

At step 4 (see Figure 4), since {510, 526} are comparable with 500 and {1977, 2009} are much greater than 500, the results lead to the conclusion that the existing method produces type I errors by leading to false positive detection of outliers in the samples while the proposed method does not.

10. Going Further with the Outlier Analysis

What if "596" is removed from the sample? The following table provides mirror-like results for this scenario (Table 7).

Table 7. Outlier analysis results for {568, 570, 570, 570, 572, 572, 572, 578, 584} dataset.

Step	Results (for $\alpha = 5\%$)							
1	$\mu = 572.889$; $\sigma = 4.725$ (MLE) $\rightarrow$ CPs = {0.1504, 0.2705, 0.2705, 0.2705, 0.4254, 0.4254, 0.4254, 0.8603, 0.9907}							
2	**Statistic**	**AD**	**KS**	**CM**	**KV**	**WU**	**H1**	**FCS**
	Value	0.935	1.057	0.174	1.535	0.155	4.678	9.715
	$\alpha_{Statistic}$	0.389	0.167	0.327	0.082	0.088	0.394	0.137
3	$x_{crit}(5\%) = 572.89 \pm 2.215 \cdot 5.011$; $p_{extreme}(5\%) = 0.5 \pm \text{InvCDF}_{\text{"g1"}}(1\text{-}0.05; 9)$; $x_{extreme}(5\%) = \{559.822, 585.956\}$							
4	**Number of samples containing outliers**	**Existing method (Grubbs)**	**Proposed method (g1)**					
	First run	2341 (23.41%)	563 (5.63%)					
	Second run	2333 (23.33%)	543 (5.43%)					

As can be observed in Table 7, the data is not in good agreement with normality (α_{FCS} in Table 6 is 7%, while in Table 7 it is 16%) and there is no change in the accuracy of the classification ({563, 543} comparable with 500, {2341, 2333} is much greater than 500; the existing method produces type I errors by leading to false positive detection of outliers in the samples, while the proposed method does not). When comparing the results given in Table 6 with the results given in Table 7 it should be noted that both tests (Grubbs and the newly proposed g1) produce somewhat confusing results (see Table 8 for side-by-side outcomes).

Table 8. Side-by-side comparison of the analysis of the samples.

Sample	{568, 570, 570, 570, 572, 572, 572, 578, 584, 596}	{568, 570, 570, 570, 572, 572, 572, 578, 584}
At 5% risk being in error can the hypothesis that the sample was drawn from a normal distribution be rejected?	No (α_{FCS} = 7%)	No (α_{FCS} = 15.8%)
Grubbs confidence interval for 'no outliers' at 5% risk being in error	(555.27, 595.13) 596 is detected as being outlier	(561.79, 583.99) 584 is detected as being outlier
g1 confidence interval for 'no outliers' at 5% risk being in error	(552.08, 598.32) no outliers	(559.82, 585.96) no outliers

Table 8 highlights the fact that based on the {568, 570, 570, 570, 572, 572, 572, 578, 584} sample, the g1 test may be interpreted as identifying 596 as being an outlier. This is not quite true because the g1 test was not intended to be used in this way. That is, 596 is outside of the dataset, so at the time of constructing the confidence intervals for the extreme values, the information regarding its observation was missing.

Another trial was done, this time with **601** replacing **596** in the initial dataset (Table 9).

Table 9. Outlier analysis results for the {568, 570, 570, 570, 572, 572, 572, 578, 584, **601**} dataset.

Step	Results (for α = 5%)							
1	From the CM method: μ = 575.7; σ = 10.067; from MLE method: μ = 575.7; σ = 9.550							
	Statistic	**AD**	**KS**	**CM**	**KV**	**WU**	**H1**	**FCS**
2	Value	1.267	1.109	0.225	1.774	0.198	5.411	13.652
	$\alpha_{Statistic}$	0.241	0.132	0.226	0.018	0.035	0.254	0.034
3	Grubbs confidence interval for 'no outliers' at 5% risk being in error: (552.647,598.753); **601** is an outlier; g1 confidence interval for 'no outliers' at 5% risk being in error: (548.963, 602.437); no outliers							

In a further trial, **604** replaced **596** in the initial dataset (Table 10).

Table 10. Outlier analysis results for the {568, 570, 570, 570, 572, 572, 572, 578, 584, 604} dataset.

Step	Results (for α = 5%)							
1	From the CM method: μ = 576.0; σ = 10.914; from MLE method: μ = 576.0; σ = 10.354							
	Statistic	**AD**	**KS**	**CM**	**KV**	**WU**	**H1**	**FCS**
2	Value	1.348	1.108	0.238	1.803	0.209	5.481	14.468
	$\alpha_{Statistic}$	0.216	0.133	0.206	0.015	0.028	0.215	0.025
3	Grubbs confidence interval for 'no outliers' at 5% risk being in error: (551.00, 601.00); 604 is an outlier; g1 confidence interval for 'no outliers' at 5% risk being in error: (547.01, 604.99); no outliers							

The conclusion is simple (see the results in the Tables 6, 7, 9 and 10): A test will hardly ever detect an outlier for a small sample; it is more likely to reject the hypothesis of the sample drawn from the distribution itself!

The same trick was used on a bigger sample and the results are shown in Table 11 (the dataset is from Table 4).

Table 11. Outlier analysis results for Table 4 dataset under the assumption of normal distribution.

Step	Results (for $\alpha = 5\%$)							
1	Table 5 Dataset; Normal distribution → CM: $\mu = 6.481$; $\sigma = 0.831$; MLE: $\mu = 6.481$; $\sigma = 0.829$							
2	Statistic	AD	KS	CM	KV	WU	H1	FCS
	Value	0.439	0.484	0.049	0.952	0.047	104.2	1.276
	$\alpha_{\text{Statistic}}$	0.812	0.965	0.886	0.852	0.743	0.641	0.973
3	Grubbs confidence interval for 'no outliers' at 5% risk being in error: (3.492, 9.470); 9.603 is an outlier g1 confidence interval for 'no outliers' at 5% risk being in error: (3.444, 9.517); 9.603 is an outlier							
4	Number of samples containing outliers	Existing method (Grubbs)	Proposed method (g1)					
	First run	637 (6.37%)	511 (5.11%)					
	Second run	630 (6.3%)	481 (4.81%)					

On one hand, as the results in Table 11 prove, the proposed method correctly identifies the confidence interval for the extreme values, while the existing method does not.

On the other hand, the results in Table 11 also show that the likelihood of identifying the outliers increases with the sample size, making it perfectly possible to identify outliers with the proposed method, although this is not the case in small samples. It is possible to detect the outliers in small samples as well, but not when the parameters of the distribution are derived from the sample data—only when the parameters of the distribution are known a priori or determined from other samples (the results given in Tables 6–10 are proof of this).

11. Further Discussion

The obtained expression for CDF of "g1" (Equation (11)) reveals the domain of a random variable distributed by the "g1" statistic ([0, 0.5]), which is consistent with the definition of "g1" (Equations (9) and (10)).

Independently of the shape of the theoretical distribution being tested (the generic case is defined by Equation (5)), as defined by Equations (9) and (10), the newly proposed statistic "g1" defines a symmetric confidence interval for the extreme values in samples in the probability space (Equation (14)). Later, this symmetric confidence interval may be changed back into an asymmetrical one when it is expressed in the domain of the theoretical distribution being tested (Equation (15)). It should be recognized that "g1" uses a symmetrization strategy to obtain the confidence interval for the extreme values in samples.

It might seem that the literature on robust statistics was ignored in this work, however, this is not entirely true. In fact, a whole pool of robust statistics was used extensively in the study (see Equation (8)), introduced as a tool in Table 5 and involved in the later calculations (Tables 6, 7 and 9, Tables 10 and 11). Also, it should be noted that the substitution of the mean by the median is not a new idea; it is well known in the field of robust statistics (for example, Watson U^2 [29], the $WU_{\text{Statistic}}$ in Equation (8), uses it).

A short literature survey provides several of examples of current real applications that require the proposed method. Thus, in signal processing, non-stationary, non-Gaussian, spiky signals are usually regarded as outliers and thus discarded (see [35–38] as typical cases). In this context, it should be noted that Mood's median test is preferred to the Kruskal-Wallis test when outliers are present [39]. The identification of outliers is also recognized as an issue in the validation of protein structures, and the current methods are revised in [40]. Other examples can be found in [41].

In the wider context, an alternate window-based strategy has been proposed in which outliers are detected in each window by the Tukey method and labeled so that they can be excluded from the realization of the process points to be used for model identification [42]. A contingency-based strategy proposes maximization of true positive (TP) values and minimization of false negative (FN) and false positive (FP) values [43]. Finally, another distribution testing procedure has been proposed in [44].

12. Conclusions

A new method for detecting outliers was proposed in this paper. The method is applicable to any continuous distribution at any risk being in error. It was proved that the method correctly detects the outliers. For a normal distribution at 5% risk being in error, it was also shown that the proposed method outperforms the classical Grubbs test for detecting the outliers.

Acknowledgments: Thanks to my colleague S.D. Bolboacă and for our fruitful discussions during the development stage of the study, which helped and motivated the author to complete the study.

References

1. Gauss, C.F. *Theoria Motus Corporum Coelestium*; (Translated in 1857 as "Theory of Motion of the Heavenly Bodies Moving about the Sun in Conic Sections" by C. H. Davis. Little, Brown: Boston. Reprinted in 1963 by Dover: New York); Perthes et Besser: Hamburg, Germany, 1809; pp. 249–259.
2. Tippett, L.H.C. The extreme individuals and the range of samples taken from a normal population. *Biometrika* **1925**, *17*, 151–164. [CrossRef]
3. Fisher, R.A.; Tippett, L.H.C. Limiting forms of the frequency distribution of the largest and smallest member of a sample. *Proc. Camb. Philos. Soc.* **1928**, *24*, 180–190. [CrossRef]
4. Thompson, W.R. On a criterion for the rejection of observations and the distribution of the ratio of the deviation to the sample standard deviation. *Ann. Math. Stat.* **1935**, *6*, 214–219. [CrossRef]
5. Pearson, E.; Sekar, C.C. The efficiency of the statistical tools and a criterion for the rejection of outlying observations. *Biometrika* **1936**, *28*, 308–320. [CrossRef]
6. Grubbs, F.E. Sample criteria for testing outlying observations. *Ann. Math. Stat.* **1950**, *21*, 27–58. [CrossRef]
7. Grubbs, F.E. Procedures for detecting outlying observations in samples. *Technometrics* **1969**, *11*, 1–21. [CrossRef]
8. Nooghabi, M.; Nooghabi, H.; Nasiri, P. Detecting outliers in gamma distribution. *Commun. Stat. Theory Methods* **2010**, *39*, 698–706. [CrossRef]
9. Kumar, N.; Lalitha, S. Testing for upper outliers in gamma sample. *Commun. Stat. Theory Methods* **2012**, *41*, 820–828. [CrossRef]
10. Lucini, M.; Frery, A. Comments on "Detecting Outliers in Gamma Distribution" by M. Jabbari Nooghabi et al. (2010). *Commun. Stat. Theory Methods* **2017**, *46*, 5223–5227. [CrossRef]
11. Hartley, H. The range in random samples. *Biometrika* **1942**, *32*, 334–348. [CrossRef]
12. Bardet, J.-M.; Dimby, S.-F. A new non-parametric detector of univariate outliers for distributions with unbounded support. *Extremes* **2017**, *20*, 751–775. [CrossRef]
13. Gosset, W. The probable error of a mean. *Biometrika* **1908**, *6*, 1–25.
14. Jäntschi, L.; Bolboacă, S.-D. Computation of probability associated with Anderson-Darling statistic. *Mathematics* **2018**, *6*, 88. [CrossRef]
15. Fisher, R. On an Absolute Criterion for Fitting Frequency Curves. *Messenger Math.* **1912**, *41*, 155–160.
16. Fisher, R. Questions and answers #14. *Am. Stat.* **1948**, *2*, 30–31.
17. Bolboacă, S.D.; Jäntschi, L.; Sestraș, A.F.; Sestraș, R.E.; Pamfil, D.C. Supplementary material of 'Pearson-Fisher chi-square statistic revisited'. *Information* **2011**, *2*, 528–545. [CrossRef]
18. Jäntschi, L.; Bolboacă, S.D. Performances of Shannon's Entropy Statistic in Assessment of Distribution of Data. *Ovidius Univ. Ann. Chem.* **2017**, *28*, 30–42. [CrossRef]
19. Davis, P.; Rabinowitz, P. *Methods of Numerical Integration*; Academic Press: New York, NY, USA, 1975; pp. 51–198.
20. Pearson, K. Note on Francis Gallon's problem. *Biometrika* **1902**, *1*, 390–399.
21. Cramér, H. On the composition of elementary errors. *Scand. Actuar. J.* **1928**, *1*, 13–74. [CrossRef]
22. Von Mises, R.E. *Wahrscheinlichkeit, Statistik und Wahrheit*; Julius Springer: Berlin, Germany, 1928; pp. 100–138.
23. Kolmogorov, A. Sulla determinazione empirica di una legge di distribuzione. *Giornale dell'Istituto Italiano degli Attuari* **1933**, *4*, 83–91.
24. Kolmogorov, A. Confidence Limits for an Unknown Distribution Function. *Ann. Math. Stat.* **1941**, *12*, 461–463. [CrossRef]

25. Smirnov, N. Table for estimating the goodness of fit of empirical distributions. *Ann. Math. Stat.* **1948**, *19*, 279–281. [CrossRef]
26. Anderson, T.W.; Darling, D.A. Asymptotic theory of certain "goodness-of-fit" criteria based on stochastic processes. *Ann. Math. Stat.* **1952**, *23*, 193–212. [CrossRef]
27. Anderson, T.W.; Darling, D.A. A Test of Goodness-of-Fit. *J. Am. Stat. Assoc.* **1954**, *49*, 765–769. [CrossRef]
28. Kuiper, N.H. Tests concerning random points on a circle. *Proc. Koninklijke Nederlandse Akademie van Wetenschappen Series A* **1960**, *63*, 38–47. [CrossRef]
29. Watson, G.S. Goodness-Of-Fit Tests on a Circle. *Biometrika* **1961**, *48*, 109–114. [CrossRef]
30. Metropolis, N.; Ulam, S. The Monte Carlo Method. *J. Am. Stat. Assoc.* **1949**, *44*, 335–341. [CrossRef] [PubMed]
31. Fisher, R.A. On the mathematical foundations of theoretical statistics. *Philos. Trans. R. Soc. A* **1922**, *222*, 309–368. [CrossRef]
32. Jäntschi, L. Distribution fitting 1. Parameters estimation under assumption of agreement between observation and model. *Bull. UASVM Hortic.* **2009**, *66*, 684–690.
33. Jäntschi, L.; Bolboacă, S.D. Distribution fitting 2. Pearson-Fisher, Kolmogorov-Smirnov, Anderson-Darling, Wilks-Shapiro, Kramer-von-Misses and Jarque-Bera statistics. *Bull. UASVM Hortic.* **2009**, *66*, 691–697.
34. Bolboacă, S.D.; Jäntschi, L. Distribution fitting 3. Analysis under normality assumption. *Bull. UASVM Hortic.* **2009**, *66*, 698–705.
35. Liu, K.; Chen, Y.Q.; Domański, P.D.; Zhang, X. A novel method for control performance assessment with fractional order signal processing and its application to semiconductor manufacturing. *Algorithms* **2018**, *11*, 90. [CrossRef]
36. Paiva, J.S.; Ribeiro, R.S.R.; Cunha, J.P.S.; Rosa, C.C.; Jorge, P.A.S. Single particle differentiation through 2D optical fiber trapping and back-scattered signal statistical analysis: An exploratory approach. *Sensors* **2018**, *18*, 710. [CrossRef] [PubMed]
37. Teunissen, P.J.G.; Imparato, D.; Tiberius, C.C.J.M. Does RAIM with correct exclusion produce unbiased positions? *Sensors* **2017**, *17*, 1508. [CrossRef] [PubMed]
38. Pan, Z.; Liu, L.; Qiu, X.; Lei, B. Fast vessel detection in Gaofen-3 SAR images with ultrafine strip-map mode. *Sensors* **2017**, *17*, 1578. [CrossRef] [PubMed]
39. Vergura, S.; Carpentieri, M. Statistics to detect low-intensity anomalies in PV systems. *Energies* **2018**, *11*, 30. [CrossRef]
40. Chen, L.; He, J.; Sazzed, S.; Walker, R. An investigation of atomic structures derived from X-ray crystallography and cryo-electron microscopy using distal blocks of side-chains. *Molecules* **2018**, *23*, 610. [CrossRef] [PubMed]
41. Bolboacă, S.D.; Jäntschi, L. The effect of leverage and influential on structure-activity relationships. *Comb. Chem. High Throughput Screen.* **2013**, *16*, 288–297. [CrossRef] [PubMed]
42. Faes, L.; Porta, A.; Nollo, G.; Javorka, M. Information decomposition in multivariate systems: Definitions, implementation and application to cardiovascular networks. *Entropy* **2017**, *19*, 5. [CrossRef]
43. Li, G.; Wang, J.; Liang, J.; Yue, C. Application of sliding nest window control chart in data stream anomaly detection. *Symmetry* **2018**, *10*, 113. [CrossRef]
44. Paolella, M.S. Stable-GARCH models for financial returns: Fast estimation and tests for stability. *Econometrics* **2016**, *4*, 25. [CrossRef]

14

One-Dimensional Optimal System for 2D Rotating Ideal Gas

Andronikos Paliathanasis

Institute of Systems Science, Durban University of Technology, P.O. Box 1334, Durban 4000, South Africa; anpaliat@phys.uoa.gr

Abstract: We derive the one-dimensional optimal system for a system of three partial differential equations, which describe the two-dimensional rotating ideal gas with polytropic parameter $\gamma > 2$. The Lie symmetries and the one-dimensional optimal system are determined for the nonrotating and rotating systems. We compare the results, and we find that when there is no Coriolis force, the system admits eight Lie point symmetries, while the rotating system admits seven Lie point symmetries. Consequently, the two systems are not algebraic equivalent as in the case of $\gamma = 2$, which was found by previous studies. For the one-dimensional optimal system, we determine all the Lie invariants, while we demonstrate our results by reducing the system of partial differential equations into a system of first-order ordinary differential equations, which can be solved by quadratures.

Keywords: lie symmetries; invariants; shallow water; similarity solutions; optimal system

1. Introduction

A powerful mathematical treatment for the determination of exact solutions for nonlinear differential equations is the Lie symmetry analysis [1–3]. Specifically, Lie point symmetries help us in the simplification of differential equations by means of similarity transformations, which reduce the differential equation. The reduction process is based on the existence of functions that are invariant under a specific group of point transformations. When someone uses these invariants as new dependent and independent variables, the differential equation is reduced. The reduction process differs between ordinary differential equations (ODEs) and partial differential equations (PDEs). For ODEs, Lie point symmetries are applied to reduce the order of ODE by one; while on PDEs, Lie point symmetries are applied to reduce by one the number of independent variables, while the order of the PDEs remains the same. The solutions that are found with the application of those invariant functions are called similarity solutions. Some applications on the determination of similarity solutions for nonlinear differential equations can be found in [4–9] and the references therein.

A common characteristic in the reduction process is that the Lie point symmetries are not preserved during the reduction; hence, we can say that the symmetries can be lost. That is not an accurate statement, because symmetries are not "destroyed" or "created" under point transformations, but the "nature" of the symmetry changes. In addition, Lie symmetries can be used to construct new similarity solutions for a given differential equation by applying the adjoint representation of the Lie group [10].

It is possible that a given differential equation admits more than one similarity solution when the given differential equation admits a "large" number of Lie point symmetries. Hence, in order for someone to classify a differential equation according to the admitted similarity solutions, all the inequivalent Lie subalgebras of the admitted Lie symmetries should be determined.

The first group classification problem was carried out by Ovsiannikov [11], who demonstrated the construction of the one-dimensional optimal system for the Lie algebra. Since then, the classification of the one-dimensional optimal system has become a main tool for the study of nonlinear differential equations [12–15].

In this work, we focus on the classification of the one-dimensional optimal system for the two-dimensional rotating ideal gas system described by the following system of PDEs [16–18]:

$$h_t + (hu)_x + (hv)_y = 0, \tag{1}$$

$$u_t + uu_x + vu_y + h^{\gamma-2}h_x - fv = 0, \tag{2}$$

$$v_t + uv_x + vv_y + h^{\gamma-2}h_y + fu = 0. \tag{3}$$

where u and v are the velocity components in the x and y directions, respectively, h is the density of the ideal gas, f is the Coriolis parameter, and γ is the polytropic parameter of the fluid. Usually, γ is assumed to be $\gamma = 2$ where Equations (1)–(3) reduce to the shallow water system. However, in this work, we consider that $\gamma > 2$. In this work, polytropic index γ is defined as $\frac{C_p}{C_v} = \gamma - 1$.

Shallow water equations describe the flow of a fluid under a pressure surface. There are various physical phenomena that are described by the shallow water system with emphasis on atmospheric and oceanic phenomena [19–21]. Hence, the existence of the Coriolis force becomes critical in the description of the physical phenomena.

In the case of $\gamma = 2$, the complete symmetry analysis of the system (1)–(3) is presented in [22]. It was found that for $\gamma = 2$, the given system of PDEs is invariant under a nine-dimensional Lie algebra. The same Lie algebra, but in a different representation, is also admitted by the nonrotating system, i.e., $f = 0$. One of the main results of [22] is that the transformation that relates the two representations of the admitted Lie algebras for the rotating and nonrotating system transforms the rotating system (1)–(3) into the nonrotating one. For other applications of Lie symmetries on shallow water equations, we refer the reader to [23–28].

For the case of an ideal gas [17], i.e., parameter $\gamma > 1$ from our analysis, it follows that this property is lost. The nonrotating system and the rotating one are invariant under a different number of Lie symmetries and consequently under different Lie algebras. For each of the Lie algebras, we have the one-dimensional optimal system and all the Lie invariants. The results are presented in tables. We demonstrate the application of the Lie invariants by determining some similarity solutions for the system (1)–(3) for $\gamma > 2$. The paper is structured as follows.

In Section 2, we briefly discuss the theory of Lie symmetries for differential equations and the adjoint representation. The nonrotating system (1)–(3) is studied in Section 3. Specifically, we determine the Lie points symmetries, which form an eight-dimensional Lie algebra. The commutators and the adjoint representation are presented. We make use of these results, and we perform, a classification of the one-dimensional optimal system. We found that in total, there are twenty-three one-dimensional independent Lie symmetries and possible reductions, and the corresponding invariants are determined and presented in tables. In Section 4, we perform the same analysis for the rotating system. There, we find that the admitted Lie symmetries form a seven-dimensional Lie algebra, while there are twenty independent one-dimensional Lie algebras. We demonstrate the results by reducing the system of PDEs (1)–(3) into an integrable system of three first-order ODEs, the solution of which is given by quadratures. In Section 5, we discuss our results and draw our conclusions. Finally, in Appendix A, we present the tables, which include the results of our analysis.

2. Lie Symmetry Analysis

Let $H^A\left(x^i, \Phi^A, \Phi_i^A, \ldots\right) = 0$ be a system of partial differential equations (PDEs) where Φ^A denotes the dependent variables and x^i are the independent variables. At this point, it is important to mention that we make use of the Einstein summation convention. By definition, under the action of the infinitesimal one-parameter point transformation (1PPT):

$$\bar{x}^i = x^i\left(x^j, \Phi^B; \varepsilon\right), \ \bar{\Phi}^A = \Phi^A\left(x^j, \Phi^B; \varepsilon\right), \tag{4}$$

which connects two different points $P\left(x^{j},\Phi^{B}\right)\rightarrow Q\left(\bar{x}^{j},\bar{\Phi}^{B},\varepsilon\right)$, the differential equation $H^{A}=0$ remains invariant if and only if $\bar{H}^{A}=H^{A}$, that is [2]:

$$\lim_{\varepsilon\rightarrow 0}\frac{\bar{H}^{A}\left(\bar{y}^{i},\bar{u}^{A},...;\varepsilon\right)-H^{A}\left(y^{i},u^{A},...\right)}{\varepsilon}=0. \tag{5}$$

The latter condition means that the $\Phi^{A}\left(P\right)$ and $\Phi^{A}\left(Q\right)$ are connected through the transformation.

The lhs of Expression (5) defines the Lie derivative of H^{A} along the vector field X of the one-parameter point transformation (4), in which X is defined as:

$$X=\frac{\partial\bar{x}^{i}}{\partial\varepsilon}\partial_{i}+\frac{\partial\bar{\Phi}}{\partial\varepsilon}\partial_{A}.$$

Thus, Condition (5) is equivalent to the following expression: [2]

$$\mathcal{L}_{X}\left(H^{A}\right)=0, \tag{6}$$

where $\mathcal{L}$ denotes the Lie derivative with respect to the vector field $X^{[n]}$, which is the nth-extension of generator X of the transformation (4) in the jet space $\left\{x^{i},\Phi^{A},\Phi^{A}_{,i},\Phi^{A}_{,ij},...\right\}$ given by the expression [2]:

$$X^{[n]}=X+\eta^{[1]}\partial_{\Phi^{A}_{i}}+...+\eta^{[n]}\partial_{\Phi^{A}_{i_{i}i_{j}...i_{n}}}, \tag{7}$$

in which:

$$\eta^{[n]}=D_{i}\eta^{[n-1]}-u_{i_{1}i_{2}...i_{n-1}}D_{i}\left(\frac{\partial\bar{x}^{j}}{\partial\varepsilon}\right)\ ,\ i\succeq 1\ ,\ \eta^{[0]}=\left(\frac{\partial\bar{\Phi}^{A}}{\partial\varepsilon}\right). \tag{8}$$

Condition (6) provides a system of PDEs whose solution determines the components of the X, consequently the infinitesimal transformation. The vector fields X, which satisfy condition (6), are called Lie symmetries for the differential equation $H^{A}=0$. The Lie symmetries for a given differential equation form a Lie algebra.

Lie symmetries can be used in different ways [2] in order to study a differential equation. However, their direct application is on the determination of the so-called similarity solutions. The steps that we follow to determine a similarity solution are based on the determination and application of the Lie invariant functions.

Let X be a Lie symmetry for a given differential equation $H^{A}=0$, then the differential equation $X\left(F\right)=0$, where F is a function, provides the Lie invariants where by replacing in the differential equation $H^{A}=0$, we reduce the number of the independent variables (in the case of PDEs) or the order of the differential equation (in the case of ordinary differential equations (ODEs)).

Optimal System

Consider the n-dimensional Lie algebra G_{n} with elements $X_{1},\ X_{2},\ ...\ X_{n}$. Then, we shall say that the two vector fields [2]:

$$Z=\sum_{i=1}^{n}a_{i}X_{i}\ ,\ W=\sum_{i=1}^{n}b_{i}X_{i}\ ,\ \ a_{i},\ b_{i}\text{ are constants.} \tag{9}$$

are equivalent iff there:

$$\mathbf{W}=lim_{j=i}^{n}Ad\left(\exp\left(\varepsilon_{i}X_{i}\right)\right)\mathbf{Z} \tag{10}$$

or:

$$W=cZ\ ,\ c=const. \tag{11}$$

where the operator [2]:

$$Ad\left(\exp\left(\varepsilon X_i\right)\right) X_j = X_j - \varepsilon\left[X_i, X_j\right] + \frac{1}{2}\varepsilon^2\left[X_i,\left[X_i, X_j\right]\right] + \ldots \tag{12}$$

is called the adjoint representation.

Therefore, in order to perform a complete classification for the similarity solutions of a given differential equation, we should determine all the one-dimensional independent symmetry vectors of the Lie algebra G_n.

We continue our analysis by calculating the Lie point symmetries for the system (1)–(3) for the case where the system is rotating ($f \neq 0$) and nonrotating ($f = 0$).

3. Symmetries and the Optimal System for Nonrotating Shallow Water

We start our analysis by applying the symmetry condition (6) for the Coriolis free system (1)–(3) with $f = 0$. We found that the system of PDEs admits eight Lie point symmetries, as are presented in the following [11]:

$$\begin{aligned}
X_1 &= \partial_t\,,\ X_2 = \partial_x\,,\ X_3 = \partial_y\,,\\
X_4 &= t\partial_x + \partial_u\,,\ X_5 = t\partial_y + \partial_v\,,\\
X_6 &= y\partial_x - x\partial_y + v\partial_u - u\partial_v,\\
X_7 &= t\partial_t + x\partial_x + y\partial_y,\\
X_8 &= (\gamma - 1)\left(x\partial_x + y\partial_y + u\partial_u + v\partial_v\right) + 2h\partial_h.
\end{aligned}$$

The commutators of the Lie symmetries and the adjoint representation are presented in Table 1 and Table A1, respectively.

Table 1. Commutators of the admitted Lie point symmetries for the nonrotating 2D shallow water.

[,]	X_1	X_2	X_3	X_4	X_5	X_6	X_7	X_8
X_1	0	0	0	X_2	X_3	0	$-(\gamma-1)X_1$	0
X_2	0	0	0	0	0	$-X_3$	0	$(\gamma-1)X_2$
X_3	0	0	0	0	0	X_2	0	$(\gamma-1)X_3$
X_4	$-X_2$	0	0	0	0	$-X_5$	$(\gamma-1)X_4$	$(\gamma-1)X_4$
X_5	$-X_3$	0	0	0	0	X_4	$(\gamma-1)X_5$	$(\gamma-1)X_5$
X_6	0	X_3	$-X_2$	X_5	$-X_4$	0	0	0
X_7	$(\gamma-1)X_1$	0	0	$-(\gamma-1)X_4$	$-(\gamma-1)X_5$	0	0	0
X_8	0	$-(\gamma-1)X_2$	$-(\gamma-1)X_3$	$-(\gamma-1)X_4$	$-(\gamma-1)X_5$	0	0	0

We continue by determining the one-dimensional optimal system. Let us consider the generic symmetry vector:

$$Z^8 = a_1X_1 + a_2X_2 + a_3X_3 + a_4X_4 + a_5X_5 + a_6X_6 + a_7X_7 + a_8X_8$$

From Table A1, we see that by applying the following adjoint representations:

$$Z'^8 = Ad\left(\exp\left(\varepsilon_5X_5\right)\right) Ad\left(\exp\left(\varepsilon_4X_4\right)\right) Ad\left(\exp\left(\varepsilon_3X_3\right)\right) Ad\left(\exp\left(\varepsilon_2X_2\right)\right) Ad\left(\exp\left(\varepsilon_1X_1\right)\right) Z^8$$

parameters ε_1, ε_2, ε_3, ε_4, and ε_5 can be determined such that:

$$Z'^8 = a_6'X_6 + a_7'X_7 + a_8'X_8$$

Parameters a_6, a_7, and a_8 are the relative invariants of the full adjoint action. Indeed, in order to determine the relative invariants, we solve the following system of partial differential equations [1]:

$$\Delta\left(\phi\left(a_i\right)\right) = C_{ij}^k a^i \frac{\partial}{\partial a_j}$$

where C_{ij}^k are the structure constants of the admitted Lie algebra as presented in Table 1. Consequently, in order to derive all the possible one-dimensional Lie symmetries, we should study various cases were none of the invariants are zero, one of the invariants is zero, two of the invariants are zero, or all the invariants are zero.

Hence, for the first three cases, infer the following one-dimensional independent Lie algebras:

$$X_6\ ,\ X_7\ ,\ X_8\ ,\ \xi_{(67)} = X_6 + \alpha X_7\ ,\ \xi_{(68)} = X_6 + \alpha X_8$$

$$\xi_{(78)} = X_7 + \alpha X_8\ ,\ \xi_{(678)} = X_6 + \alpha X_7 + \beta X_8.$$

We apply the same procedure for the rest of the possible linear combinations of the symmetry vectors, and we find the one-dimensional-dependent Lie algebras:

$$X_1,\ X_2\ ,\ X_3\ ,\ X_4\ ,\ X_5\ ,\ \xi_{(12)} = X_1 + \alpha X_2\ ,\ \xi_{(13)} = X_1 + \alpha X_3\ ,\ \xi_{(23)} = X_2 + \alpha X_3\ ,\ \xi_{(14)} = X_1 + \alpha X_4\ ,$$

$$\xi_{(15)} = X_1 + \alpha X_5\ ,\ \xi_{(16)} = X_1 + \alpha X_6,\ \xi_{(34)} = X_3 + \alpha X_4\ ,\ \xi_{(25)} = X_2 + \alpha X_5\ \xi_{(45)} = X_4 + \alpha X_5\ ,$$

$$\xi_{(123)} = X_1 + \alpha X_2 + \beta X_3\ \xi_{(145)} = X_1 + \alpha X_4 + \beta X_5\ ,\ \xi_{(125)} = X_1 + \alpha X_2 + \beta X_5\ ,\ \xi_{(134)} = X_1 + \alpha X_3 + \beta X_4,$$

in which α and β are constants.

Therefore, by applying one of the above Lie symmetry vectors, we find all the possible reductions from a system of $1+2$ PDEs to a system of $1+1$ PDEs. The reduced system will not admit all the remaining Lie symmetries. The Lie symmetries that survive under a reduction process are given as described in the following example.

Let a PDE admit the Lie point symmetries Γ_1, Γ_2, which are such that $[\Gamma_1, \Gamma_2] = C_{12}^1 X_1$, with $C_{12}^1 \neq 0$. Reduction with the symmetry vector Γ_1 leads to a reduced differential equation, which admits Γ_2 as the Lie symmetry. On the other hand, reduction of the mother equation with respect to the Lie symmetry Γ_2 leads to a different reduced differential equation, which does not admit as a Lie point symmetry the vector field Γ_1. In case the two Lie symmetries form an Abelian Lie algebra, i.e., $C_{12}^1 = 0$, then under any reduction process, symmetries are preserved by any reduction.

We found that the optimal system admits twenty-three one-dimensional Lie symmetries and possible independent reductions. All the possible twenty-three Lie invariants are presented in Tables A2 and A3.

An application of the Lie invariants is presented below.

Application of ξ_{145}

Let us now demonstrate the results of Tables A2 and A3 by the Lie invariants of the symmetry vector ξ_{145} and construct the similarity solution for the system.

The application of ξ_{145} in the nonrotating system (1)–(3) reduces the PDEs in the following system:

$$(hu)_z + (hv)_w = 0 \tag{13}$$

$$\alpha + uu_z + vu_w + h^{\gamma-2}h_z = 0 \tag{14}$$

$$\beta + uv_z + vv_w + h^{\gamma-2}h_w = 0 \tag{15}$$

where $z = x - \frac{\alpha}{2}t^2$ and $w = y - \frac{\beta}{2}t^2$.

System (13)–(15) admits the Lie point symmetries:

$$\partial_z\,,\ \partial_w\,,\ z\partial_z + w\partial_w + \frac{2}{\gamma-1}h\partial_h + u\partial_u + v\partial_v \tag{16}$$

Reduction with the symmetry vector $\partial_z + c\partial_w$ provides the following system of first-order ODEs:

$$Fh_\sigma = (c\alpha - \beta)\,h^2, \tag{17}$$

$$Fv_\sigma = \frac{(\alpha - c\beta)\,ch^\gamma - \alpha h\,(v-cu)^2}{v-cu}, \tag{18}$$

$$Fh_\sigma = \frac{(\alpha - c\beta)\,cu^\gamma - \beta h\,(v-cu)^2}{v-cu}. \tag{19}$$

where $F = (1+c^2)\,h^\gamma - h\,(v-cu)^2$ and $\sigma = z + cw$.

By performing the change of variable $d\sigma = fd\tau$, function f can be removed from the above system. For $h(\tau) = 0$, the system (17)–(19) admits a solution $u = u_0$, $v = v_0$, which is a critical point. The latter special solutions are always unstable when $\alpha c > \beta$.

We proceed with our analysis by considering the rotating system.

4. Symmetries and Optimal System for Rotating Shallow Water

For the rotating system ($f \neq 0$), the Lie symmetries are:

$$\begin{aligned}
Y_1 &= \partial_t\,,\ Y_2 = \partial_x\,,\ Y_3 = \partial_y\,,\\
Y_4 &= y\partial_x - x\partial_y + v\partial_u - u\partial_v\,,\\
Y_5 &= \sin(ft)\,\partial_x + \cos(ft)\,\partial_y + f\,(\cos(ft)\,\partial_u - \sin(ft)\,\partial_v)\\
Y_6 &= \cos(ft)\,\partial_x - \sin(ft)\,\partial_y - f\,(\sin(ft)\,\partial_u + \cos(ft)\,\partial_v)\\
Y_7 &= (\gamma-1)\,(x\partial_x + y\partial_y + u\partial_u + v\partial_v) + 2h\partial_h
\end{aligned}$$

The commutators and the adjoint representation are given in Table 2 and Table A4. The Lie symmetries for the rotating system form a smaller dimension Lie algebra than the non-rotating system. That is not the case when $\gamma = 2$, where the two Lie algebras have the same dimension and are equivalent under point transformation [22]. Therefore, for $\gamma > 2$, the Coriolis force cannot be eliminated by a point transformation as in the $\gamma = 2$ case.

Table 2. Commutators of the admitted Lie point symmetries for the rotating 2D shallow water.

$[\,,\,]$	$\mathbf{Y}_1$	$\mathbf{Y}_2$	$\mathbf{Y}_3$	$\mathbf{Y}_4$	$\mathbf{Y}_5$	$\mathbf{Y}_6$	$\mathbf{Y}_7$
$\mathbf{Y}_1$	0	0	0	0	fY_6	$-fY_5$	0
$\mathbf{Y}_2$	0	0	0	$-Y_3$	0	0	$(\gamma-1)\,Y_2$
$\mathbf{Y}_3$	0	0	0	Y_2	0	0	$(\gamma-1)\,Y_3$
$\mathbf{Y}_4$	0	Y_3	$-Y_2$	0	$-Y_6$	Y_5	0
$\mathbf{Y}_5$	$-fY_6$	0	0	Y_6	0	0	$(\gamma-1)\,Y_5$
$\mathbf{Y}_6$	fY_5	0	0	$-Y_5$	0	0	$(\gamma-1)\,Y_6$
$\mathbf{Y}_7$	0	$-(\gamma-1)\,Y_2$	$-(\gamma-1)\,Y_3$	0	$-(\gamma-1)\,Y_5$	$-(\gamma-1)\,Y_6$	0

As for the admitted Lie symmetries admitted by the given system of PDEs with or without the Coriolis terms for $\gamma > 2$, we remark that the rotating and the nonrotating systems have a common Lie subalgebra of one-parameter point transformations consisting of the symmetry vectors Y_1, Y_2, Y_3, Y_4, and Y_7 or for the nonrotating system X_1, X_2, X_3, X_6, and X_8.

We proceed with the determination of the one-dimensional optimal system and the invariant functions. Specifically, the relative invariants for the adjoint representation are calculated to be a_1 , a_7 and a_8. From Table 2 and Table A4, we can find the one-dimensional optimal system, which is:

$$Y_1,\, Y_2,\, Y_3,\, Y_4,\, Y_5,\, Y_6,\, Y_7,\, \chi_{12} = Y_1 + \alpha Y_2,\, \chi_{13} = Y_1 + \alpha Y_3,$$

$$\chi_{14} = Y_1 + \alpha Y_4\,,\, \chi_{15} = Y_1 + \alpha Y_5,\;\; \chi_{16} = Y_1 + \alpha Y_6,\, \chi_{17} = Y_1 + \alpha Y_7,$$

$$\chi_{23} = Y_2 + \alpha Y_3\,,\;\; \chi_{45} = Y_4 + \alpha Y_5,\, \chi_{46} = Y_4 + \alpha Y_6,\, \chi_{56} = Y_5 + \alpha Y_6$$

$$\chi_{47} = Y_4 + \alpha Y_6\,,\, \chi_{123} = Y_1 + \alpha Y_2 + \beta Y_3,\, \chi_{147} = Y_1 + \alpha Y_4 + \beta Y_7.$$

The Lie invariants, which correspond to all the above one-dimensional Lie algebras, are presented in Tables A5 and A6.

Let us demonstrate the application of the Lie invariants by the following, from which we can see that the Lie invariants reduce the nonlinear field equations into a system of integrable first-order ODEs, which can be solved with quadratures.

4.1. Application of χ_{12}

We consider the travel-wave similarity solution in the x-plane provided by the symmetry vector χ_{12} and the vector field Y_3. The resulting equations are described by the following system of first order ODEs:

$$v_z = f\frac{u}{\alpha - u} \tag{20}$$

$$\bar{F}u_z = f\,(\alpha - u)\,vh \tag{21}$$

$$\bar{F}h_z = fvh^2 \tag{22}$$

where $\bar{F} = h^\gamma - (a-u)^2\,h$ and $z = t - \alpha x$. Because we performed reduction with a subalgebra admitted by the nonrotating system, by setting $f = 0$ in (20)–(22), we get the similarity solution for the nonrotating system, where in this case, it is found to be $h\,(z) = h_0,\; u\,(z) = u_0$ and $v\,(z) = v_0$.

We perform the substitution $dz = \frac{\bar{F}}{fv}d\tau$, and the latter system is simplified as follows:

$$\frac{v}{\bar{F}}v_\tau = \frac{u}{\alpha - u} \tag{23}$$

$$u_\tau = (\alpha - u)\,h \tag{24}$$

$$h_\tau = h^2 \tag{25}$$

from which we get the solution:

$$h\,(\tau) = (h_0 - \tau)^{-1}\;,\; u\,(\tau) = \alpha + u_0 - \frac{u_0}{h_0}\tau \tag{26}$$

and:

$$v\,(t)^2 = 2\int \frac{\left(a + u_0 - \frac{u_0}{h_0}\tau\right)}{\frac{u_0}{h_0}\,(h_0 - \tau)}\left((h_0 - \tau)^{-\gamma} + \left(\frac{u_0}{h_0}\right)^2\tau - \frac{(u_0)^2}{h_0}\right)d\tau. \tag{27}$$

4.2. Application of χ_{23}

Consider now the reduction with the symmetry vector fields χ_{23}. The resulting system of 1 + 1 differential equations admits five Lie point symmetries, and they are:

$$\partial_t,\, \partial_w\,,\, (\sin(ft) + \alpha\cos(ft))\,\partial_w + f\,(\sin(ft)\,\partial_u + \cos(ft)\,\partial_v)$$
$$(\alpha\sin(ft) - \cos(ft))\,\partial_w - f\,(\cos(ft)\,\partial_u - \sin(ft)\,\partial_v)\;,\; (\gamma - 1)\,(\partial_w + u\partial_u + v\partial_v) + 2h\partial_h.$$

where $w = y - \alpha x$. For simplicity of our calculations, let us assume $\gamma = 3$.

Reduction with the scaling symmetry provides the following system of first order ODEs:

$$H_t = 2H(\alpha U - V), \tag{28}$$

$$U_t = \alpha H^2 + u(\alpha U - V) + fV, \tag{29}$$

$$V_t = -H^2 - v(\alpha U - V) - fU, \tag{30}$$

where $h = wH$, $u = wU$, and $v = wU$. The latter system is integrable and can be solved with quadratures.

Reducing with respect to the symmetry vector $(\alpha \sin(ft) - \cos(ft))\,\partial_w - f(\cos(ft)\,\partial_u - \sin(ft)\,\partial_v)$, we find the reduced system:

$$\frac{H_t}{H} = -\frac{\alpha\cos(ft) + \sin(ft)}{\cos(ft) - \alpha\sin(ft)}, \tag{31}$$

$$U_t = -\alpha f\frac{\sin(ft)\,V - \cos(ft)\,U}{\cos(ft) - \alpha\sin(ft)}, \tag{32}$$

$$V_t = -f\frac{\sin(ft)\,V - \cos(ft)\,U}{\cos(ft) - \alpha\sin(ft)}, \tag{33}$$

where now:

$$h = H(t), \tag{34}$$

$$u = \frac{\cos(ft)}{\cos(ft) - \alpha\sin(ft)} fw + U(t), \tag{35}$$

$$v = -\frac{\sin(ft)}{\cos(ft) - \alpha\sin(ft)} fw + V(t). \tag{36}$$

System (31)–(33) is integrable, and the solution is expressed in terms of quadratures.

5. Conclusions

In this work, we determined the one-dimensional optimal system for the two-dimensional ideal gas equations. The nonrotating system was found to be invariant under an eight-dimensional group of one-parameter point transformations. and there were twenty-three independent one-dimensional Lie algebras. One the other hand, when the Coriolis force was introduced, the dynamical admitted seven Lie point symmetries and twenty one-dimensional Lie algebras.

For all the independent Lie algebras, we determined all the invariant functions, which corresponded to all the independent similarity solutions.

In a future work, we plan to classify all the independent one-dimensional Lie algebras, which lead to analytic forms for the similarity solutions.

Acknowledgments: The author wants to thank Kevin Duffy for gracious hospitality and P.G.L.Leach for a fruitful discussion on the subject.

Appendix A

In this Appendix, we present the Tables A1–A6, which are referenced in the main article.

Table A1. Adjoint representation of the admitted Lie point symmetries for the nonrotating 2D shallow water.

$Ad\left(e^{(\varepsilon X_i)}\right) X_j$	X_1	X_2	X_3	X_4	X_5	X_6	X_7	X_8
X_1	X_1	X_2	X_3	$X_4 - \varepsilon X_2$	$X_5 - \varepsilon X_3$	X_6	$X_7 + \varepsilon(\gamma - 1) X_1$	X_8
X_2	X_1	X_2	X_3	X_4	X_5	$X_6 + \varepsilon X_3$	X_7	$X_8 - \varepsilon(\gamma - 1) X_2$
X_3	X_1	X_2	X_3	X_4	X_5	$X_6 - \varepsilon X_2$	X_7	$X_8 - \varepsilon(\gamma - 1) X_3$
X_4	$X_1 + \varepsilon X_2$	X_2	X_3	X_4	X_5	$X_6 + \varepsilon X_5$	$X_7 - \varepsilon(\gamma - 1) X_4$	$X_8 - \varepsilon(\gamma - 1) X_4$
X_5	$X_1 + \varepsilon X_3$	X_2	X_3	X_4	X_5	$X_6 - \varepsilon X_4$	$X_7 - \varepsilon(\gamma - 1) X_5$	$X_8 - \varepsilon(\gamma - 1) X_5$
X_6	X_1	$X_2 \cos\varepsilon - X_3 \sin\varepsilon$	$X_2 \sin\varepsilon + X_3 \cos\varepsilon$	$X_4 \cos\varepsilon - X_5 \sin\varepsilon$	$X_4 \sin\varepsilon + X_5 \cos\varepsilon$	X_6	X_7	X_8
X_7	$e^{-(\gamma-1)\varepsilon} X_1$	X_2	X_3	$e^{-(\gamma-1)\varepsilon} X_4$	$e^{-(\gamma-1)\varepsilon} X_5$	X_6	X_7	X_8
X_8	X_1	$e^{(\gamma-1)\varepsilon} X_3$	$e^{(\gamma-1)\varepsilon} X_4$	$e^{-(\gamma-1)\varepsilon} X_4$	$e^{-(\gamma-1)\varepsilon} X_5$	X_6	X_7	X_8

Table A2. Lie invariants for the optimal system of the nonrotating system.

Symmetry	Invariants
X_1	$x,\ y,\ h(x,y),\ u(x,y),\ v(x,y)$
X_2	$t,\ y,\ h(t,y),\ u(t,y),\ v(t,y)$
X_3	$t,\ x,\ h(t,x),\ u(t,x),\ v(t,x)$
X_4	$t,\ y,\ h(t,y),\ \frac{x}{t} + U(t,y),\ v(t,y)$
X_5	$t,\ x,\ h(t,x),\ u(t,x),\ \frac{y}{t} + V(t,x)$
X_6	$t,\ x^2 + y^2,\ h(t, x^2 + y^2),\ \frac{xU(t,x^2+y^2) + yV(t,x^2+y^2)}{\sqrt{x^2+y^2}},\ \frac{yU(t,x^2+y^2) - xV(t,x^2+y^2)}{\sqrt{x^2+y^2}}$
X_7	$\frac{x}{t},\ \frac{y}{t},\ h\left(\frac{x}{t}, \frac{y}{t}\right),\ u\left(\frac{x}{t}, \frac{y}{t}\right),\ v\left(\frac{x}{t}, \frac{y}{t}\right)$
X_8	$H(x,y)\, t^{\frac{2}{1-\gamma}},\ U(x,y)\, t^{-1},\ V(x,y)\, t^{-1}$
$\xi_{(12)}$	$x - \alpha t,\ y,\ h(x - \alpha t, y),\ u(x - \alpha t, y),\ v(x - \alpha t, y)$
$\xi_{(13)}$	$x,\ y - \alpha t,\ h(x, y - \alpha t),\ u(x, y - \alpha t),\ v(x, y - \alpha t)$
$\xi_{(14)}$	$x - \frac{\alpha}{2}t^2,\ y,\ h\left(x - \frac{\alpha}{2}t^2, y\right),\ u\left(x - \frac{\alpha}{2}t^2, y\right),\ v\left(x - \frac{\alpha}{2}t^2, y\right)$
$\xi_{(15)}$	$x,\ y - \frac{\alpha}{2}t^2,\ h\left(x, y - \frac{\alpha}{2}t^2\right),\ u\left(x, y - \frac{\alpha}{2}t^2\right),\ v\left(x, y - \frac{\alpha}{2}t^2\right)$

Table A3. Lie invariants for the optimal system of the nonrotating system.

Symmetry	Invariants
$\xi_{(16)}$	$t,\ e^{-\alpha t}\left(x^2+y^2\right),\ u\left(t, e^{-\alpha t}x^2+y^2\right)\cos(\alpha t)+v\left(t, e^{-\alpha t}x^2+y^2\right)\sin(\alpha t)$ $\frac{y}{x},\ h\left(e^{-\alpha t}x^2+y^2, \frac{y}{x}\right),\ u\left(t, e^{-\alpha t}x^2+y^2\right)\sin(\alpha t)-v\left(t, e^{-\alpha t}x^2+y^2\right)\cos(\alpha t)$
$\xi_{(23)}$	$t,\ x-\alpha y,\ h(t, x-\alpha y),\ u(t, x-\alpha y), v(t, x-\alpha y)$
$\xi_{(34)}$	$t,\ y-\frac{x}{\alpha t},\ h\left(t, y-\frac{x}{\alpha t}\right),\ u\left(t, y-\frac{x}{\alpha t}\right),\ v\left(t, y-\frac{x}{\alpha t}\right)$
$\xi_{(25)}$	$t,\ y-\alpha t x,\ h(t, y-\alpha t x),\ u(t, y-\alpha t x),\ v(t, y-\alpha t x)$
$\xi_{(45)}$	$t,\ y-\alpha x,\ h(t, y-\alpha x),\ \alpha\frac{x}{t}+U(t, y-\alpha x),\ \alpha\frac{x}{t}+V(t, y-\alpha x)$
$\xi_{(123)}$	$t-\alpha x,\ t-\beta y,\ h(t-\alpha x, t-\beta y),\ u(t-\alpha x, t-\beta y), v(t-\alpha x, t-\beta y)$
$\xi_{(145)}$	$x-\frac{\alpha}{2}t^2,\ y-\frac{\beta}{2}t^2,\ h\left(x-\frac{\alpha}{2}t^2, y-\frac{\beta}{2}t^2\right),\ \alpha t+U\left(x-\frac{\alpha}{2}t^2,\ y-\frac{\beta}{2}t^2\right),\ \beta t+V\left(x-\frac{\alpha}{2}t^2,\ y-\frac{\beta}{2}t^2\right)$
$\xi_{(125)}$	$x-\alpha t,\ y-\frac{\beta}{2}t^2,\ h\left(x-\alpha t, y-\frac{\beta}{2}t^2\right),\ u\left(x-\alpha t, y-\frac{\beta}{2}t^2\right),\ \beta t+V\left(x-\alpha t, y-\frac{\beta}{2}t^2\right)$
$\xi_{(134)}$	$x-\frac{\beta}{2}t^2,\ y-\alpha t,\ h\left(x-\frac{\beta}{2}t^2, y-\alpha t\right),\ \beta t+U\left(x-\frac{\beta}{2}t^2, y-\alpha t\right),\ V\left(x-\frac{\beta}{2}t^2, y-\alpha t\right)$
$\xi_{(67)}$	$\frac{\ln t}{\alpha}, w=\frac{t^{-\frac{\alpha+\sqrt{\alpha(\alpha-4)-4}}{2\alpha}}}{2\sqrt{\alpha(\alpha-4)-4}}\left(x-\left(\alpha+\sqrt{\alpha(\alpha-4)-4}\right)y\right),\ z=\frac{t^{-\frac{\alpha+\sqrt{\alpha(\alpha-4)-4}}{2\alpha}}}{2\sqrt{\alpha(\alpha-4)-4}}\left(x+\left(\alpha+\sqrt{\alpha(\alpha-4)-4}\right)y\right)$ $h(w,z),\ U(w,z)\sin\left(\frac{\ln t}{\alpha}\right)+V(w,z)\sin\left(\frac{\ln t}{\alpha}\right),\ U(w,z)\cos\left(\frac{\ln t}{\alpha}\right)-V(w,z)\sin\left(\frac{\ln t}{\alpha}\right)$
$\xi_{(68)}$	$t,\ x^2+y^2,\ x^{-\frac{2}{\gamma-1}}h\left(t, x^2+y^2\right),\ \frac{U\left(t,x^2+y^2\right)\cos\left(\frac{\ln x}{\alpha}\right)+V\left(t,x^2+y^2\right)\sin\left(\frac{\ln x}{\alpha}\right)}{x},\ \frac{U\left(t,x^2+y^2\right)\sin\left(\frac{\ln x}{\alpha}\right)-V\left(t,x^2+y^2\right)\cos\left(\frac{\ln x}{\alpha}\right)}{x}$
$\xi_{(78)}$	$w=xt^{-\frac{(\gamma-1)}{\alpha(\gamma-1)-2}},\ z=yt^{-\frac{(\gamma-1)}{\alpha(\gamma-1)-2}},\ t^{-\frac{2\alpha}{\alpha(\gamma-1)-2}}h(w,z), t^{-\frac{(\gamma-1)\alpha}{\alpha(\gamma-1)-2}}u(w,z),\ t^{-\frac{(\gamma-1)\alpha}{\alpha(\gamma-1)-2}}v(w,z)$
$\xi_{(678)}$	$t,\ t^{-1-\beta}\left(x^2+y^2\right),\ t^{-\beta}\left(U\left(t,x^2+y^2\right)\sin(\alpha t)+V\left(t,x^2+y^2\right)\cos(\alpha t)\right)$ $t^{-\frac{2\beta}{\gamma-1}}H\left(t,x^2+y^2\right),\ t^{-\beta}\left(U\left(t,x^2+y^2\right)\cos(\alpha t)-V\left(t,x^2+y^2\right)\sin(\alpha t)\right)$

Table A4. Adjoint representation of the admitted Lie point symmetries for the rotating 2D shallow water.

$Ad\left(e^{(\varepsilon Y_i)}\right)Y_j$	$\mathbf{Y}_1$	$\mathbf{Y}_2$	$\mathbf{Y}_3$	$\mathbf{Y}_4$	$\mathbf{Y}_5$	$\mathbf{Y}_6$	$\mathbf{Y}_7$
$\mathbf{Y}_1$	Y_1	Y_2	Y_3	Y_4	$Y_5\cos(f\varepsilon)-Y_6\sin(f\varepsilon)$	$Y_5\sin(f\varepsilon)+Y_6\cos(f\varepsilon)$	Y_7
$\mathbf{Y}_2$	Y_1	Y_2	Y_3	$Y_4+\varepsilon Y_3$	Y_5	Y_6	$Y_7-\varepsilon(\gamma-1)Y_2$
$\mathbf{Y}_3$	Y_1	Y_2	Y_3	$Y_4-\varepsilon Y_2$	Y_5	Y_6	$Y_7-\varepsilon(\gamma-1)Y_3$
$\mathbf{Y}_4$	Y_1	$Y_2\cos\varepsilon-Y_3\sin\varepsilon$	$Y_2\sin\varepsilon+Y_3\cos\varepsilon$	Y_4	$Y_5\cos\varepsilon+Y_6\sin\varepsilon$	$Y_6\cos\varepsilon-Y_5\sin\varepsilon$	Y_7
$\mathbf{Y}_5$	$Y_1+f\varepsilon Y_6$	Y_2	Y_3	$Y_4-\varepsilon Y_6$	Y_5	Y_6	$Y_7-\varepsilon(\gamma-1)Y_5$
$\mathbf{Y}_6$	$Y_1-f\varepsilon Y_5$	Y_2	Y_3	$Y_4+\varepsilon Y_5$	Y_5	Y_6	$Y_7-\varepsilon(\gamma-1)Y_6$
$\mathbf{Y}_7$	Y_1	$e^{(\gamma-1)\varepsilon}Y_2$	$e^{(\gamma-1)\varepsilon}Y_3$	Y_4	$e^{(\gamma-1)\varepsilon}Y_5$	$e^{(\gamma-1)\varepsilon}Y_6$	Y_7

Table A5. Lie invariants for the optimal system of the rotating system.

Symmetry	Invariants
$\mathbf{Y}_1$	$x,\ y,\ h\left(x,y\right),\ u\left(x,y\right),\ v\left(x,y\right)$
$\mathbf{Y}_2$	$t,\ y,\ h\left(t,y\right),\ u\left(t,y\right),\ v\left(t,y\right)$
$\mathbf{Y}_3$	$t,\ x,\ h\left(t,x\right),\ u\left(t,x\right),\ v\left(t,x\right)$
$\mathbf{Y}_4$	$t,\ x^2+y^2,\ h\left(t,x^2+y^2\right),\ \frac{xU\left(t,x^2+y^2\right)+yV\left(t,x^2+y^2\right)}{\sqrt{x^2+y^2}},\ \frac{yU\left(t,x^2+y^2\right)-xV\left(t,x^2+y^2\right)}{\sqrt{x^2+y^2}}$
$\mathbf{Y}_5$	$t,\ x\cot\left(ft\right)-y,\ h\left(t,x\cot\left(ft\right)-y\right),\ fx\cot\left(ft\right)+U\left(t,x\cot\left(ft\right)-y\right),\ -fx+V\left(t,x\cot\left(ft\right)-y\right)$
$\mathbf{Y}_6$	$t,\ x\tan\left(ft\right)+y,\ h\left(t,x\tan\left(ft\right)+y\right),\ -fx\tan\left(ft\right)+U\left(t,x\tan\left(ft\right)+y\right),\ -fx+V\left(t,x\tan\left(ft\right)+y\right)$
$\mathbf{Y}_7$	$\frac{x}{t},\ \frac{y}{t},\ h\left(\frac{x}{t},\frac{y}{t}\right),\ u\left(\frac{x}{t},\frac{y}{t}\right),\ v\left(\frac{x}{t},\frac{y}{t}\right)$
$\mathcal{X}_{(12)}$	$x-\alpha t,\ y,\ h\left(x-\alpha t,y\right),\ u\left(x-\alpha t,y\right),\ v\left(x-\alpha t,y\right)$
$\mathcal{X}_{(13)}$	$x,\ y-\alpha t,\ h\left(x,y-\alpha t\right),\ u\left(x,y-\alpha t\right),\ v\left(x,y-\alpha t\right)$
$\mathcal{X}_{(14)}$	$t,\ e^{-\alpha t}\left(x^2+y^2\right),\ u\left(t,e^{-\alpha t}x^2+y^2\right)\cos\left(\alpha t\right)+v\left(t,e^{-\alpha t}x^2+y^2\right)\sin\left(\alpha t\right)$ $\frac{y}{x},\ h\left(e^{-\alpha t}x^2+y^2,\frac{y}{x}\right),\ u\left(t,e^{-\alpha t}x^2+y^2\right)\sin\left(\alpha t\right)-v\left(t,e^{-\alpha t}x^2+y^2\right)\cos\left(\alpha t\right)$

Table A6. Lie invariants for the optimal system of the rotating system.

Symmetry	Invariants
$X_{(15)}$	$x+\frac{\alpha}{f}\cos(ft),\ y-\frac{\alpha}{f}\sin(ft),\ h\left(x+\frac{\alpha}{f}\cos(ft), y-\frac{\alpha}{f}\sin(ft)\right),$ $\alpha\sin(ft)+U\left(x+\frac{\alpha}{f}\cos(ft), y-\frac{\alpha}{f}\sin(ft)\right),\ \alpha\cos(ft)+V\left(x+\frac{\alpha}{f}\cos(ft), y-\frac{\alpha}{f}\sin(ft)\right)$
$X_{(16)}$	$x-\frac{\alpha}{f}\sin(ft),\ y-\frac{\alpha}{f}\cos(ft),\ h\left(x-\frac{\alpha}{f}\sin(ft), y-\frac{\alpha}{f}\cos(ft)\right),$ $\alpha\cos(ft)+U\left(x-\frac{\alpha}{f}\sin(ft), y-\frac{\alpha}{f}\cos(ft)\right),\ -\alpha\sin(ft)+V\left(x-\frac{\alpha}{f}\sin(ft), y-\frac{\alpha}{f}\cos(ft)\right)$
$X_{(17)}$	$xe^{-\alpha t},\ ye^{-\alpha t},\ e^{\frac{2\alpha}{\gamma-1}t}h\left(xe^{-\alpha t}, ye^{-\alpha t}\right),\ e^{\alpha t}u\left(xe^{-\alpha t}, ye^{-\alpha t}\right),\ e^{\alpha t}v\left(xe^{-\alpha t}, ye^{-\alpha t}\right)$
$X_{(23)}$	$t,\ x-\alpha y,\ h(t, x-\alpha y),\ u(t, x-\alpha y),\ v(t, x-\alpha y)$
$X_{(45)}$	$t,\ w=\left(x^2+y^2-2x\cos(ft)+2y\sin(ft)\right),\ \frac{U(t,w)+f\sin(ft)}{V(t,w)+f\cos(ft)},\ \frac{U(t,w)^2+V(t,w)^2}{2}+f\left(U(t,w)\sin(ft)+V(t,w)\cos(ft)\right)$
$X_{(46)}$	$t,\ w=\left(x^2+y^2-2x\sin(ft)-2y\cos(ft)\right),\ \frac{U(t,w)-f\cos(ft)}{V(t,w)+f\sin(ft)},\ \frac{U(t,w)^2+V(t,w)^2}{2}+f\left(V(t,w)\sin(ft)-U(t,w)\cos(ft)\right)$
$X_{(56)}$	$t,\ z=y-\frac{x(\cos(ft)-\alpha\sin(ft))}{\sin(ft)+\alpha\cos(ft)},\ h(t,z),\ f\frac{x(\cos(ft)-\alpha\sin(ft))}{\sin(ft)+\alpha\cos(ft)}+U(t,z),\ -x+V(t,z)$
$X_{(47)}$	$t,\ x^2+y^2,\ x^{-\frac{2}{\gamma-1}}h\left(t, x^2+y^2\right),\ \frac{U\left(t,x^2+y^2\right)\cos\left(\frac{\ln x}{\alpha}\right)+V\left(t,x^2+y^2\right)\sin\left(\frac{\ln x}{\alpha}\right)}{x},\ \frac{U\left(t,x^2+y^2\right)\sin\left(\frac{\ln x}{\alpha}\right)-V\left(t,x^2+y^2\right)\cos\left(\frac{\ln x}{\alpha}\right)}{x}$
$X_{(123)}$	$t-\alpha x,\ t-\beta y,\ h(t-\alpha x, t-\beta y),\ u(t-\alpha x, t-\beta y), v(t-\alpha x, t-\beta y)$
$X_{(147)}$	$z=e^{-t(\gamma-1)}(x\cos t-y\sin t),\ w=e^{-t(\gamma-1)}(y\cos t+x\sin t),$ $e^{-t(\gamma-1)}h(z,w),\ e^{-t(\gamma-1)}\left(U(z,w)\cos t-V(z,w)\sin t\right),\ e^{-t(\gamma-1)}\left(U(z,w)\sin t+V(z,w)\cos t\right)$

References

1. Olver, P.J. *Applications of Lie Groups to Differential Equations*; Springer: New York, NY, USA, 1993.
2. Bluman, G.W.; Kumei, S. *Symmetries and Differential Equations*; Springer: New York, NY, USA, 1989.
3. Ibragimov, N.H. *CRC Handbook of Lie Group Analysis of Differential Equations, Volume I: Symmetries, Exact Solutions, and Conservation Laws*; CRS Press LLC: Boca Raton, FL, USA, 2000.
4. Webb, G.M. Lie symmetries of a coupled nonlinear Burgers-heat equation system. *J. Phys A Math. Gen.* **1990**, *23*, 3885. [CrossRef]
5. Tsamparlis, M.; Paliathanasis, A.; Karpathopoulos, L. Dynamics of ionization wave splitting and merging of atmospheric-pressure plasmas in branched dielectric tubes and channels. *J. Phys. A Math. Theor.* **2012**, *45*, 275201. [CrossRef]
6. Azad, H.; Mustafa, M.T. Group classification, optimal system and optimal reductions of a class of Klein Gordon equations. *Commun. Nonlinear Sci. Numer. Simul.* **2010**, *15*, 1132. [CrossRef]
7. Tsamparlis, M.; Paliathanasis, A. Symmetry analysis of the Klein–Gordon equation in Bianchi I spacetimes. *Int. J. Geom. Methods Mod. Phys.* **2005**, *12*, 155003.
8. Meleshko, S.V.; Shapeev, V.P. Nonisentropic solutions of simple wave type of the gas dynamics equations. *J. Nonlinear Math. Phys.* **2011**, *18*, 195. [CrossRef]
9. Halder, A.; Paliathanasis, A.; Leach, P.G.L. Noether's Theorem and Symmetry. *Symmetry* **2018**, *10*, 744. [CrossRef]
10. Jamal, S.; Leach, P.G.L.; Paliathanasis, A. Nonlocal Representation of the sl (2,R) Algebra for the Chazy equation. *Quaest. Math.* **2019**, *42*, 125. [CrossRef]
11. Ovsiannikov, L.V. *Group Analysis of Differential Equations*; Academic Press: New York, NY, USA, 1982.
12. Chou, K.S.; Qu, C.Z. Optimal systems and group classification of (1 + 2)-dimensional heat equation. *Acta Appl. Math.* **2004**, *83*, 257. [CrossRef]
13. Galas, F.; Richter, E.W. Exact similarity solutions of ideal MHD equations for plane motions. *Phys. D* **1991**, *50*, 297. [CrossRef]
14. Coggeshalla, S.V.; Meyer-ter-Vehn, J. Group-invariant solutions and optimal systems for multidimensional hydrodynamics. *J. Math. Phys.* **1992**, *33*, 3585. [CrossRef]
15. Hu, X.; Li, Y.; Chen, Y. A direct algorithm of one-dimensional optimal system for the group invariant solutions. *J. Math. Phys.* **2015**, *56*, 053504. [CrossRef]
16. Vallis, G.K. *Atmospheric and Oceanic Fluid Dynamics: Fundamentals and Large-Scale Circulation*; Cambridge University Press: Cambridge, UK, 2006.
17. Courant, R.; Friedrichs, K.O. *Supersonic Flow and Shock Waves*; Interscience Publishers: Geneva, Switzerland, 1948.
18. Kevorkian, J. *Partial Differential Equations: Analytical Solutions Techniques*; Chapman and Hall: New York, NY, USA, 1990.
19. Caleffi, V.; Valiani, A.; Zanni, A. Finite volume method for simulating extreme flood events in natural channels. *J. Hydraul. Res.* **2003**, *41*, 167. [CrossRef]
20. Akkermans, R.A.D.; Kamp, L.P.J.; Clercx, H.J.H.; van Heijst, G.J.F. Three-dimensional flow in electromagnetically driven shallow two-layer fluids. *Phys. Rev. E* **2010**, *82*, 026314. [CrossRef] [PubMed]
21. Kim, D.H.; Cho, Y.S.; Yi, Y.K. Propagation and run-up of nearshore tsunamis with HLLC approximate Riemann solver. *Ocean Eng.* **2007**, *34*, 1164. [CrossRef]
22. Chesnokov, A.A. Symmetries and exact solutions of the rotating shallow-water equations. *Eur. J. Appl. Math.* **2009**, *20*, 461. [CrossRef]
23. Xin, X.; Zhang, L.; Xia, Y.; Liu, H. Nonlocal symmetries and exact solutions of the (2 + 1)-dimensional generalized variable coefficient shallow water wave equation. *Appl. Math. Lett.* **2019**, *94*, 112. [CrossRef]
24. Szatmari, S.; Bihlo, A. Symmetry analysis of a system of modified shallow-water equations. *Comm. Nonlinear Sci. Num. Simul.* **2014**, *19*, 530. [CrossRef]
25. Chesnokov, A.A. Symmetries and exact solutions of the shallow water equations for a two-dimensional shear flow. *J. Appl. Mech. Tech. Phys.* **2008**, *49*, 737. [CrossRef]
26. Liu, J.-G.; Zeng, Z.-F.; He, Y.; Ai, G.-P. A Class of Exact Solution of $(3+1)$-Dimensional Generalized Shallow Water Equation System. *Int. J. Nonlinear Sci. Num. Simul.* **2014**, *19*, 37. [CrossRef]

27. Pandey, M. Lie Symmetries and Exact Solutions of Shallow Water Equations with Variable Bottom. *Int. J. Nonlinear Sci. Num. Simul.* **2015**, *16*, 93. [CrossRef]
28. Paliathanasis, A. Lie symmetries and similarity solutions for rotating shallow water. *Preprint arXiv* **2019**, arXiv:1906.00689.

15

On a Class of Optimal Fourth Order Multiple Root Solvers without using Derivatives

Janak Raj Sharma [1,*], Sunil Kumar [1] and Lorentz Jäntschi [2,3,*]

1 Department of Mathematics, Sant Longowal Institute of Engineering and Technology, Longowal, Sangrur 148106, India; sfageria1988@gmail.com
2 Department of Physics and Chemistry, Technical University of Cluj-Napoca, 400114 Cluj-Napoca, Romania
3 Institute of Doctoral Studies, Babeş-Bolyai University, 400084 Cluj-Napoca, Romania
* Correspondence: jrshira@yahoo.co.in (J.R.S.); lorentz.jantschi@gmail.com (L.J.)

Abstract: Many optimal order multiple root techniques involving derivatives have been proposed in literature. On the contrary, optimal order multiple root techniques without derivatives are almost nonexistent. With this as a motivational factor, here we develop a family of optimal fourth-order derivative-free iterative schemes for computing multiple roots. The procedure is based on two steps of which the first is Traub–Steffensen iteration and second is Traub–Steffensen-like iteration. Theoretical results proved for particular cases of the family are symmetric to each other. This feature leads us to prove the general result that shows the fourth-order convergence. Efficacy is demonstrated on different test problems that verifies the efficient convergent nature of the new methods. Moreover, the comparison of performance has proven the presented derivative-free techniques as good competitors to the existing optimal fourth-order methods that use derivatives.

Keywords: iterative function; multiple root; composite method; derivative-free method; optimal convergence

MSC: 65H05; 41A25; 49M15

---

1. Introduction

We consider derivative-free methods for finding the multiple root (say, α) with multiplicity m of a nonlinear equation $f(t) = 0$, i.e., $f^{(j)}(\alpha) = 0$, $j = 0, 1, 2, \ldots, m-1$ and $f^{(m)}(\alpha) \neq 0$.

Several higher order methods, with or without the use of modified Newton's method [1]

$$t_{k+1} = t_k - m\frac{f(t_k)}{f'(t_k)}, \tag{1}$$

have been derived and analyzed in literature (see, for example, [2–15] and references cited therein). In such methods, one requires determining the derivatives of either first order or both first and second order. Contrary to this, higher-order derivative-free methods to compute multiple roots are yet to be investigated. These methods are important in the problems where derivative f' is complicated to process or is costly to evaluate. The basic derivative-free method is the Traub–Steffensen method [16], which uses the approximation

$$f'(t_k) \simeq \frac{f(t_k + \beta f(t_k)) - f(t_k)}{\beta f(t_k)}, \quad \beta \in \mathbb{R} - \{0\},$$

or

$$f'(t_k) \simeq f[s_k, t_k],$$

for the derivative f' in the classical Newton method in Equation (1). Here, $s_k = t_k + \beta f(t_k)$ and $f[s,t] = \frac{f(s)-f(t)}{s-t}$ is a divided difference of first order. In this way, the modified Newton method in Equation (1) transforms to the modified Traub–Steffensen derivative free method

$$t_{k+1} = t_k - m\frac{f(t_k)}{f[s_k,t_k]}. \tag{2}$$

The modified Traub–Steffensen method in Equation (2) is a noticeable improvement over the Newton method, because it preserves the convergence of order two without using any derivative.

In this work, we aim to design derivative-free multiple root methods of high efficient quality, i.e., the methods of higher convergence order that use the computations as small as we please. Proceeding in this way, we introduce a class of derivative-free fourth-order methods that require three new pieces of information of the function f per iteration, and hence possess optimal fourth-order convergence in the terminology of Kung–Traub conjecture [17]. This conjecture states that multi-point iterative functions without memory based on n function evaluations may attain the convergence order 2^{n-1}, which is maximum. The methods achieving this convergence order are usually called optimal methods. The new iterative scheme uses the modified Traub–Steffensen iteration in Equation (2) in the first step and Traub–Steffensen-like iteration in the second step. The methods are examined numerically on many practical problems of different kind. The comparison of performance with existing techniques requiring derivative evaluations verifies the efficient character of the new methods in terms of accuracy and executed CPU time.

The rest of the paper is summarized as follows. In Section 2, the scheme of fourth-order method is proposed and its convergence order is studied for particular cases. The main result for the general case is studied in Section 3. Numerical tests to demonstrate applicability and efficiency of the methods are presented in Section 4. In this section, a comparison of performance with already established methods is also shown. In Section 5, a conclusion of the main points is drawn.

2. Formulation of Method

To compute a multiple root with multiplicity $m \geq 1$, consider the following two-step iterative scheme:

$$\begin{aligned} z_k &= t_k - m\frac{f(t_k)}{f[s_k,t_k]}, \\ t_{k+1} &= z_k - H(x_k,y_k)\frac{f(t_k)}{f[s_k,t_k]}, \end{aligned} \tag{3}$$

where $x_k = \sqrt[m]{\frac{f(z_k)}{f(t_k)}}$, $y_k = \sqrt[m]{\frac{f(z_k)}{f(s_k)}}$ and $H : \mathbb{C}^2 \to \mathbb{C}$ is analytic in a neighborhood of $(0,0)$. Notice that this is a two-step scheme with first step as the Traub–Steffensen iteration in Equation (2) and the next step as the Traub–Steffensen-like iteration. The second step is weighted by the factor $H(x,y)$, thus we can call it weight factor or more appropriately weight function.

In the sequel, we study the convergence results of proposed iterative scheme in Equation (3). For clarity, the results are obtained separately for different cases based on the multiplicity m. Firstly, for the case $m = 1$, the following theorem is proved:

Theorem 1. *Assume that $f : \mathbb{C} \to \mathbb{C}$ is an analytic function in a domain containing a multiple zero (say, α) with multiplicity $m = 1$. Suppose that the initial point t_0 is close enough to α, then the convergence order of Equation (3) is at least 4, provided that $H_{00} = 0$, $H_{10} = 1$, $H_{01} = 0$, $H_{20} = 2$, $H_{11} = 11$ and $H_{02} = 0$, where $H_{ij} = \frac{\partial^{i+j}}{\partial x^i \partial y^j} H(x_k,y_k)|_{(x_k=0,y_k=0)}$, for $0 \leq i,j \leq 2$.*

Proof. Assume that the error at kth stage is $e_k = t_k - \alpha$. Using the Taylor's expansion of $f(t_k)$ about α and keeping into mind that $f(\alpha) = 0$ and $f'(\alpha) \neq 0$, we have

$$f(t_k) = f'(\alpha)e_k(1 + A_1e_k + A_2e_k^2 + A_3e_k^3 + A_4e_k^4 + \cdots), \tag{4}$$

where $A_n = \frac{1}{(1+n)!}\frac{f^{(1+n)}(\alpha)}{f'(\alpha)}$ for $n \in \mathbb{N}$.

Similarly we have the Taylor's expansion of $f(s_k)$ about α

$$f(s_k) = f'(\alpha)e_{s_k}(1 + A_1e_{s_k} + A_2e_{s_k}^2 + A_3e_{s_k}^3 + A_4e_{s_k}^4 + \cdots), \tag{5}$$

where $e_{s_k} = s_k - \alpha = e_k + \beta f'(\alpha)e_k(1 + A_1e_k + A_2e_k^2 + A_3e_k^3 + A_4e_k^4 + \cdots)$.

Then, the first step of Equation (3) yields

$$\begin{aligned} e_{z_k} &= z_k - \alpha \\ &= (1+\beta f'(\alpha))A_1e_k^2 - ((2+2\beta f'(\alpha)+(\beta f'(\alpha))^2)A_1^2 - (2+3\beta f'(\alpha)+(\beta f'(\alpha))^2)A_2)e_k^3 + ((4+5\beta f'(\alpha) \\ &\quad + 3(\beta f'(\alpha))^2 + (\beta f'(\alpha))^3)A_1^3 - (7+10\beta f'(\alpha)+7(\beta f'(\alpha))^2+2(\beta f'(\alpha))^3)A_1A_2 + (3+6\beta f'(\alpha) \\ &\quad + 4(\beta f'(\alpha))^2 + (\beta f'(\alpha))^3)A_3)e_k^4 + O(e_k^5). \end{aligned} \tag{6}$$

Expanding $f(z_k)$ about α, it follows that

$$f(z_k) = f'(\alpha)e_{z_k}(1 + A_1e_{z_k} + A_2e_{z_k}^2 + A_3e_{z_k}^3 + \cdots). \tag{7}$$

Using Equations (4), (5) and (7) in x_k and y_k, after some simple calculations, we have

$$\begin{aligned} x_k &= (1+\beta f'(\alpha))A_1e_k - ((3+3\beta f'(\alpha)+(\beta f'(\alpha))^2)A_1^2 - (2+3\beta f'(\alpha)+(\beta f'(\alpha))^2)A_2)e_k^2 + ((8+10\beta f'(\alpha) \\ &\quad + 5(\beta f'(\alpha))^2 + (\beta f'(\alpha))^3)A_1^3 - 2(5+7\beta f'(\alpha)+4(\beta f'(\alpha))^2+(\beta f'(\alpha))^3)A_1A_2 + (3+6\beta f'(\alpha) \\ &\quad + 4(\beta f'(\alpha))^2 + (\beta f'(\alpha))^3)A_3)e_k^3 + O(e_k^4) \end{aligned} \tag{8}$$

and

$$\begin{aligned} y_k &= A_1e_k - ((3+2\beta f'(\alpha))A_1^2 - (2+\beta f'(\alpha))A_2)e_k^2 + ((8+8\beta f'(\alpha)+3(\beta f'(\alpha))^2)A_1^3 \\ &\quad - (10+11\beta f'(\alpha)+4(\beta f'(\alpha))^2)A_1A_2 + (3+3\beta f'(\alpha)+(\beta f'(\alpha))^2)A_3)e_k^3 + O(e_k^4). \end{aligned} \tag{9}$$

Developing $H(x_k, y_k)$ by Taylor series in the neighborhood of origin $(0,0)$,

$$H(x_k, y_k) \approx H_{00} + x_kH_{10} + y_kH_{01} + \frac{1}{2}x_k^2H_{20} + x_ky_kH_{11} + \frac{1}{2}y_k^2H_{02}. \tag{10}$$

Inserting Equations (4)–(10) into the second step of Equation (3), and then some simple calculations yield

$$\begin{aligned} e_{k+1} = &- H_{00}\,e_k + (H_{00} - H_{01} + \beta f'(\alpha)H_{00} - (-1+H_{10})(1+\beta f'(\alpha)))A_1e_k^2 - \frac{1}{2}((4+H_{02}-8H_{10}+2H_{11} \\ &+ H_{20} + 4\beta f'(\alpha) - 10\beta f'(\alpha)H_{10} + 2\beta f'(\alpha)H_{11} + 2\beta f'(\alpha)H_{20} + 2(\beta f'(\alpha))^2 - 4(\beta f'(\alpha))^2H_{10} \\ &+ (\beta f'(\alpha))^2H_{20} - 2H_{01}(4+3\beta f'(\alpha)) + 2H_{00}(2+2\beta f'(\alpha)+(\beta f'(\alpha))^2))A_1^2 - 2(2+\beta f'(\alpha))(H_{00} \\ &- H_{01} + \beta f'(\alpha)H_{00} - (-1+H_{10})(1+\beta f'(\alpha)))A_2)e_k^3 + \delta e_k^4 + O(e_k^5), \end{aligned} \tag{11}$$

where $\delta = \delta(\beta, A_1, A_2, A_3, H_{00}, H_{10}, H_{01}, H_{20}, H_{11}, H_{02})$. Here, expression of δ is not being produced explicitly since it is very lengthy.

It is clear from Equation (11) that we would obtain at least fourth-order convergence if we set coefficients of e_k, e_k^2 and e_k^3 simultaneously equal to zero. Then, solving the resulting equations, one gets

$$H_{00} = 0,\ H_{10} = 1,\ H_{01} = 0,\ H_{20} = 2,\ H_{11} = 1,\ H_{02} = 0. \tag{12}$$

As a result, the error equation is given by

$$e_{k+1} = (1 + \beta f'(\alpha))A_1((5 + 5\beta f'(\alpha) + (\beta f'(\alpha))^2)A_1^2 - (1 + \beta f'(\alpha))A_2)e_k^4 + O(e_k^5). \tag{13}$$

Thus, the theorem is proved. □

Next, we show the conditions for $m = 2$ by the following theorem:

Theorem 2. *Using the hypotheses of Theorem 1, the order of convergence of the scheme in Equation* (3) *for the case* $m = 2$ *is at least 4, if* $H_{00} = 0$, $H_{10} = 1$, $H_{01} = 1$, *and* $H_{20} = 8 - H_{02} - 2H_{11}$, *wherein* $\{|H_{11}|, |H_{02}|\} < \infty$.

Proof. Assume that the error at kth stage is $e_k = t_k - \alpha$. Using the Taylor's expansion of $f(t_k)$ about α and keeping in mind that $f(\alpha) = 0$, $f'(\alpha) = 0$, and $f^{(2)}(\alpha) \neq 0$, we have

$$f(t_k) = \frac{f^{(2)}(\alpha)}{2!} e_k^2 (1 + B_1 e_k + B_2 e_k^2 + B_3 e_k^3 + B_4 e_k^4 + \cdots), \tag{14}$$

where $B_n = \frac{2!}{(2+n)!} \frac{f^{(2+n)}(\alpha)}{f^{(2)}(\alpha)}$ for $n \in \mathbb{N}$.

Similarly, we have the Taylor's expansion of $f(s_k)$ about α

$$f(s_k) = \frac{f^{(2)}(\alpha)}{2!} e_{s_k}^2 (1 + B_1 e_{s_k} + B_2 e_{s_k}^2 + B_3 e_{s_k}^3 + B_4 e_{s_k}^4 + \cdots), \tag{15}$$

where $e_{s_k} = s_k - \alpha = e_k + \frac{\beta f^{(2)}(\alpha)}{2!} e_k^2 (1 + B_1 e_k + B_2 e_k^2 + B_3 e_k^3 + B_4 e_k^4 + \cdots)$.

Then, the first step of Equation (3) yields

$$\begin{aligned} e_{z_k} &= z_k - \alpha \\ &= \frac{1}{2}\Big(\frac{\beta f^{(2)}(\alpha)}{2} + B_1\Big)e_k^2 - \frac{1}{16}((\beta f^{(2)}(\alpha))^2 - 8\beta f^{(2)}(\alpha)B_1 + 12B_1^2 - 16B_2)e_k^3 + \frac{1}{64}((\beta f^{(2)}(\alpha))^3 \\ &\quad - 20\beta f^{(2)}(\alpha)B_1^2 + 72B_1^3 + 64\beta f^{(2)}(\alpha)B_2 - 10B_1((\beta f^{(2)}(\alpha))^2 + 16B_2) + 96B_3)e_k^4 + O(e_k^5). \end{aligned} \tag{16}$$

Expanding $f(z_k)$ about α, it follows that

$$f(z_k) = \frac{f^{(2)}(\alpha)}{2!} e_{z_k}^2 (1 + B_1 e_{z_k} + B_2 e_{z_k}^2 + B_3 e_{z_k}^3 + B_4 e_{z_k}^4 + \cdots). \tag{17}$$

Using Equations (14), (15) and (17) in x_k and y_k, after some simple calculations, we have

$$\begin{aligned} x_k &= \frac{1}{2}\Big(\frac{\beta f^{(2)}(\alpha)}{2} + B_1\Big)e_k - \frac{1}{16}((\beta f^{(2)}(\alpha))^2 - 6\beta f^{(2)}(\alpha)B_1 + 16(B_1^2 - B_2))e_k^2 + \frac{1}{64}((\beta f^{(2)}(\alpha))^3 \\ &\quad - 22\beta f^{(2)}(\alpha)B_1^2 + 4(29B_1^3 + 14\beta f^{(2)}(\alpha)B_2) - 2B_1(3(\beta f^{(2)}(\alpha))^2 + 104B_2) + 96B_3)e_k^3 + O(e_k^4) \end{aligned} \tag{18}$$

and

$$\begin{aligned} y_k &= \frac{1}{2}\Big(\frac{\beta f^{(2)}(\alpha)}{2} + B_1\Big)e_k - \frac{1}{16}(3(\beta f^{(2)}(\alpha))^2 - 2\beta f^{(2)}(\alpha)B_1 + 16(B_1^2 - B_2))e_k^2 + \frac{1}{64}(7(\beta f^{(2)}(\alpha))^3 \\ &\quad + 24\beta f^{(2)}(\alpha)B_2 - 14\beta f^{(2)}(\alpha)B_1^2 + 116B_1^3 - 2B_1(11(\beta f^{(2)}(\alpha))^2 + 104B_2) + 96B_3)e_k^3 + O(e_k^4). \end{aligned} \tag{19}$$

Developing by Taylor series the weight function $H(x_k, y_k)$ in the neighborhood of origin $(0,0)$,

$$H(x_k, y_k) \approx H_{00} + x_k H_{10} + y_k H_{01} + \frac{1}{2}x_k^2 H_{20} + x_k y_k H_{11} + \frac{1}{2}y_k^2 H_{02}. \tag{20}$$

Inserting Equations (14)–(20) intothe second step of Equation (3), and then some simple calculations yield

$$\begin{aligned} e_{k+1} = & -\frac{H_{00}}{2}e_k + \frac{1}{4}(2 + H_{00} - H_{01} - H_{10})\Big(\frac{\beta f^{(2)}(\alpha)}{2} + B_1\Big)e_k^2 - \frac{1}{64}\Big((\beta f^{(2)}(\alpha))^2(4 + 2H_{00} - 8H_{01} + H_{02} \\ & - 4H_{10} + 2H_{11} + H_{20}) + 4\beta f^{(2)}(\alpha)(-8 - 4H_{00} - H_{01} + H_{02} + H_{10} + 2H_{11} + H_{20})B_1 + 4(12 + 6H_{00} \\ & - 10H_{01} + H_{02} - 10H_{10} + 2H_{11} + H_{20})B_1^2 - 32(2 + H_{00} - H_{01} - H_{10})B_2\Big)e_k^3 + \phi e_k^4 + O(e_k^5), \end{aligned} \tag{21}$$

where $\phi = \phi(\beta, B_1, B_2, B_3, H_{00}, H_{10}, H_{01}, H_{20}, H_{11}, H_{02})$. Here, expression of ϕ is not being produced explicitly since it is very lengthy.

It is clear from Equation (21) that we would obtain at least fourth-order convergence if we set coefficients of e_k, e_k^2 and e_k^3 simultaneously equal to zero. Then, solving the resulting equations, one gets

$$H_{00} = 0, \quad H_{10} = 1, \quad H_{01} = 1, \quad H_{20} = 8 - H_{02} - 2H_{11}. \tag{22}$$

As a result, the error equation is given by

$$e_{k+1} = \frac{1}{32}\Big(\frac{\beta f^{(2)}(\alpha)}{2} + B_1\Big)\big((2\beta f^{(2)}(\alpha)(3 + H_{02} + H_{11})B_1 + 22B_1^2 + ((\beta f^{(2)}(\alpha))^2(H_{02} + H_{11}) - 8B_2)\big)e_k^4 + O(e_k^5).$$

Thus, the theorem is proved. □

Below, we state the theorems (without proof) for the cases $m = 3, 4, 5$ as the proof is similar to the above proved theorems.

Theorem 3. *Using the hypotheses of Theorem 1, the order of convergence of scheme in Equation* (3) *for the case* $m = 3$ *is at least 4, if* $H_{00} = 0$, $H_{10} = 3 - H_{01}$, *and* $H_{20} = 12 - H_{02} - 2H_{11}$, *where* $\{|H_{01}|, |H_{02}|, |H_{11}|\} < \infty$. *Moreover, the scheme satisfies error equation*

$$e_{k+1} = \frac{1}{54}(\beta f^{(3)}(\alpha)(-3 + H_{01})C_1 + 12C_1^3 - 6C_1C_2)e_k^4 + O(e_k^5),$$

where $C_n = \frac{3!}{(3+n)!}\frac{f^{(3+n)}(\alpha)}{f^{(3)}(\alpha)}$ *for* $n \in \mathbb{N}$.

Theorem 4. *Using the hypotheses of Theorem 1, the order of convergence of scheme in Equation* (3) *for the case* $m = 4$ *is at least 4, if* $H_{00} = 0$, $H_{10} = 4 - H_{01}$, *and* $H_{20} = 16 - H_{02} - 2H_{11}$, *where* $\{|H_{01}|, |H_{02}|, |H_{11}|\} < \infty$. *Moreover, the scheme satisfies error equation*

$$e_{k+1} = \frac{1}{128}(13D_1^3 - 8D_1D_2)e_k^4 + O(e_k^5),$$

where $D_n = \frac{4!}{(4+n)!}\frac{f^{(4+n)}(\alpha)}{f^{(4)}(\alpha)}$ *for* $n \in \mathbb{N}$.

Theorem 5. *Using the hypotheses of Theorem 1, the order of convergence of scheme in Equation (3) for the case $m = 5$ is at least 4, if $H_{00} = 0$, $H_{10} = 5 - H_{01}$, and $H_{20} = 20 - H_{02} - 2H_{11}$, where $\{|H_{01}|, |H_{02}|, |H_{11}|\} < \infty$. Moreover, the scheme satisfies error equation*

$$e_{k+1} = \frac{1}{125}(7E_1^3 - 5E_1E_2)e_k^4 + O(e_k^5),$$

where $E_n = \frac{5!}{(5+n)!}\frac{f^{(5+n)}(\alpha)}{f^{(5)}(\alpha)}$ for $n \in \mathbb{N}$.

Remark 1. *We can observe from the above results that the number of conditions on H_{ij} is 6, 4, 3, 3, 3 corresponding to cases $m = 1, 2, 3, 4, 5$ to attain the fourth-order convergence of the method in Equation (3). The cases $m = 3, 4, 5$ satisfy the common conditions, $H_{00} = 0$, $H_{10} = m - H_{01}$, and $H_{20} = 4m - H_{02} - 2H_{11}$. Nevertheless, their error equations differ from each other as the parameter β does not appear in the equations for $m = 4, 5$. It has been seen that when $m \geq 4$ the conditions on H_{ij} are always three in number and the error equation in each such case does not contain β term. This type of symmetry in the results helps us to prove the general result, which is presented in next section.*

3. Main Result

For the multiplicity $m \geq 4$, we prove the order of convergence of the scheme in Equation (3) by the following theorem:

Theorem 6. *Assume that the function $f : \mathbb{C} \to \mathbb{C}$ is an analytic in a domain containing zero α having multiplicity $m \geq 4$. Further, suppose that the initial estimation t_0 is close enough to α. Then, the convergence of the iteration scheme in Equation (3) is of order four, provided that $H_{00} = 0$, $H_{10} = m - H_{01}$, and $H_{20} = 4m - H_{02} - 2H_{11}$, wherein $\{|H_{01}|, |H_{11}|, |H_{02}|\} < \infty$. Moreover, the error in the scheme is given by*

$$e_{k+1} = \frac{1}{2m^3}((9+m)K_1^3 - 2mK_1K_2)e_k^4 + O(e_k^5).$$

Proof. Taking into account that $f^{(j)}(\alpha) = 0$, $j = 0, 1, 2, \ldots, m-1$ and $f^m(\alpha) \neq 0$, then, developing $f(t_k)$ about α in the Taylor's series,

$$f(t_k) = \frac{f^m(\alpha)}{m!}e_k^m(1 + K_1e_k + K_2e_k^2 + K_3e_k^3 + K_4e_k^4 + \cdots), \tag{23}$$

where $K_n = \frac{m!}{(m+n)!}\frac{f^{(m+n)}(\alpha)}{f^{(m)}(\alpha)}$ for $n \in \mathbb{N}$.

In addition, from the expansion of $f(s_k)$ about α, it follows that

$$f(s_k) = \frac{f^m(\alpha)}{m!}e_{s_k}^m(1 + K_1e_{s_k} + K_2e_{s_k}^2 + K_3e_{s_k}^3 + K_4e_{s_k}^4 + \cdots), \tag{24}$$

where $e_{s_k} = s_k - \alpha = e_k + \frac{\beta f^m(\alpha)}{m!}e_k^m(1 + K_1e_k + K_2e_k^2 + K_3e_k^3 + K_4e_k^4 + \cdots)$. From the first step of Equation (3),

$$\begin{aligned} e_{z_k} &= z_k - \alpha \\ &= \frac{K_1}{m}e_k^2 + \frac{1}{m^2}(2mK_2 - (1+m)K_1^2)e_k^3 + \frac{1}{m^3}((1+m)^2K_1^3 - m(4+3m)K_1K_2 + 3m^2K_3)e_k^4 + O(e_k^5). \end{aligned} \tag{25}$$

Expansion of $f(z_k)$ around α yields

$$f(z_k) = \frac{f^m(\alpha)}{m!}e_{z_k}^2(1 + K_1e_{z_k} + K_2e_{z_k}^2 + K_3e_{z_k}^3 + K_4e_{z_k}^4 + \cdots). \tag{26}$$

Using Equations (23), (24) and (26) in the expressions of x_k and y_k, we have that

$$x_k = \frac{K_1}{m}e_k + \frac{1}{m^2}(2mK_2 - (2+m)K_1^2)e_k^2 + \frac{1}{2m^3}((7+7m+2m^2)K_1^3 - 2m(7+3m)K_1K_2 + 6m^2K_3)e_k^3 + O(e_k^4) \quad (27)$$

and

$$y_k = \frac{K_1}{m}e_k + \frac{1}{m^2}(mK_2 - (2+m)K_1^2)e_k^2 + \frac{1}{2m^3}((6+7m+2m^2)K_1^3 - 2m(6+3m)K_1K_2 + 6m^2K_3)e_k^3 + O(e_k^4). \quad (28)$$

Developing $H(x_k, y_k)$ in Taylor's series in the neighborhood of origin $(0,0)$,

$$H(x_k, y_k) \approx H_{00} + x_kH_{10} + y_kH_{01} + \frac{1}{2}x_k^2H_{20} + x_ky_kH_{11} + \frac{1}{2}y_k^2H_{02}. \quad (29)$$

Inserting Equations (23)–(29) into the second step of Equation (3), it follows that

$$\begin{aligned} e_{k+1} = & -\frac{H_{00}}{m}e_k + \frac{1}{m^2}(H_{00} - H_{01} - H_{10} + m)K_1e_k^2 - \frac{1}{2m^3}((H_{02} - 6H_{10} + 2H_{11} + H_{20} + 2m - 2mH_{10} + 2m^2 \\ & + 2(1+m)H_{00} - 2(3+m)H_{01})K_1^2 - 4m(H_{00} - H_{01} - H_{10} + m)K_2)e_k^3 + \frac{1}{2m^4}((5H_{02} - 13H_{10} + 10H_{11} \\ & + 5H_{20} + 2m + 2mH_{02} - 11mH_{10} + 4mH_{11} + 2mH_{20} + 4m^2 - 2m^2H_{10} + 2m^3 + 2(1+m)^2H_{00} \\ & - (13+11m+2m^2)H_{01})K_1^3 - 2m(2H_{02} - 11H_{10} + 4H_{11} + 2H_{20} + 4m - 3mH_{10} + 3m^2 + (4+3m)H_{00} \\ & - (11+3m)H_{01})K_1K_2 + 6m^2(H_{00} - H_{01} - H_{10} + m)K_3)e_k^4 + O(e_k^5). \end{aligned} \quad (30)$$

It is clear that we can obtain at least fourth-order convergence if the coefficients of e_k, e_k^2, and e_k^3 vanish. On solving the resulting equations, we get

$$H_{00} = 0, \;\; H_{10} = m - H_{01}, \;\; H_{20} = 4m - \; H_{02} - 2\,H_{11}. \quad (31)$$

Then, error of Equation (30) is given by

$$e_{k+1} = \frac{1}{2m^3}((9+m)K_1^3 - 2mK_1K_2)e_k^4 + O(e_k^5). \quad (32)$$

Thus, the theorem is proved. □

Remark 2. *The proposed scheme in Equation* (3) *reaches fourth-order convergence provided that the conditions of Theorems 1–3 and 6 are satisfied. This convergence rate is achieved by using only three function evaluations, viz.* $f(t_n)$, $f(s_n)$, *and* $f(z_n)$, *per iteration. Therefore, the scheme in Equation* (3) *is optimal by the Kung–Traub hypothesis [17].*

Remark 3. *It is important to note that parameter* β, *which is used in* s_k, *appears only in the error equations of the cases* $m = 1, 2, 3$ *but not for* $m \geq 4$. *For* $m \geq 4$, *we have observed that this parameter appears in the coefficients of* e_k^5 *and higher order. However, we do not need such terms to show the required fourth-order convergence.*

Some Special Cases

We can generate many iterative schemes as the special cases of the family in Equation (3) based on the forms of function $H(x, y)$ that satisfy the conditions of Theorems 1, 2 and 6. However, we restrict ourselves to the choices of low-degree polynomials or simple rational functions. These choices should be such that the resulting methods may converge to the root with order four for $m \geq 1$. Accordingly, the following simple forms are considered:

(1) Let us choose the function

$$H(x_k, y_k) = x_k + m\,x_k^2 + (m-1)y_k + m\,x_k\,y_k,$$

which satisfies the conditions of Theorems 1, 2 and 6. Then, the corresponding fourth-order iterative scheme is given by

$$t_{k+1} = z_k - \left(x_k + m\,x_k^2 + (m-1)y_k + m\,x_k\,y_k\right)\frac{f(t_k)}{f[s_k, t_k]}. \tag{33}$$

(2) Next, consider the rational function

$$H(x_k, y_k) = -\frac{x_k + m\,x_k^2 - (m-1)y_k(m\,y_k - 1)}{m\,y_k - 1},$$

satisfying the conditions of Theorems 1, 2 and 6. Then, corresponding fourth-order iterative scheme is given by

$$t_{k+1} = z_k + \frac{x_k + m\,x_k^2 - (m-1)y_k(m\,y_k - 1)}{m\,y_k - 1}\,\frac{f(t_k)}{f[s_k, t_k]}. \tag{34}$$

(3) Consider another rational function satisfying the conditions of Theorems 1, 2 and 6, which is given by

$$H(x_k, y_k) = \frac{x_k - y_k + m\,y_k + 2m\,x_k\,y_k - m^2\,x_k\,y_k}{1 - m\,x_k + x_k^2}.$$

The corresponding fourth-order iterative scheme is given by

$$t_{k+1} = z_k - \frac{x_k - y_k + m\,y_k + 2m\,x_k\,y_k - m^2\,x_k\,y_k}{1 - m\,x_k + x_k^2}\,\frac{f(t_k)}{f[s_k, t_k]}. \tag{35}$$

For each of the above cases, $z_k = t_k - m\frac{f(t_k)}{f[s_k,t_k]}$. For future reference, the proposed methods in Equations (33)–(35) are denoted by NM1, NM2, and NM3, respectively.

4. Numerical Results

To validate the theoretical results proven in previous sections, the special cases NM1, NM2, and NM3 of new family were tested numerically by implementing them on some nonlinear equations. Moreover, their comparison was also performed with some existing optimal fourth-order methods that use derivatives in the formulas. We considered, for example, the methods by Li et al. [7,8], Sharma and Sharma [9], Zhou et al. [10], Soleymani et al. [12], and Kansal et al. [14]. The methods are expressed as follows:

Li–Liao–Cheng method (LLC):

$$z_k = t_k - \frac{2m}{m+2}\frac{f(t_k)}{f'(t_k)},$$

$$t_{k+1} = t_k - \frac{m(m-2)\left(\frac{m}{m+2}\right)^{-m} f'(z_k) - m^2 f'(t_k)}{f'(t_k) - \left(\frac{m}{m+2}\right)^{-m} f'(z_k)}\,\frac{f(t_k)}{2f'(t_k)}.$$

Li–Cheng–Neta method (LCN):

$$z_k = t_k - \frac{2m}{m+2}\frac{f(t_k)}{f'(t_k)},$$
$$t_{k+1} = t_k - \alpha_1\frac{f(t_k)}{f'(z_k)} - \frac{f(t_k)}{\alpha_2 f'(t_k) + \alpha_3 f'(z_k)},$$

where

$$\alpha_1 = -\frac{1}{2}\frac{\left(\frac{m}{m+2}\right)^m m(m^4 + 4m^3 - 16m - 16)}{m^3 - 4m + 8},$$
$$\alpha_2 = -\frac{(m^3 - 4m + 8)^2}{m(m^4 + 4m^3 - 4m^2 - 16m + 16)(m^2 + 2m - 4)},$$
$$\alpha_3 = \frac{m^2(m^3 - 4m + 8)}{\left(\frac{m}{m+2}\right)^m (m^4 + 4m^3 - 4m^2 - 16m + 16)(m^2 + 2m - 4)}.$$

Sharma–Sharma method (SS):

$$z_k = t_k - \frac{2m}{m+2}\frac{f(t_k)}{f'(t_k)},$$
$$t_{k+1} = t_k - \frac{m}{8}\Big[(m^3 - 4m + 8) - (m+2)^2\Big(\frac{m}{m+2}\Big)^m \frac{f'(t_k)}{f'(z_k)}$$
$$\times \Big(2(m-1) - (m+2)\Big(\frac{m}{m+2}\Big)^m \frac{f'(t_k)}{f'(z_k)}\Big)\Big]\frac{f(t_k)}{f'(t_k)}.$$

Zhou–Chen–Song method (ZCS):

$$z_k = t_k - \frac{2m}{m+2}\frac{f(t_k)}{f'(t_k)},$$
$$t_{k+1} = t_k - \frac{m}{8}\Big[m^3\Big(\frac{m+2}{m}\Big)^{2m}\Big(\frac{f'(z_k)}{f'(t_k)}\Big)^2 - 2m^2(m+3)\Big(\frac{m+2}{m}\Big)^m \frac{f'(z_k)}{f'(t_k)}$$
$$+ (m^3 + 6m^2 + 8m + 8)\Big]\frac{f(t_k)}{f'(t_k)}.$$

Soleymani–Babajee–Lotfi method (SBL):

$$z_k = t_k - \frac{2m}{m+2}\frac{f(t_k)}{f'(t_k)},$$
$$t_{k+1} = t_k - \frac{f'(z_k)f(t_k)}{q_1(f'(z_k))^2 + q_2 f'(z_k)f'(t_k) + q_3(f'(t_k))^2},$$

where

$$q_1 = \frac{1}{16}m^{3-m}(2+m)^m,$$
$$q_2 = \frac{8 - m(2+m)(-2+m^2)}{8m},$$
$$q_3 = \frac{1}{16}(-2+m)m^{-1+m}(2+m)^{3-m}.$$

Kansal–Kanwar–Bhatia method (KKB):

$$z_k = t_k - \frac{2m}{m+2}\frac{f(t_k)}{f'(t_k)},$$

$$t_{k+1} = t_k - \frac{m}{4}f(t_k)\left(1 + \frac{m^4 p^{-2m}\left(-\frac{f'(z_k)}{f'(t_k)} + p^{-1+m}\right)^2(-1+p^m)}{8(2p^m + m(-1+p^m))}\right)$$
$$\times \left(\frac{4 - 2m + m^2(-1+p^{-m})}{f'(t_k)} - \frac{p^{-m}(2p^m + m(-1+p^m))^2}{f'(t_k) - f'(z_k)}\right),$$

where $p = \frac{m}{m+2}$.

Computational work was compiled in the programming package of Mathematica software [18] in a PC with Intel(R) Pentium(R) CPU B960 @ 2.20 GHz, 2.20 GHz (32-bit Operating System) Microsoft Windows 7 Professional and 4 GB RAM. Performance of the new methods was tested by choosing value of the parameter $\beta = 0.01$. The tabulated results obtained by the methods for each problem include: (a) the number of iterations (k) required to obtain the solution using the stopping criterion $|t_{k+1} - t_k| + |f(t_k)| < 10^{-100}$; (b) the estimated error $|t_{k+1} - t_k|$ in the first three iterations; (c) the calculated convergence order (CCO); and (d) the elapsed time (CPU time in seconds) in execution of a program, which was measured by the command "TimeUsed[ ]". The calculated convergence order (CCO) to confirm the theoretical convergence order was calculated by the formula (see [19])

$$\text{CCO} = \frac{\log|(t_{k+2} - \alpha)/(t_{k+1} - \alpha)|}{\log|(t_{k+1} - \alpha)/(t_k - \alpha)|}, \quad \text{for each } k = 1, 2, \ldots \tag{36}$$

The following numerical examples were chosen for experimentation:

Example 1. *Planck law of radiation to calculate the energy density in an isothermal black body [20] is stated as*

$$\phi(\lambda) = \frac{8\pi ch\lambda^{-5}}{e^{ch/\lambda kT} - 1}. \tag{37}$$

where λ is wavelength of the radiation, c is speed of light, T is absolute temperature of the black body, k is Boltzmann's constant, and h is Planck's constant. The problem is to determine the wavelength λ corresponding to maximum energy density $\phi(\lambda)$. Thus, Equation (37) leads to

$$\phi'(\lambda) = \left(\frac{8\pi ch\lambda^{-6}}{e^{ch/\lambda kT} - 1}\right)\left(\frac{(ch/\lambda kT)e^{ch/\lambda kT}}{e^{ch/\lambda kT} - 1} - 5\right) = A.B. \ (say)$$

Note that a maxima for ϕ will occur when $B = 0$, that is when

$$\frac{(ch/\lambda kT)e^{ch/\lambda kT}}{e^{ch/\lambda kT} - 1} = 5.$$

Setting $t = ch/\lambda kT$, the above equation assumes the form

$$1 - \frac{t}{5} = e^{-t}. \tag{38}$$

Define

$$f_1(t) = e^{-t} - 1 + \frac{t}{5}. \tag{39}$$

The root $t = 0$ is trivial and thus is not taken for discussion. Observe that for $t = 5$ the left-hand side of Equation (38) is zero and the right-hand side is $e^{-5} \approx 6.74 \times 10^{-3}$. Thus, we guess that another root might occur *somewhere near to* $t = 5$. *In fact, the expected root of Equation* (39) *is given by* $\alpha \approx 4.96511423174427630369$ *with* $t_0 = 5.5$. *Then, the wavelength of radiation* (λ) *corresponding to maximum energy density is*

$$\lambda \approx \frac{ch}{4.96511423174427630369(kT)}.$$

The results so obtained are shown in Table 1.

Example 2. *Consider the van der Waals equation (see [15])*

$$\left(P + \frac{a_1 n^2}{V^2}\right)(V - na_2) = nRT,$$

that explains the nature of a real gas by adding two parameters a_1 *and* a_2 *in the ideal gas equation. To find the volume* V *in terms of rest of the parameters, one requires solving the equation*

$$PV^3 - (na_2P + nRT)V^2 + a_1n^2V - a_1a_2n^2 = 0.$$

One can find values of n, P, *and* T, *for a given a set of values of* a_1 *and* a_2 *of a particular gas, so that the equation has three roots. Using a particular set of values, we have the function*

$$f_2(t) = t^3 - 5.22t^2 + 9.0825t - 5.2675,$$

that has three roots from which one is simple zero $\alpha = 1.72$ *and other one is a multiple zero* $\alpha = 1.75$ *of multiplicity two. However, our desired zero is* $\alpha = 1.75$. *The methods are tested for initial guess* $t_0 = 2.5$. *Computed results are given in Table 2.*

Example 3. *Next, we assume a standard nonlinear test function which is defined as*

$$f_3(t) = \left[\tan^{-1}\left(\frac{\sqrt{5}}{2}\right) - \tan^{-1}(\sqrt{t^2-1}) + \sqrt{6}\left(\tan^{-1}\left(\sqrt{\frac{t^2-1}{6}}\right) - \tan^{-1}\left(\frac{1}{2}\sqrt{\frac{5}{6}}\right)\right) - \frac{11}{63}\right]^3.$$

The function f_3 *has multiple zero at* $\alpha = 1.8411027704926161\ldots$ *of multiplicity three. We select initial approximation* $t_0 = 1.6$ *to obtain zero of this function. Numerical results are exhibited in Table 3.*

Example 4. *Lastly, we consider another standard test function, which is defined as*

$$f_4(t) = t(t^2+1)(2e^{t^2+1} + t^2 - 1)\cosh^2\left(\frac{\pi t}{2}\right).$$

The function f_4 *has multiple zero at* i *of multiplicity four. We choose the initial approximation* $x_0 = 1.2i$ *for obtaining the zero of the function. Numerical results are displayed in Table 4.*

Table 1. Numerical results for Example 1.

Method	k	$\|t_2 - t_1\|$	$\|t_3 - t_2\|$	$\|t_4 - t_3\|$	CCO	CPU-Time
LLC	4	1.51×10^{-5}	1.47×10^{-23}	1.30×10^{-95}	4.000	0.4993
LCN	4	1.55×10^{-5}	1.73×10^{-23}	2.65×10^{-95}	4.000	0.5302
SS	4	1.52×10^{-5}	1.51×10^{-23}	1.47×10^{-95}	4.000	0.6390
ZCS	4	1.57×10^{-5}	1.87×10^{-23}	3.75×10^{-95}	4.000	0.6404
SBL	4	1.50×10^{-5}	1.43×10^{-23}	1.19×10^{-95}	4.000	0.8112
KKB	fails	-	-	-	-	-
NM1	3	5.59×10^{-6}	1.35×10^{-25}	0	4.000	0.3344
NM2	3	5.27×10^{-6}	9.80×10^{-26}	0	4.000	0.3726
NM3	3	5.43×10^{-6}	1.16×10^{-25}	0	4.000	0.3475

Table 2. Numerical results for Example 2.

Method	k	$\lvert t_2 - t_1 \rvert$	$\lvert t_3 - t_2 \rvert$	$\lvert t_4 - t_3 \rvert$	CCO	CPU-Time
LLC	6	9.09×10^{-2}	8.03×10^{-3}	2.33×10^{-5}	4.000	0.0780
LCN	6	9.09×10^{-2}	8.03×10^{-3}	2.33×10^{-5}	4.000	0.0784
SS	6	9.26×10^{-2}	8.58×10^{-3}	3.11×10^{-5}	4.000	0.0945
ZCS	6	9.62×10^{-2}	9.84×10^{-3}	5.64×10^{-5}	4.000	0.0792
SBL	6	9.09×10^{-2}	8.03×10^{-3}	2.33×10^{-5}	4.000	0.0797
KKB	6	8.97×10^{-2}	7.62×10^{-3}	1.68×10^{-5}	4.000	0.0934
NM1	6	9.91×10^{-2}	1.08×10^{-2}	8.79×10^{-5}	4.000	0.0752
NM2	6	8.06×10^{-2}	5.08×10^{-3}	2.81×10^{-5}	4.000	0.0684
NM3	6	8.78×10^{-2}	7.02×10^{-3}	1.31×10^{-5}	4.000	0.0788

Table 3. Numerical results for Example 3.

Method	k	$\lvert t_2 - t_1 \rvert$	$\lvert t_3 - t_2 \rvert$	$\lvert t_4 - t_3 \rvert$	CCO	CPU-Time
LLC	4	1.11×10^{-4}	9.02×10^{-19}	3.91×10^{-75}	4.000	2.5743
LCN	4	1.11×10^{-4}	8.93×10^{-19}	3.72×10^{-75}	4.000	2.6364
SS	4	1.11×10^{-4}	8.71×10^{-19}	3.29×10^{-75}	4.000	2.8718
ZCS	4	1.11×10^{-4}	8.16×10^{-19}	2.38×10^{-75}	4.000	2.8863
SBL	4	1.11×10^{-4}	8.63×10^{-19}	3.15×10^{-75}	4.000	3.2605
KKB	4	1.11×10^{-4}	9.80×10^{-19}	5.87×10^{-75}	4.000	2.9011
NM1	4	2.31×10^{-5}	4.04×10^{-21}	3.78×10^{-84}	4.000	2.2935
NM2	4	2.07×10^{-5}	1.32×10^{-21}	2.18×10^{-86}	4.000	2.5287
NM3	4	2.11×10^{-5}	1.66×10^{-21}	6.36×10^{-86}	4.000	2.4964

Table 4. Numerical results for Example 4.

Method	k	$\lvert t_2 - t_1 \rvert$	$\lvert t_3 - t_2 \rvert$	$\lvert t_4 - t_3 \rvert$	CCO	CPU-Time
LLC	4	2.64×10^{-4}	2.13×10^{-15}	9.11×10^{-60}	4.000	1.7382
LCN	4	2.64×10^{-4}	2.14×10^{-15}	9.39×10^{-60}	4.000	2.4035
SS	4	2.64×10^{-4}	2.18×10^{-15}	1.01×10^{-59}	4.000	2.5431
ZCS	4	2.65×10^{-4}	2.24×10^{-15}	1.14×10^{-59}	4.000	2.6213
SBL	4	2.66×10^{-4}	2.28×10^{-15}	1.23×10^{-59}	4.000	3.2610
KKB	4	2.61×10^{-4}	2.00×10^{-15}	6.83×10^{-60}	4.000	2.6524
NM1	4	1.43×10^{-4}	1.29×10^{-16}	8.61×10^{-65}	4.000	0.5522
NM2	4	4.86×10^{-5}	5.98×10^{-20}	1.36×10^{-79}	4.000	0.6996
NM3	4	6.12×10^{-5}	6.69×10^{-19}	9.54×10^{-75}	4.000	0.6837

From the computed results shown in Tables 1–4, we can observe a good convergence behavior of the proposed methods similar to those of existing methods. The reason for good convergence is the increase in accuracy of the successive approximations per iteration, as is evident from numerical results. This also points to the stable nature of methods. It is also clear that the approximations to the solutions by the new methods have accuracies greater than or equal to those computed by existing methods. We display the value 0 of $|t_{k+1} - t_k|$ at the stage when stopping criterion $|t_{k+1} - t_k| + |f(t_k)| < 10^{-100}$ has been satisfied. From the calculation of computational order of convergence shown in the penultimate column in each table, we verify the theoretical fourth-order of convergence.

The efficient nature of presented methods can be observed by the fact that the amount of CPU time consumed by the methods is less than the time taken by existing methods (result confirmed by similar numerical experiments on many other different problems). The methods requiring repeated evaluations of the roots (such as the ones tackled in [21–24]), also may benefit greatly from the use of proposed methods (NM1–NM3, Equations (33)–(35)).

5. Conclusions

In this paper, we propose a family of fourth-order derivative-free numerical methods for obtaining multiple roots of nonlinear equations. Analysis of the convergence was carried out, which proved the order four under standard assumptions of the function whose zeros we are looking for. In addition, our designed scheme also satisfies the Kung–Traub hypothesis of optimal order of convergence. Some special cases are established. These are employed to solve nonlinear equations including those arising in practical problems. The new methods are compared with existing techniques of same order. Testing of the numerical results shows that the presented derivative-free methods are good competitors to the existing optimal fourth-order techniques that require derivative evaluations in the algorithm. We conclude the work with a remark that derivative-free methods are good alternatives to Newton-type schemes in the cases when derivatives are expensive to compute or difficult to obtain.

Author Contributions: Methodology, J.R.S.; writing, review and editing, J.R.S.; investigation, S.K.; data curation, S.K.; conceptualization, L.J.; formal analysis, L.J.

References

1. Schröder, E. Über unendlich viele Algorithmen zur Auflösung der Gleichungen. *Math. Ann.* **1870**, *2*, 317–365. [CrossRef]
2. Hansen E.; Patrick, M. A family of root finding methods. *Numer. Math.* **1977**, *27*, 257–269. [CrossRef]
3. Victory, H.D.; Neta, B. A higher order method for multiple zeros of nonlinear functions. *Int. J. Comput. Math.* **1983**, *12*, 329–335. [CrossRef]
4. Dong, C. A family of multipoint iterative functions for finding multiple roots of equations. *Int. J. Comput. Math.* **1987**, *21*, 363–367. [CrossRef]
5. Osada, N. An optimal multiple root-finding method of order three. *J. Comput. Appl. Math.* **1994**, *51*, 131–133. [CrossRef]
6. Neta, B. New third order nonlinear solvers for multiple roots. *App. Math. Comput.* **2008**, *202*, 162–170. [CrossRef]
7. Li, S.; Liao, X.; Cheng, L. A new fourth-order iterative method for finding multiple roots of nonlinear equations. *Appl. Math. Comput.* **2009**, *215*, 1288–1292.
8. Li, S.G.; Cheng, L.Z.; Neta, B. Some fourth-order nonlinear solvers with closed formulae for multiple roots. *Comput Math. Appl.* **2010**, *59*, 126–135. [CrossRef]
9. Sharma, J.R.; Sharma, R. Modified Jarratt method for computing multiple roots. *Appl. Math. Comput.* **2010**, *217*, 878–881. [CrossRef]
10. Zhou, X.; Chen, X.; Song, Y. Constructing higher-order methods for obtaining the multiple roots of nonlinear equations. *J. Comput. Appl. Math.* **2011**, *235*, 4199–4206. [CrossRef]
11. Sharifi, M.; Babajee, D.K.R.; Soleymani, F. Finding the solution of nonlinear equations by a class of optimal methods. *Comput. Math. Appl.* **2012**, *63*, 764–774. [CrossRef]
12. Soleymani, F.; Babajee, D.K.R.; Lotfi, T. On a numerical technique for finding multiple zeros and its dynamics. *J. Egypt. Math. Soc.* **2013**, *21*, 346–353. [CrossRef]
13. Geum, Y.H.; Kim Y.I.; Neta, B. A class of two-point sixth-order multiple-zero finders of modified double-Newton type and their dynamics. *Appl. Math. Comput.* **2015**, *270*, 387–400. [CrossRef]
14. Kansal, M.; Kanwar, V.; Bhatia, S. On some optimal multiple root-finding methods and their dynamics. *Appl. Appl. Math.* **2015**, *10*, 349–367.
15. Behl, R.; Zafar, F.; Alshormani, A.S.; Junjua, M.U.D.; Yasmin, N. An optimal eighth-order scheme for multiple zeros of unvariate functions. *Int. J. Comput. Meth.* **2019**, *16*, 1843002. [CrossRef]
16. Traub, J.F. *Iterative Methods for the Solution of Equations*; Chelsea Publishing Company: New York, NY, USA, 1982.
17. Kung, H.T.; Traub, J.F. Optimal order of one-point and multipoint iteration. *J. Assoc. Comput. Mach.* **1974**, *21*, 643–651. [CrossRef]

18. Wolfram, S. *The Mathematica Book*, 5th ed.; Wolfram Media: Bengaluru, India, 2003.
19. Weerakoon, S.; Fernando, T.G.I. A variant of Newton's method with accelerated third-order convergence. *Appl. Math. Lett.* **2000**, *13*, 87–93. [CrossRef]
20. Bradie, B. *A Friendly Introduction to Numerical Analysis*; Pearson Education Inc.: New Delhi, India, 2006.
21. Jäntschi L.; Bolboacă S.D. Conformational study of C_{24} cyclic polyyne clusters. *Int. J. Quantum Chem.* **2018**, *118*, e25614. [CrossRef]
22. Azad H.; Anaya K.; Al-Dweik A. Y.; Mustafa M. T. Invariant solutions of the wave equation on static spherically symmetric spacetimes admitting G7 isometry algebra. *Symmetry* **2018**, *10*, 665. [CrossRef]
23. Matko, V.; Brezovec, B. Improved data center energy efficiency and availability with multilayer node event processing. *Energies* **2018**, *11*, 2478. [CrossRef]
24. Jäntschi L. The eigenproblem translated for alignment of molecules. *Symmetry* **2019**, *11*, 1027. [CrossRef]

Permissions

All chapters in this book were first published by MDPI; hereby published with permission under the Creative Commons Attribution License or equivalent. Every chapter published in this book has been scrutinized by our experts. Their significance has been extensively debated. The topics covered herein carry significant findings which will fuel the growth of the discipline. They may even be implemented as practical applications or may be referred to as a beginning point for another development.

The contributors of this book come from diverse backgrounds, making this book a truly international effort. This book will bring forth new frontiers with its revolutionizing research information and detailed analysis of the nascent developments around the world.

We would like to thank all the contributing authors for lending their expertise to make the book truly unique. They have played a crucial role in the development of this book. Without their invaluable contributions this book wouldn't have been possible. They have made vital efforts to compile up to date information on the varied aspects of this subject to make this book a valuable addition to the collection of many professionals and students.

This book was conceptualized with the vision of imparting up-to-date information and advanced data in this field. To ensure the same, a matchless editorial board was set up. Every individual on the board went through rigorous rounds of assessment to prove their worth. After which they invested a large part of their time researching and compiling the most relevant data for our readers.

The editorial board has been involved in producing this book since its inception. They have spent rigorous hours researching and exploring the diverse topics which have resulted in the successful publishing of this book. They have passed on their knowledge of decades through this book. To expedite this challenging task, the publisher supported the team at every step. A small team of assistant editors was also appointed to further simplify the editing procedure and attain best results for the readers.

Apart from the editorial board, the designing team has also invested a significant amount of their time in understanding the subject and creating the most relevant covers. They scrutinized every image to scout for the most suitable representation of the subject and create an appropriate cover for the book.

The publishing team has been an ardent support to the editorial, designing and production team. Their endless efforts to recruit the best for this project, has resulted in the accomplishment of this book. They are a veteran in the field of academics and their pool of knowledge is as vast as their experience in printing. Their expertise and guidance has proved useful at every step. Their uncompromising quality standards have made this book an exceptional effort. Their encouragement from time to time has been an inspiration for everyone.

The publisher and the editorial board hope that this book will prove to be a valuable piece of knowledge for researchers, students, practitioners and scholars across the globe.

List of Contributors

Yermek Aldabergenov
Department of Physics, Faculty of Science, Chulalongkorn University, Thanon Phayathai, Pathumwan, Bangkok 10330, Thailand
Institute of Experimental and Theoretical Physics, Al-Farabi Kazakh National University, 71 Al-Farabi Avenue, Almaty 050040, Kazakhstan

Sergei V. Ketov
Department of Physics, Tokyo Metropolitan University, Minami-ohsawa 1-1, Hachioji-shi, Tokyo 192-0397, Japan
Research School of High Energy Physics, Tomsk Polytechnic University, 30 Lenin Avenue, Tomsk 634050, Russia
Kavli Institute for the Physics and Mathematics of the Universe (IPMU), The University of Tokyo, Chiba 277-8568, Japan

Andrew James Bruce
Mathematics Research Unit, University of Luxembourg, Maison du Nombre 6, Avenue de la Fonte, L-4364 Esch-sur-Alzette, Luxembourg

Irina Rozgacheva
All-Russian Institute for Scientific and Technical Information of Russian Academy of Sciences (VINITI RAS), Usievich 20, Moscow 125190, Russia

Benedek Nagy and Khaled Abuhmaidan
Department of Mathematics, Faculty of Arts and Sciences, Eastern Mediterranean University, via Mersin 10, Turkey, Famagusta 99450, North Cyprus

Yuanxiang Dong and Chenjing Hou
School of Management Science and Engineering, Shanxi University of Finance and Economics, Taiyuan 030006, China

Yuchen Pan
Business School, Southwest University of Political Science & Law, Chongqing 401120, China

Ke Gong
School of Economics and Management, Chongqing Jiaotong University, Chongqing 400074, China

Roger Tovar-Falón and Guillermo Martínez-Flórez
Departamento de Matemáticas y Estadística, Facultad de Ciencias Básicas, Universidad de Córdoba, Montería 230027, Colombia

Heleno Bolfarine
Departamento de Estatística, IME, Universidade de São Paulo, São Paulo 1010, Brazil

Mujahid Abbas
Department of Mathematics, Government College University, Lahore 54000, Pakistan
Department of Medical Research, China Medical University No. 91, Hsueh-Shih Road, Taichung 400, Taiwan

Hira Iqbal
Department of Sciences and Humanities, National University of Computer and Emerging Sciences, Lahore Campus, Lahore 54000, Pakistan

Manuel De la Sen
Institute of Research and Development of Processes, University of the Basque Country, Campus of Leioa (Bizkaia), P.O. Box 644, Bilbao, Barrio Sarriena, 48940 Leioa, Spain

Deepak Kumar
Department of Mathematics, Sant Longowal Institute of Engineering and Technology, Longowal, Sangrur 148106, India

Flavia-Corina Mitroi-Symeonidis
Academy of Economic Studies, Department of Applied Mathematics, Calea Dorobantilor 15-17, Sector 1, RO-010552 Bucharest, Romania
Police Academy "Alexandru Ioan Cuza", Fire Officers Faculty, Str. Morarilor 3, Sector 2, RO-022451 Bucharest, Romania

Ion Anghel
Police Academy "Alexandru Ioan Cuza", Fire Officers Faculty, Str. Morarilor 3, Sector 2, RO-022451 Bucharest, Romania

Nicuşor Minculete
Faculty of Mathematics and Computer Science, Transilvania University of Bras, ov, Str. Iuliu Maniu50, 500091 Bras, ov, Romania

Adrian Holhoş and Daniela Roşca
Department of Mathematics, Technical University of Cluj-Napoca, str. Memorandumului 28, RO 400114 Cluj-Napoca, Romania

Pablo V. Negrón-Marrero and Melissa López-Serrano
Department of Mathematics, University of Puerto Rico, Humacao, PR 00791-4300, USA

Jared Lynskey, Kyi Thar, Thant Zin Oo and Choong Seon Hong
Department of Computer Science and Engineering, Kyung Hee University, Yongin-si, Gyeonggi-do 17104, Korea

Andronikos Paliathanasis
Institute of Systems Science, Durban University of Technology, Durban 4000, South Africa

Janak Raj Sharma and Sunil Kumar
Department of Mathematics, Sant Longowal Institute of Engineering and Technology, Longowal, Sangrur 148106, India

Lorentz Jäntschi
Institute of Doctoral Studies, Babeş-Bolyai University, 400084 Cluj-Napoca, Romania
Department of Physics and Chemistry, Technical University of Cluj-Napoca, Cluj-Napoca 400114, Romania
Chemical Doctoral School, Babeş-Bolyai University, 400084 Cluj-Napoca, Romania

Index

www.ingramcontent.com/pod-product-compliance
Lightning Source LLC
LaVergne TN
LVHW081031210926
848485LV00004B/995